AF845293

Regional Cooperation for Water Quality Improvement in Southwestern Pennsylvania

Committee on Water Quality Improvement for the Pittsburgh Region

Water Science and Technology Board

Division on Earth and Life Studies

NATIONAL RESEARCH COUNCIL
OF THE NATIONAL ACADEMIES

THE NATIONAL ACADEMIES PRESS
Washington, D.C.
www.nap.edu

THE NATIONAL ACADEMIES PRESS 500 Fifth Street, N.W. Washington, DC 20001

NOTICE: The project that is the subject of this report was approved by the Governing Board of the National Research Council, whose members are drawn from the councils of the National Academy of Sciences, the National Academy of Engineering, and the Institute of Medicine. The members of the committee responsible for the report were chosen for their special competences and with regard for appropriate balance.

Support for this study was provided by a grant from the Allegheny Conference on Community Development. Any opinions, findings, conclusions, or recommendations expressed in this publication are those of the author(s) and do not necessarily reflect the views of the organization that provided support for the project.

International Standard Book Number 0-309-09524-7 (Book)
International Standard Book Number 0-309-54724-5 (PDF)
Library of Congress Control Number 2005922546

Additional copies of this report are available from the National Academies Press, 500 Fifth Street, N.W., Lockbox 285, Washington, DC 20055; (800) 624-6242 or (202) 334-3313 (in the Washington metropolitan area); Internet, http://www.nap.edu.

Cover Photo: The Pittsburgh Point where the Allegheny and Monongahela Rivers converge to form the Ohio River. Photograph courtesy of Jessica Hess. Copyright 2005 by Jessica Hess. All rights reserved.

Copyright 2005 by the National Academy of Sciences. All rights reserved.

Printed in the United States of America.

THE NATIONAL ACADEMIES

Advisers to the Nation on Science, Engineering, and Medicine

The **National Academy of Sciences** is a private, nonprofit, self-perpetuating society of distinguished scholars engaged in scientific and engineering research, dedicated to the furtherance of science and technology and to their use for the general welfare. Upon the authority of the charter granted to it by the Congress in 1863, the Academy has a mandate that requires it to advise the federal government on scientific and technical matters. Dr. Bruce M. Alberts is president of the National Academy of Sciences.

The **National Academy of Engineering** was established in 1964, under the charter of the National Academy of Sciences, as a parallel organization of outstanding engineers. It is autonomous in its administration and in the selection of its members, sharing with the National Academy of Sciences the responsibility for advising the federal government. The National Academy of Engineering also sponsors engineering programs aimed at meeting national needs, encourages education and research, and recognizes the superior achievements of engineers. Dr. Wm. A. Wulf is president of the National Academy of Engineering.

The **Institute of Medicine** was established in 1970 by the National Academy of Sciences to secure the services of eminent members of appropriate professions in the examination of policy matters pertaining to the health of the public. The Institute acts under the responsibility given to the National Academy of Sciences by its congressional charter to be an adviser to the federal government and, upon its own initiative, to identify issues of medical care, research, and education. Dr. Harvey V. Fineberg is president of the Institute of Medicine.

The **National Research Council** was organized by the National Academy of Sciences in 1916 to associate the broad community of science and technology with the Academy's purposes of furthering knowledge and advising the federal government. Functioning in accordance with general policies determined by the Academy, the Council has become the principal operating agency of both the National Academy of Sciences and the National Academy of Engineering in providing services to the government, the public, and the scientific and engineering communities. The Council is administered jointly by both Academies and the Institute of Medicine. Dr. Bruce M. Alberts and Dr. Wm. A. Wulf are chair and vice chair, respectively, of the National Research Council

www.national-academies.org

COMMITTEE ON WATER QUALITY IMPROVEMENT FOR THE PITTSBURGH REGION*

JEROME B. GILBERT, *Chair*, J. Gilbert Inc., Orinda, California
BRIAN J. HILL, Pennsylvania Environmental Council, Meadville†
JEFFREY M. LAURIA, Malcolm Pirnie, Inc., Columbus, Ohio
GARY S. LOGSDON, Black & Veatch Corporation (retired), Cincinnati, Ohio
PERRY L. MCCARTY, Stanford University, Stanford, California
PATRICIA MILLER, Tetra Tech, Inc., Cincinnati, Ohio
DAVID H. MOREAU, University of North Carolina, Chapel Hill
NELSON P. MOYER, The CADMUS Group, Inc., Iowa City, Iowa
RUTHERFORD H. PLATT, University of Massachusetts, Amherst
STUART S. SCHWARTZ, Cleveland State University, Cleveland, Ohio
JAMES S. SHORTLE, Pennsylvania State University, University Park
JOEL A. TARR, Carnegie Mellon University, Pittsburgh, Pennsylvania
JEANNE M. VANBRIESEN, Carnegie Mellon University, Pittsburgh, Pennsylvania
PAUL F. ZIEMKIEWICZ, West Virginia University, Morgantown

NRC Staff

MARK C. GIBSON, Study Director
DOROTHY K. WEIR, Senior Project Assistant

---

* The activities of this committee were overseen and supported by the National Research Council's Water Science and Technology Board (see Appendix D for listing). Biographical information on committee members and staff is contained in Appendix E.

† Mr. Hill resigned from the committee in May 2004 after accepting a position in the Policy Office of the Governor, Harrisburg, Pennsylvania.

Preface

The City of Pittsburgh is located in western Pennsylvania and sits astride the Monongahela and Allegheny Rivers, at the junction of the Ohio River. These rivers have played critical roles in the city's history. Although utilized by various Native American tribes, the region moved into a different stage when it became a place of strategic significance in mid-eighteenth century struggles between the French and British empires, and soon after between European colonists and native tribes. Later, in the nineteenth century, the region became a "gateway" to the Ohio valley and the American West. Although it initially thrived as a commercial center, the region's resources and locational advantages moved it rapidly toward industrial production, especially of iron products.

By the time of the Civil War, Pittsburgh had become a prosperous industrial city with a population of more than 50,000, surrounded by other industrial towns. Cheap energy from the easily exploited Pittsburgh coal seam and the activities of many notable entrepreneurs, inventors, and venture capitalists made Pittsburgh one of the nation's leading manufacturing centers by the turn of the twentieth century. For the first quarter or so of the 1900s, the city and region enjoyed a booming industrial economy, including creation of an extensive urban infrastructure and vibrant cultural institutions. Industrial growth, however, came at a high environmental cost, with degradation of air, water, and land resources.

After its boom industrial years, the region suffered a number of severe shocks, including the great flood of 1936, the economic impacts of the Depression, and further deterioration of its environment and infrastructure because of intensive wartime manufacturing demands during World War II. At the end of that war however, under the leadership of Richard King Mellon, Mayor David Lawrence, and the Allegheny Conference on Community Development, the city embarked on its first so-called "Renaissance." This included redevelopment of downtown Pittsburgh, creation of a new highway system, and environmental improvements such as smoke control and construction of a major sewage treatment plant serving most of Allegheny County. In addition, city elites and politicians successfully lobbied for the construction of several U.S. Army Corps of Engineers flood control dams in the upstream watersheds that have significantly reduced flood risks on the three major rivers.

Between 1978 and 1983, however, the steel industry—the prime component of the region's industrial economy—folded, and more than 100,000 manufacturing jobs were lost. In spite of the collapse of steel, the ever-resilient region embarked in the 1980s on "Renaissance II," concentrating on both downtown and neighborhood improvements. Regional adjustments and new endeavors continued through the 1990s, as Pittsburgh reinvented itself, creating an economy featuring high technology, medical research, institutions of higher education, and other enterprises to replace the heavy industry of its past.

The "reinvented" Pittsburgh region recognizes the importance of clean water and other natural resources, but it must confront myriad issues and problems across southwestern Pennsylvania and therefore can ill afford to approach such problems inefficiently. As part of a proactive effort to strategically address the region's water quality and related problems, this National Research Council (NRC) study was commissioned by the nonprofit Allegheny Conference on Community Development (ACCD), an institution involving leaders from industry, government, and academia that has for decades brought intellectual and political power to address economic and other issues of the region.

To undertake the study that resulted in this report, the NRC's Water Science and Technology Board (WSTB) formed the Committee on Water Quality Improvement for the Pittsburgh Region. The committee carried out an independent assessment of the wastewater and water quality problems of the region and has made recommendations on how these issues and needs can best be addressed. The charge to our committee (see statement of task, Box ES-1) was based on regional needs, as identified by project sponsors and other regional experts and interests and negotiated with the WSTB. At the outset, these needs were considered and the committee surveyed available data to see if they would support detailed answers to the questions posed in the statement of task. In general, the committee's analyses were constrained by data limitations—ranging from concentrations and sources of main stem river bacteria to on-site waste disposal conditions—that did not allow it to provide specific technical recommendations, and such a level of prescription is not characteristic of the NRC in any case. However, an assessment of the data and information that do exist allowed the committee to recommend a comprehensive watershed-based approach and strategy for the region. We believe our report should help serve as a basis for developing a water quality improvement investment strategy to be pursued by the multiple jurisdictions on a cooperative basis. The committee also hopes that this report will be of interest to the U.S. Environmental Protection Agency and to other urban areas where a regional cooperative approach to water quality management would be beneficial.

The committee consisted of 14 volunteer experts in environmental and hydrologic engineering, public health and aquatic microbiology, watershed management, urban and regional planning, history, public policy, law, and economics. The committee was constituted to help generate multidisciplinary strategies for addressing regional wastewater and water quality problems and included members with experience in southwestern Pennsylvania and others with relevant expertise from throughout the nation. The committee consulted with the study sponsors, the public, and members of a "resource panel" that included representatives of a wide variety of local, regional, and state organizations concerned with the region's water quality (see Appendix A). That panel of regional experts was formed by the ACCD to assist the study and, especially, to respond to the committee's requests for information. This report's conclusions and recommendations are based on a review of relevant technical literature; information gathered at seven committee meetings; a public stakeholder workshop held at Carnegie Mellon University on July 8, 2002, in conjunction with the first committee meeting; and the collective expertise of committee members.

I would like to thank the members of this committee for dedicating their time and expertise in addressing the water quality problems of southwestern Pennsylvania. The committee was guided in the generation of this report by Stephen D. Parker, director of the WSTB, and Mark C. Gibson, study director and WSTB senior staff officer. Mark set the pace and agenda for the study, helped the committee maintain focus on the study tasks, and ensured compliance with NRC policies. Assisting Mark and Steve in these efforts was Dorothy K. Weir, who as our project assistant was responsible for meeting logistics, research assistance, report preparation, and editorial tasks. The committee members had the benefit of information from a wide range of the members of the resource panel mentioned above and other concerned residents of the region—especially that gained at the public stakeholder workshop. It is particularly indebted to Jared L. Cohon, president of Carnegie Mellon University who chaired the Western Division of the Pennsylvania Economy League's Southwestern Pennsylvania Water and Sewer

Infrastructure Project Steering Committee that produced the 2002 report *Investing in Clean Water* and who initially pursued this NRC study. He, along with ACCD officials Harold Miller, Jan Lauer, and Joshua Donner, and John W. Schombert, Executive Director of the Three Rivers Wet Weather Demonstration Program, deserve particular credit and appreciation for their participation and considerable assistance throughout this study. The committee also thanks Jiayi Li and Sherie Mershon of Pennsylvania State University and Carnegie Mellon University, respectively, for their graphics and research contributions to this report.

This report was reviewed in draft form by individuals chosen for their diverse perspectives and technical expertise in accordance with procedures approved by the NRC's Report Review Committee. The purpose of this independent review is to provide candid and critical comments that will assist the institution in making its published report as sound as possible and to ensure that the report meets institutional standards for objectivity, evidence, and responsiveness to the study charge. The review comments and draft manuscript remain confidential to protect the integrity of the deliberative process. We wish to thank the following individuals for their review of this report: Daniel P. Loucks, Cornell University; William V. Luneburg, Jr., University of Pittsburgh School of Law; James M. McElfish, Jr., Environmental Law Institute; William J. Miller, consulting engineer, Berkeley, California; Max J. Pfeffer, Cornell University; Larry A. Roesner, Colorado State University; Mary W. Stoertz, Ohio University; and Marylynn V. Yates, University of California, Riverside.

Although the reviewers listed above have provided many constructive comments and suggestions, they were not asked to endorse the conclusions or recommendations, nor did they see the final draft of the report before its release. The review of this report was overseen by Patrick R. Atkins of the Aluminum Company of America, New York, N.Y. Appointed by the NRC, he was responsible for making certain that an independent examination of the report was carefully carried out in accordance with institutional procedures and that all review comments were carefully considered. Responsibility for the final content of this report rests entirely with the authoring committee and the NRC.

Jerome B. Gilbert, *Chair*

Contents

Executive Summary

BACKGROUND

The City of Pittsburgh is located in southwestern Pennsylvania where the Allegheny and Monongahela Rivers meet to form the Ohio River (see Figures ES-1 and ES-2). These "Three Rivers" have been central to the history, economy, and identity of the region. Pittsburgh initially thrived as a commercial and transportation center in the mid-nineteenth century but soon transformed to a region characterized by a growing industrial sector with a specialty in metals production and major mining activity. This growth, however, came at a high environmental cost in terms of polluted air and water, which afflicted the Pittsburgh region for most of the twentieth century. By the 1980s, new laws, advances in technology, and the significant decline of the steel industry-based economy combined to reduce industrial air pollution. Water quality problems in the region, although lessened, have persisted.

Drainage from abandoned coal mines—typically a highly acidic solution bearing a large load of iron, either dissolved or precipitated as ferric hydroxide—is the source of significant residual water pollution in certain streams in southwestern Pennsylvania and can produce biologically "dead" waters. However, beginning in the late 1960s, state and federal legislation requiring the treatment of polluted water prior to discharge to local streams has reduced (but not eliminated) the widespread ecological impacts of this drainage.

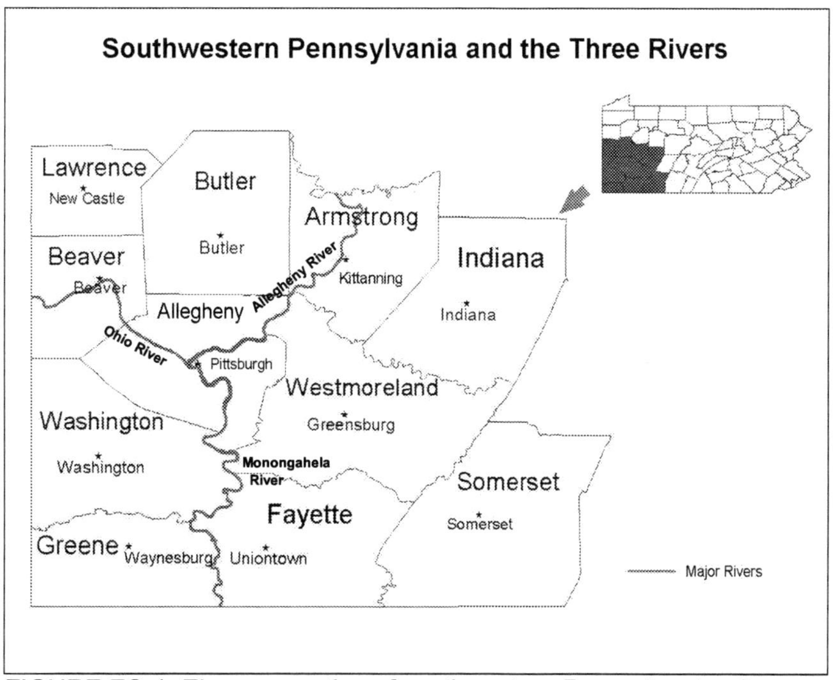

FIGURE ES-1 Eleven counties of southwestern Pennsylvania. See Figure ES-2 for a map of Allegheny County and the City of Pittsburgh.

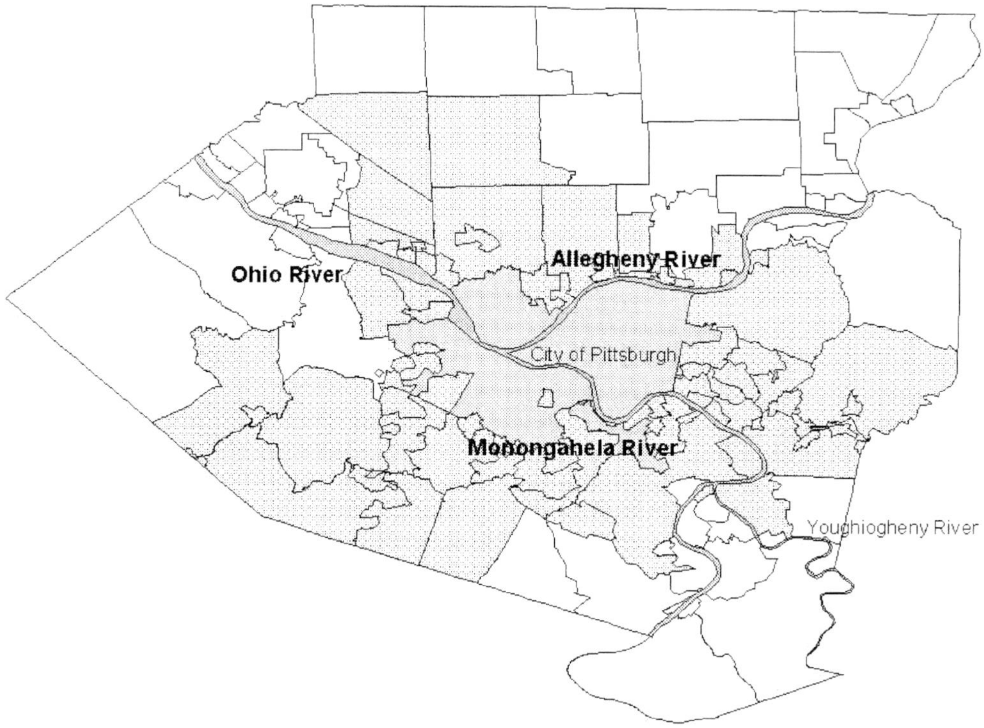

FIGURE ES-2 The Allegheny, Monongahela, and Ohio Rivers in Allegheny County in southwestern Pennsylvania; shaded areas include the 83 Allegheny County communities serviced by the Allegheny County Sanitary Authority (ALCOSAN), including the City of Pittsburgh.

Since the late 1950s, the development of sewage treatment plants throughout the region—the largest of which is operated by the Allegheny County Sanitary Authority (ALCOSAN) and serves the City of Pittsburgh and 82 other communities in Allegheny County (see Figure ES-2)—has alleviated downstream pollution in the Ohio River from the municipal sewers that previously discharged directly to local waterways. Yet releases of untreated sewage and surface runoff, especially on wet weather days and due to failing sewers, continue to degrade the quality of waters and impair their value for habitat, recreation, and water supply. Sewage-related water quality problems also persist in dry weather because of aging and deteriorating on-site sewage treatment and disposal ("septic") systems and sewage pipes that may be a significant source of contamination to groundwater supplies. These problems threaten the region's public health, environment, economy, and image. For example, there has been a steady rise in the last decade in the number of days of the summer recreational season that the Allegheny County Health Department (ACHD) has issued river advisories (i.e., when rainfall in the region is great enough to potentially cause sewer overflows and lead to excessive levels of bacterial indicator organisms[1]) that recommend restricted recreational contact exposure. Indeed, the City of Pittsburgh, ALCOSAN, and other communities in the region face extensive and costly regulatory

[1] Because it is impractical to test waters for all possible pathogenic microorganisms, the microbial quality of water is often assessed through the use of indicator microorganisms (usually bacteria). Although such fecal indicator bacteria are generally not pathogenic, they provide estimates of the amount of feces and, indirectly, the presence and quantity of fecal pathogens in water (see Chapter 3 for further information).

action under the federal Clean Water Act for both combined sewer overflows (CSOs) and sanitary sewer overflows (SSOs) resulting from wet weather conditions.

The costs of water and sewer infrastructure improvements necessary to address the release of untreated or inadequately treated sewage into the region's surface waters are uncertain; however, based on investments made in other cities and on national studies, significant investment is expected to be needed to rehabilitate and upgrade aging municipal infrastructure and replace failing on-site systems. Meeting such costs is expected to be difficult given the economic climate of the region. Although the region's economic base has shifted in the last two decades from mining and traditional manufacturing to other sectors, many communities in southwestern Pennsylvania continue to experience significant economic weakness as reflected in population decline (in both the central city and its metropolitan area), unemployment rates, and other indicators such as poverty level and income (see Chapter 2 for further information). Unlike the 1970s and 1980s, little federal and state assistance is available for the development and expansion of major water supply and wastewater facilities, and even that is likely to be restricted to extreme situations and poor communities. The costs of these improvements must be considered in the context of the potential costs of inaction, which would include adverse impacts on public health, the environment, and economic growth, and possible further federal and state regulatory action or private lawsuits by concerned parties.

This report was written by the Committee on Water Quality Improvement for the Pittsburgh Region overseen by the National Research Council (NRC)'s Water Science and Technology Board. The committee was formed in 2002 at the request of the Allegheny Conference on Community Development (ACCD) to conduct an independent assessment of the wastewater and water quality problems of the Pittsburgh area in southwestern Pennsylvania and to make recommendations on how these issues and needs of the region can best be addressed by multiple jurisdictions on a cooperative basis. The study goals combined to create a framework of guidance and recommendations to help make water quality improvement-related investments. The committee's statement of task is included in Box ES-1. This report's content, conclusions, and recommendations reflect the collective expertise and consensus of the committee. Detailed conclusions are contained in the individual chapters and in chapter summaries. The report's principal recommendations pertaining to water quality improvement in southwestern Pennsylvania are contained in Chapters 5 and 6.

WATER QUALITY AND CAUSES OF IMPAIRMENT

The condition of waterbodies across the United States is determined by comparing certain measured physical, chemical, and biological parameters within those waters to state water quality standards. Each water quality standard consists of two primary and distinct parts: (1) designated beneficial use(s) of the waterbody (e.g., aquatic life support, drinking water supply); and (2) narrative and numeric water quality criteria for biological, chemical, and physical parameters that measure attainment of designated use(s). Waterbodies can be impaired for any of their designated uses by a variety of contaminants. It is important to note that inadequacies in the type and extent of water quality data available in southwestern Pennsylvania prevented the committee from assessing the full extent of adverse effects due to pollution. Almost all of the water quality data available to the committee during this study were derived from single studies in specific areas for limited durations. Recently, several agencies have expanded water quality data collec-

BOX ES-1
Statement of Task

The NRC will establish an expert committee to undertake an assessment of the wastewater and water quality problems of the Pittsburgh, Pennsylvania area and make recommendations on how these issues and needs of the region can be best addressed by the multiple jurisdictions on a cooperative basis. The study will address several key questions, including:

1. What are the region's most pressing wastewater and water quality problems and what management and infrastructure development strategies (including consideration for relevant emerging technologies) might be pursued to most effectively address them? For example, what criteria might be applied to compare the impacts of combined sewer overflows and failing septic systems?
2. How should water quality data be used to most effectively inform priority-setting and aid decision-making for infrastructure investments in the Pittsburgh region? For example, what conclusions can be drawn about the relative contributions that sewage overflows, septic tank failures, and other point and nonpoint sources of pollution are making to surface water contamination in the region based on the water quality data that is currently being collected (or that could be collected through a special, short-term effort)?
3. What are the best approaches and cost-effective means to monitor and assess the impact of wastewater discharges on the region's water quality? What established, innovative, and emerging techniques can be used to assess or predict the public health, environmental, and economic impacts of the region's current and future wastewater discharges?
4. What is the reliability with which predictions on improvements in water quality will result from actions taken in wastewater management? What monitoring and modeling activities are appropriate to understand the links between actions and improvements in wastewater management systems for a complex watershed that includes multiple political jurisdictions and resultant water quality benefits?
5. What are the best strategies to encourage public awareness and regional cooperation between municipalities and disparate organizations to address the pervasive water quality problems? What models from other regions of the country might be applied to the Pittsburgh region? Conversely, what lessons can be learned from the Pittsburgh region and applied elsewhere?

tion in the region, although there appears to be little coordination of these activities. Therefore, it is difficult to fully identify the sources of pollution that cause these impairments, to assess the extent of adverse effects, and to prioritize remediation efforts.

Surface waters in southwestern Pennsylvania are impaired for several uses including contact recreation due to the presence of indicator microorganisms in excess of levels expected to cause human illness; fish consumption due to organic and inorganic chemicals known to bioaccumulate in fish tissue and to represent a human health risk; and aquatic habitat due to metal concentrations and low pH that alter ecosystems and can harm aquatic organisms. Statewide, the committee found that the major causes of water quality impairment in the Commonwealth of Pennsylvania are the following: (1) acid mine drainage, (2) agriculture, (3) urban and stormwater runoff, and (4) human waste handling.

Improperly managed wastewaters resulting from various human activities are degrading the microbiological water quality in the region, although available data are not sufficient to determine the relative contribution of different sources to surface and groundwaters. More specifically, wet weather biological water quality in the main stem rivers is demonstrably worse than that in dry weather, suggesting that stormwater and sewer overflows (CSOs and SSOs) may be important contributors. Furthermore, water quality in many tributaries does not meet biological standards in either wet or dry weather conditions, suggesting that failing septic systems may be important contributors. However, regional waters are not considered impaired for use as sources of drinking water because of the extensive treatment that is routinely performed on drinking water sources, particularly surface water sources. Although groundwater used for public drinking supplies generally meets water quality guidelines, private wells[2] show significant variability in terms of microbial contamination, and the effects of mining are apparent in some areas. There is no evidence that southwestern Pennsylvania has recently experienced any waterborne disease outbreak that would link impaired source water quality with human health effects. However, as with water quality data, significant gaps exist in public health monitoring, thus preventing an adequate assessment of possible endemic waterborne disease occurrences.

The contribution of agriculture to pathogen loading in rural areas of southwestern Pennsylvania could not be determined, but this is a well-known pathogen source in many parts of the nation, and many livestock management practices in southwestern Pennsylvania are likely to contribute pathogens to streams. Relative nonpoint contributions of human and nonhuman pathogen sources in both urban and rural watersheds are not known.

Acid mine drainage is a significant cause of water quality impairment in the region, predominately affecting streams and tributaries. This regional water quality issue extends beyond Pennsylvania to encompass much of the Appalachian Range. Presently, acid mine drainage is being addressed by multiple jurisdictions including federal and state programs, and continued public funding to combat this water pollution problem is essential to future water quality improvement in southwestern Pennsylvania.

WATER QUALITY IMPROVEMENT: DECISION-MAKING STRATEGIES AND TECHNICAL SOLUTIONS

From a regulatory perspective, the most important water quality problem in the region in terms of the potential for adverse human health effects is controlling microbial contamination of streams that derives from the effect of wet weather conditions on sewer systems (CSOs, SSOs, and stormwater), failing septic systems, and agricultural and urban runoff. The U.S. Environmental Protection Agency (EPA) has adopted regulations requiring CSO and SSO controls and issued consent orders through the Pennsylvania Department of Environmental Protection (PADEP) and the ACHD to many ALCOSAN partner communities (see also Figure ES-2) to address this pervasive wet weather problem through increased attention to centralized sewer systems. A similar consent decree for ALCOSAN with EPA is pending and expected to be finalized soon. The evaluation of water quality improvements related to such remedial activities will be critical. However, the implementation of solutions for identified sources of impairment does not preclude the need for additional information related to other sources and their contributions to water quality impairment in the region. To develop better understanding of sources of contamination in southwestern Pennsylvania, water quality monitoring and modeling efforts should take place concurrently with mandated remedial activities.

It is clear that the causes and nature of water quality impairments, the parties responsible, and the individuals and waterways affected differ for each of the problem contaminants in the region. A comprehensive watershed-based approach is needed to address the spectrum of water

[2] Private well-owners are not currently required by state or federal regulations to monitor for contaminants or treat their drinking water.

quality problems, including wet weather problems; such a systematic approach should recognize interrelationships among problems and the need for parties responsible for each water quality problem to share in its solution. The technical approach is embodied in what the committee calls the "Three Rivers Comprehensive Watershed Assessment and Response Plan," or CWARP.

The framework recommended for planning and implementation of CWARP consists of the following five basic steps: (I) problem identification; (II) assessment of existing conditions including quantification of loads and modeling their relationships to water quality; (III) projection of future loads and their timing, location, and impacts on streams; (IV) formulation and evaluation of alternative management strategies, including assessment of the effects of alternatives on future conditions and the preferential ordering and scheduling of various elements of the preferred strategy; and (V) adaptive implementation of elements of the strategy, relying on feedback from implementation of each element to provide the basis for continued planning of subsequent elements. This five-step CWARP process must be adapted to address planning and management needs at the following four interrelated scales: (1) river basin, (2) multicounty/metropolitan scale, (3) high-density urban areas, and (4) rural areas.

The committee recognizes that the region is not starting with a blank slate, and Step I has been largely completed for each of these scales. Substantial progress has been made on Step II, but significant gaps remain. Because the problems are largely associated with existing conditions and there is only modest growth in the region as a whole, Step III may be less important, but changes in land use that are occurring in suburban (formerly rural) areas cannot be ignored. Lastly, Steps IV and V do not appear to have been well developed in any respect and thus deserve much greater attention.

Because regional information on the biological quality (see Box 5-2 for further information) of receiving waters is scant, its collection during and in support of CWARP at the river basin scale is critical. Information collection for CWARP should include biological data to both assist in ecosystem health assessment benchmarking and to help document changes to the ecosystem that occur as a result of changing stressors. To this end, an effort should be made to expand the Ohio River component of the rejuvenated Great Rivers program of EPA's Environmental Monitoring and Assessment Program with an emphasis on the biological water quality of the main stem rivers.

At least two aspects of water management are of concern at the multicounty/metropolitan scale of CWARP. First, and at the very least, water quality planning at this scale should be sufficient to inform regional interests of the potential effects (including constraints, if any) of water quality conditions on future transportation and land development, the consequences of development on water quality where it occurs, and how those effects and consequences can and should be modified. Second, planning at this scale should also result in the identification of opportunities for economies of scale in the delivery of water and wastewater services through cooperative arrangements among local governments. The Southwestern Pennsylvania Commission (SPC) or an alternative organization should formulate regional water resource plans and integrate them with transportation and land use plans.

Several entities have estimated recently that solving wet weather problems in the urban core of the region by conventional means, using a combination of storage, conveyance, and treatment improvements, could cost several billion dollars. Investing large sums of capital based only on currently available data may not ultimately solve the most important problems or provide appropriate solutions. Although it is true that no amount of additional data and analyses would remove all uncertainty about water quality investments, it is clear that currently available information is lacking in several critical respects (e.g., how much surface water runoff from separate stormwater sewers affects water quality in receiving streams during wet weather events). Until these facts are known better, planning and implementation of cost-effective remedial measures will be impeded.

Whereas receiving water quality modeling activities appear to be extremely limited currently in the region's three main stem rivers, they should be used to estimate impacts of pollution loadings on the receiving streams and to help prioritize alternatives for pollution control. Other modeling activities required to implement CWARP in the region's urban core include sewer system routing models, dynamic sewer system modeling, dynamic stormwater modeling, and real-time sewer flow control modeling for analysis and operation. Projections of changes in the regional landscape are important in the planning and implementation of CWARP in the region's urban core. Planning studies conducted at the multicounty/metropolitan scale should be sufficient for this purpose and include projections for several land use, transportation, water supply, and wastewater parameters.

The first route to successful improvement of water quality in the region is to optimize utilization of existing infrastructure. To this end, the committee strongly recommends that all wastewater collection systems located in the watershed, particularly in the region's urban core, be fully compliant with EPA's Capacity, Management, Operations, and Maintenance (CMOM) policy or an equivalent program. Thereafter, related information, approaches, and technologies recommended in this report would be available to help guide major longterm investments in improving the region's water quality. Furthermore, ALCOSAN's draft long term control plan (LTCP) for controlling CSOs, which was drafted in 1999 and was the subject of an extensive third party review in 2001 through 2002, should be reevaluated in the context of the overall CWARP approach to reflect ongoing consent order negotiations, CMOM, and information from CWARP as it is developed in the future.

The CWARP approach is recommended as a framework for development of the LTCP and similar documents because of the circumstances (especially data limitations) that exist in southwestern Pennsylvania and, in principle, would apply in other regions of the United States with similar water quality problems and circumstances. In addition, in the development of a final LTCP, ALCOSAN and other wastewater treatment providers in southwestern Pennsylvania should evaluate the utilization of real-time control of CSOs. Storage and treatment of CSOs in nearby abandoned mine voids, which is being evaluated for the Township of Upper St. Clair, Pennsylvania, should also be evaluated. Also recommended is consideration of several innovative approaches and technologies to determine what, if any, role they may have for improving water quality in southwestern Pennsylvania—especially in the region's urban core areas.

Best management practices for septic systems should be implemented throughout the region using the CWARP framework. Although individual systems are permitted locally, and current technical standards are available to ensure proper performance, they may be ignored. Furthermore, prevention of the discharge of untreated sewage into local waterways or ditches is difficult to enforce. The region needs a coordinated, well-funded program for oversight and routine maintenance of cluster and individual septic systems. Such a program can be self-sustaining through user charges providing they are applied on a cooperative regional or county basis. Several actions to help improve water quality in the predominantly rural areas of southwestern Pennsylvania are discussed and recommended for consideration.

There are no comprehensive estimates of the economic benefits of addressing the remaining water quality problems for southwestern Pennsylvania or from proposed projects to address the region's water quality problems. Nevertheless, the region would be expected to benefit economically from measures that significantly reduce drinking water risks and enhance recreational opportunities. The CWARP process can identify a list of alternative management strategies and projects that are technically feasible and capable of addressing the region's water quality problems at a variety of scales, but the question remains: Which is the better option? A variety of economic evaluation frameworks are available; some of the more prominent are discussed in this report, including cost-effectiveness analysis, benefit-cost analysis, and multicriteria methods. In this regard, the use of cost-effectiveness as the primary method for evaluating options for achieving water quality objectives in the region is recommended and should include an analysis of incremental costs to achieve elimination of low-probability contamination events. The committee recommends the use of benefit-cost analysis in the evaluation of water quality improvement projects in the region and in helping to set priorities among them.

As the CWARP process is being planned and implemented, it is essential that it be integrated with the ongoing process of establishing total maximum daily loads (TMDLs) for impaired streams being conducted by PADEP under requirements of the federal Clean Water Act. Lastly, the recommended CWARP effort should be completed quickly to provide timely support for those water quality improvements that are required and others in the public interest. It is difficult to estimate the cost of implementing CWARP, but in the committee's judgment it should be small compared to the cost of improvements and more than offset by potential savings.

WATER QUALITY IMPROVEMENT: INSTITUTIONAL AND FINANCIAL SOLUTIONS

Water planning issues in southwestern Pennsylvania need to be addressed on a regional and holistic basis, taking into account water quality; water supply; flood hazard mitigation; aquatic and riparian habitat protection and restoration; and recreation. Moving toward regionalization will be challenging because water resource and quality management in southwestern Pennsylvania currently is highly fragmented among federal and state governments as well as 11 counties, 595 municipalities, and 492 water and sewerage providers. In choosing an appropriate organization or set of organizations to address these concerns, the following three factors should be considered: (1) water resource management functions for which improvements are necessary or desirable; (2) the level of government or private sector enterprise to which management functions should be entrusted and to which legal authority should be delegated by the legislature; and (3) the geographic scale that is appropriate to achieve efficiency by exploiting economies of scale and making significant regional interdependencies internal to the planning area. Consistent with the recommended CWARP approach, changes are necessary to address the water resource issues of southwestern Pennsylvania at the following geographic scales: (1) river basins and interstate river basins and watersheds; (2) metropolitan region scale (multicounty areas); (3) metropolitan urban areas; and (4) rural areas. In addition, information that exists today, as well as that developed under elements of CWARP, should be made readily available to the public. This would include sources of water quality problems, their significance, appropriate solutions, costs, and their social impacts.

River Basin

Some water quality problems—particularly those related to long distance transport of pathogenic organisms, heavy metals, and persistent toxic chemicals—transcend regional, state, and political boundaries. At the largest scale of river basins, water monitoring and management is the responsibility of federal agencies (particularly the EPA, U.S. Geological Survey, and the U.S. Army Corps of Engineers) and the state (PADEP). The Ohio River Valley Water Sanitation Commission (ORSANCO) also conducts water monitoring and modeling for the Ohio River basin and its tributaries; ORSANCO and the PADEP are the appropriate agencies to establish the formulation of management strategies at the river basin scale.

Metropolitan Region

The Southwestern Pennsylvania Commission (SPC) is the primary organization for transportation planning and economic development at the multicounty regional scale. Those plans can significantly affect regional land use and water-related services. Concerns about land use and associated water supplies, wastewater disposal, and stormwater management should be incorporated in planning at that scale. The SPC is probably the region's best choice for carrying out this planning function, but its present representative structure and lack of water resource expertise limit its capacity to do so. Its regional databases on land use, transportation, and economic development are its strengths relative to water resource planning.

An important step that SPC, in coordination with ORSANCO, could take to broaden representation and advance public education on regional water resources would be to establish a Three Rivers Regional Water Forum as conceptually illustrated in Figure ES-3 (see also Box 6-4). The forum should be charged with a broad mandate to assess priorities for water infrastructure planning, maintenance, and construction as those activities are related to regional transportation, land use, economic development, and current infrastructure capacity and condition. The forum should include elected and appointed officials of local governments, regional leaders in the private sector, academia, environmental organizations, and other nongovernmental organizations, and participation should be encouraged by all organizations that share some responsibility for the proposed CWARP. Although there are several options for the creation and organization of a regional water forum, an unincorporated network of public and private stakeholders established by voluntary memoranda of understanding is recommended for careful consideration. However, the participants and exact organization plan should be determined locally.

High-Density Urban Areas

As stated previously, the primary water quality management problem in the Pittsburgh region's urban core is periodic discharges of untreated wastewater from combined and separate sanitary sewers. Continued fragmented management of the sewer collection-conveyance-treatment system (i.e., maintaining the status quo) is not a satisfactory situation. Rather, planning and management of sanitary and combined sewers should be integrated with stormwater management. At least five viable organizational arrangements, represented by the

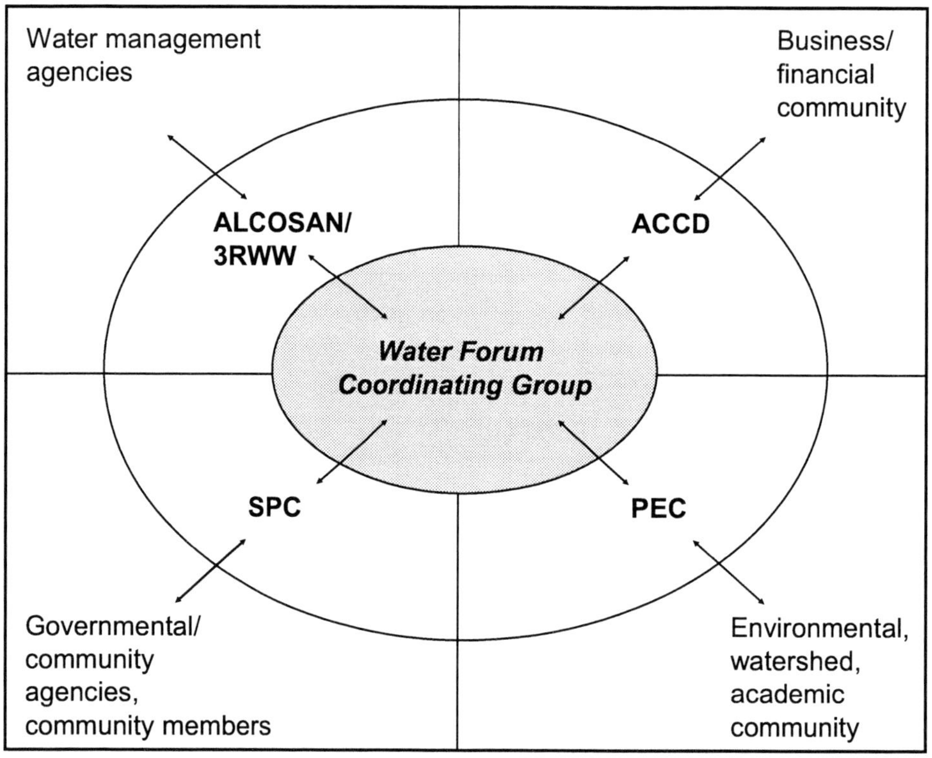

FIGURE ES-3 Concept diagram for a Three Rivers Regional Water Forum. Coordination of the forum would be provided by a group that represents major nongovernmental organizations, local, state, and federal government stakeholders; and regional academic experts, among others. NOTE: 3RWW is the 3 Rivers Wet Weather Demonstration Program; PEC is the Pennsylvania Environmental Council (see Chapter 6 for further information).

following options, could serve that purpose, including (A) merger of City of Pittsburgh and neighboring municipalities with Allegheny County; (B) establishment of county-wide management either by dedication of the systems to Allegheny County or through an administrative arrangement with Allegheny County using authority under Pennsylvania Acts 67 and 68; (C) creation of one or more special districts for sewer management; (D) expansion of the role of ALCOSAN to include sewer collection systems, with or without authority over stormwater management; and (E) continuation of the decentralized system but with performance standards and voluntary participation in a regional maintenance organization (RMO) provided on a fee-for-service basis. ALCOSAN would be encouraged to establish the RMO.

All five options are viable, and discussions on merger of services between Allegheny County and the City of Pittsburgh have already occurred. A merger of city and county government, although politically difficult, is desirable from the perspective of water quality management as it would create the potential for implementing CWARP over a significant portion of the urban core. However, such a merger would be much more effective if it included other large municipalities that generate stormwater, sanitary sewage, or combined sewage that flows through the high density core of the region. Allegheny County should take a leadership role in search of a consensus on one of the four remaining options. The committee also prefers Option

B (establishment of county-wide management) to Options C, D, or E because it captures economies of scale in planning and management, facilitates the use of a systems approach, and keeps decision making closer to politically accountable public officials.

ORSANCO, with its prior experience with similar problems in the Ohio River basin, can be of valuable assistance in reaching a consensus on all of the preceding options. The 3 Rivers Wet Weather Demonstration Program (3RWW) should be continued or expanded to conduct public education programs for stormwater and CSO management; to provide technical assistance to local governments for stormwater and CSO management; to provide education to local government on identifying and correcting illegal connections to sewer systems; and to monitor, analyze, and report on the status of stormwater and CSO management in Allegheny County.

Rural Areas

Additional steps are also needed to address water supply and wastewater disposal systematically in rural and small urban areas outside of the region's urban core. The actions recommended in Chapter 5 to address septic system deficiencies should be undertaken cooperatively by several agencies. At the state level, the Watershed Restoration Action Strategies program should be expanded to include assessment of effects of inadequate wastewater disposal on water quality. In doing so, PADEP should work closely with local governments having legal authority over such systems. The SPC could and should take strong leadership in bringing local governments together to address these issues. In addition to PADEP, SPC should request assistance from EPA and nongovernmental organizations having prior experience with programs of this kind. The Allegheny County experience in these activities should provide a sound foundation for other counties in the region.

Financing

The following actions are recommended regarding a framework for regional financing of water quality improvements in southwestern Pennsylvania:

- Develop and implement a sewer and/or water user surcharge to fund at least the first few years of planning and data gathering under CWARP or a similar program. Ideally, the charge would be in addition to wastewater/water bills throughout the basin or, as a minimum, in the region's urban core.
- Initiate a flow-based repayment system for ALCOSAN and other regional wastewater treatment providers that reflects, to the extent practicable, the actual contributions of flow into sewerage systems.
- Select one or more forms of regional governance that have the necessary legal authority and administrative expertise to finance capital improvements and operating and maintenance expenses of management programs. Such authorities should include the power to incur debt for capital projects, establish user charges, and collect revenues necessary to pay for all expenses except those financed by intergovernmental grants.
- Continue efforts to increase regional assistance through PENNVEST (Pennsylvania Infrastructure Investment Authority) and other sources of funds that can generate support for

specific programs, such as development of county-based management programs for on-site septic systems and acid mine drainage control.

- To the extent that assistance is not available, continuing studies are needed regarding the efficient application of current local taxes and user charges to cover start-up efforts identified above, with the goal of creating repayment mechanisms based on an equitable regional user charge system. Ultimately the system would generate sufficient revenues to repay debt obligations that will be necessary to fund priority facilities.

IMPLICATIONS BEYOND SOUTHWESTERN PENNSYLVANIA

During the course of this study of water quality improvement in southwestern Pennsylvania, the committee gained knowledge and insights on several matters that have broader implications and, in the committee's judgment, might be considered useful by others responsible for national efforts to protect and enhance water quality. These are discussed in Chapter 7 of this report in the following areas: (1) information systems, (2) health and ecological impacts of water quality, (3) potential federal policy conflicts with regional optimization, (4) stakeholder representation and participation, (5) paying for water quality improvements, and (6) regionalization and cooperation.

SUMMARY

As this report makes clear, water quality problems in southwestern Pennsylvania are complex and region wide. Many of southwestern Pennsylvania's current and most pressing water quality problems, such as those attributable to sewer overflows and stormwater, can be traced to historical water supply and wastewater infrastructure decisions made by individual municipalities at a time when today's population and economic and industrial climate could not have been foreseen. Other problems, such as acid mine drainage, are a legacy of the region's past heavy mining and manufacturing economy. Ongoing remediation activities and those planned to address wet weather-related problems for the mostly urban ALCOSAN service area may not be optimal (in terms of either effectiveness or economics) and, in any case, are not designed to address the full set of problems in the 11-county region or the Allegheny and Monongahela River basins. Furthermore, because of the paucity of data, it is not possible at present to make reliable predictions of water quality improvements that will result from such investments. Indeed, as stated earlier, the limited data available provide no evidence that southwestern Pennsylvania has recently experienced any waterborne disease outbreak that would link impaired source water quality with adverse human health effects.

The committee concludes that the interrelated water quality problems of southwestern Pennsylvania must be confronted on a regional basis and in a systematic way. Such an approach should improve public awareness of the issues and promote regional cooperation through the involvement of key stakeholder groups with an interest in water quality improvement. In this regard, one or more regional decision-making authorities should take responsibility for leading the development of a Comprehensive Watershed Assessment and Response Plan that would have as its principal objective the meeting of water quality standards throughout the region in the most cost-effective manner. A first step in determining effective infrastructure investment and management strategies for water quality improvement and an integral part of the CWARP is to undertake coordinated basin-wide monitoring (including biological monitoring) and modeling to estimate the amounts and relative impacts of various sources of pollutants entering the region's surface and groundwater. This is critical to ensuring that remediation efforts are appropriately targeted to the most important sources of pollution and that limited funds for remediation are spent on the highest-risk problems. The southwestern Pennsylvania experience is repeated to a greater or lesser extent around the United States, and the solutions suggested in this report relating to cooperation and regionalization are widely recognized as having national implications and benefits. Thus, the program recommended herein for water quality improvement in southwestern Pennsylvania can serve as a model for other regions. Lastly, effective implementation of this report's recommendations regarding water pollution reduction is not intended to delay current progress in regional water quality improvement.

1

Introduction

BACKGROUND

The City of Pittsburgh and surrounding counties of southwestern Pennsylvania have long suffered from air and water quality degradation. Until the 1950s, the skies were darkened by soot and smoke from bituminous coal burning by residences, businesses, and railroads, as well as industrial air pollution. Acid mine drainage, raw sewage, and untreated industrial waste routinely entered local streams and the region's three major rivers—the Allegheny, the Monongahela, and below their confluence, the Ohio (see Figure 1-1).

A smoke control law implemented in 1946, accompanied by a replacement of coal with natural gas for space heating and diesel electric for railroad locomotives in the 1950s, sharply reduced soot smoke pollution. Industrial fumes and gases, however, produced primarily by the iron and steel industry, continued to pollute the air into the 1970s when state and federal clean air laws combined with the closing of the steel mills reduced industrial air pollution (Mershon and Tarr, 2003). Water quality problems persist, however, despite the operation of a large sewage treatment plant constructed in 1958 by the Allegheny County Sanitary Authority (ALCOSAN) and other plants constructed by smaller municipalities. Combined sewer and separate sewer overflows, failing septic systems, untreated discharges from "straight pipes," stormwater, agricultural runoff, and acid mine drainage continue to degrade the quality of local streams and impair their value for habitat, recreation, and water supply.

Annually, an estimated 16 billion gallons of mixed rainwater and sewage is introduced into the region's waterways from combined sewer overflows (CSOs) and sanitary sewer overflows (SSOs) during wet weather in the ALCOSAN service area (WSIP, 2002). The ALCOSAN service area has 328 CSO structures from which untreated sewage is discharged into local streams during wet weather—more than any similar authority in the country (WSIP, 2002; see Chapter 4). Chapter 2 traces many of the region's current urban water quality problems to water supply and wastewater infrastructure decisions made by and for the City of Pittsburgh and outlying areas in the past. Currently, the City of Pittsburgh, ALCOSAN, and 82 communities served by ALCOSAN, as well as some not served by ALCOSAN, face regulatory sanctions under the federal Clean Water Act (see Box 1-1) for sewer overflows.[1]

---

[1] Consent orders were issued in October 2003 by the Allegheny County Health Department to 26 ALCOSAN-served communities to reduce their numbers of precipitation-triggered CSOs. Similarly, 55 communities with illegal sanitary sewer overflows were issued consent orders through the Pennsylvania Department of Environmental Protection in November 2003 to determine where their sewers are leaking as a first step to eliminating these discharges. The City of Pittsburgh and the Pittsburgh Water and Sewer Authority signed a consent order and agreement in February 2004 to address CSO outfalls. A draft consent decree between the U.S. Environmental Protection Agency and ALCOSAN had not yet been signed as this report was nearing completion in December 2004 (see Chapter 5 for further information).

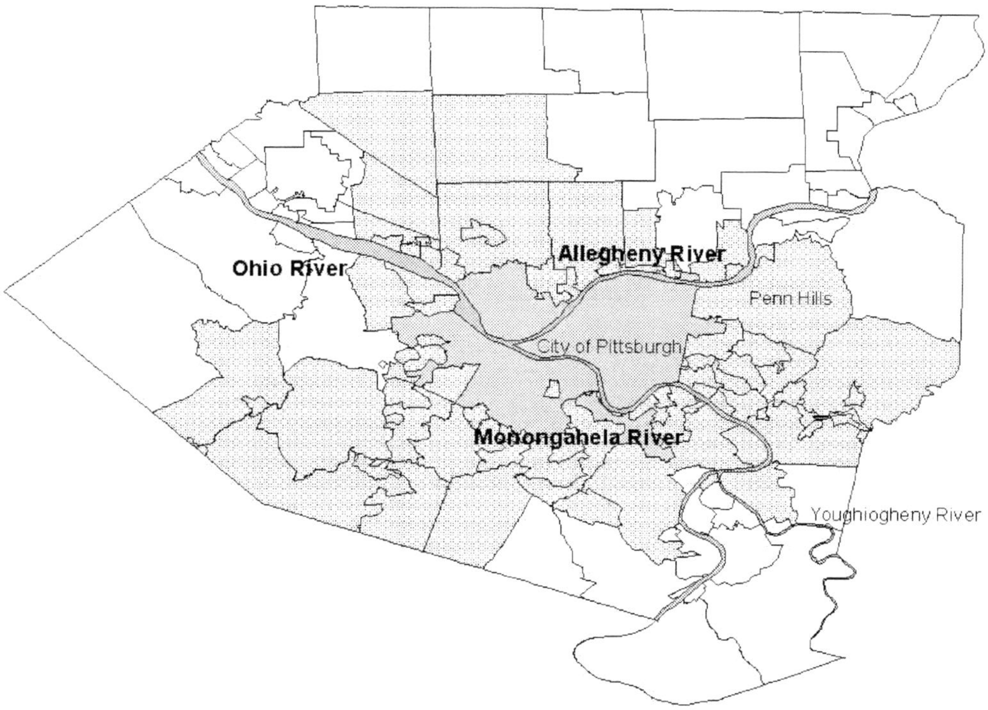

FIGURE 1-1 The Allegheny, Monongahela, and Ohio Rivers (the "Three Rivers") in Allegheny County in southwestern Pennsylvania; shaded areas includes Allegheny County communities serviced by ALCOSAN in addition to the City of Pittsburgh.

Untreated human waste, stormwater, and agricultural runoff may spread parasitic protozoa (e.g., *Giardia*, *Cryptosporidium*), enteric bacteria and viruses, and other waterborne contaminants (NRC, 2004). Such microorganisms and contaminants are public health threats particularly to children, the elderly, the immunocompromised, and other sensitive populations (Balbus et al., 2000; NRC, 2001, 2004).

For approximately the last quarter of a century, tests of water quality by the U.S. Geological Survey (USGS), the Ohio River Valley Water Sanitation Commission (ORSANCO), and other governmental, nongovernmental, and university groups have found that fecal coliform levels (bacterial indicators of fecal contamination) have repeatedly been in violation of water quality standards at certain monitoring stations on the Monongahela and Allegheny Rivers, especially during wet weather events (WSIP, 2002; see Chapter 3). Furthermore, over the last decade, the Allegheny County Health Department has issued warnings on significant numbers of days (roughly 30 to 50 days per year during the May through December recreation season) to avoid bodily contact with the water in large portions of the rivers. During dry weather, water in the main stem rivers meets recreational guidelines for indicator microorganisms. However, water in many tributaries remains contaminated by indicator organisms and pathogens even in dry weather. Dry weather sewage treatment system problems (e.g., failing on-site sewage treatment and distribution systems [OSTDSs, or "septic systems"], malfunctioning package plants), agricultural practices, and natural sources may contribute to these dry weather

BOX 1-1
Overview of the Clean Water Act

The federal Clean Water Act (33 USCA sec. 1251, et seq., referred to as the CWA in this report) provides the basic legal framework for safeguarding and restoring the quality of the nation's surface waters. The law originated in the Federal Water Pollution Control Act of 1948 as significantly expanded in the 1972 Water Quality Amendments and some 35 other amendments through 2000, and is of central importance to this report. The overall goal of the CWA in its many subprograms is to protect, restore, and enhance the "waters of the United States" (including but not limited to "navigable waters"; see more below) for the protection and propagation of fish and aquatic life and wildlife, recreational purposes, and the withdrawal of water for public water supply, agricultural, industrial and other purposes.

Among its diverse and complex provisions, the CWA authorizes the U.S. Environmental Protection Agency (EPA) to establish ambient water quality standards[1] and limits for specific classes of pollutants. The CWA also established the National Pollutant Discharge Elimination System (NPDES),[2] which regulates major industrial and sewage treatment plant discharges into the waters of the United States. Combined sewer overflows such as those in the Pittsburgh region are eligible for permits subject to various requirements to limit their environmental impacts on receiving waters, while sanitary sewer overflows are illegal. Section 404 of the CWA requires the U.S. Army Corps of Engineers to regulate public and private activities involving "dredge and fill" of navigable waters pursuant to environmental guidelines issued by EPA. Under judicial and administrative interpretation, this section underlies federal regulation of wetlands throughout the United States in conjunction with state and local wetland management programs.

Under Section 303(d) of the CWA, states and authorized tribes are required to identify surface waters that are "impaired" by pollution sources, including failing septic systems, acid mine drainage, and agricultural runoff. To remediate these ambient water quality problems, states must prepare "total maximum daily load" (TMDL) plans under which the relevant pollutants are reduced through a range of measures including "best management practices" (BMPs).

Ultimately, authority for implementation and enforcement of the CWA rests with the EPA. However, the Pennsylvania Department of Environmental Protection (PADEP) is a key partner in this process at the state level, while the Allegheny County Health Department (ACHD) exercises concurrent jurisdiction with PADEP over sewage and discharges of wastewater within Allegheny County.

---

[1] Ambient water quality standards (AWQSs) are determined by each state and consist of (1) designated beneficial uses (e.g., drinking water supply, primary contact recreation); (2) narrative and numeric criteria for biological, chemical, and physical parameters to meet designated use(s); (3) antidegradation policies to protect existing uses; and (4) general policies addressing implementation issues. State water quality standards have become the centerpiece around which most surface water quality programs revolve; for example, they serve as the benchmark for which monitoring data are compared to assess the health of waters and to list impaired waters under CWA Section 303(d) (AWQSs are discussed extensively in Chapter 3).

[2] As authorized under Section 402 of the CWA, the NPDES permitting program controls water pollution by regulating point sources (e.g., discrete conveyances such as pipes) that discharge pollutants into waters of the United States (see Chapter 3 for further information).

conditions. Furthermore, many older homes in rural areas and former coal mining towns discharge sewage directly into local streams via straight pipes and "wildcat sewers" (see Appendix C for various sewage disposal and other terms used throughout this report).

The diverse water and sewage problems of southwestern Pennsylvania are often linked hydrologically. A downstream community's poor water quality problem may result from an upstream community's overflowing sewers, straight pipes, or failing septic systems. Whereas some municipalities have taken steps to address their water and sewage problems, many others face major water quality problems. Individual efforts may bring about limited improvements but the scope of the problem is so broad that a regional approach is needed. This is a conclusion reached in several recent reports, including the April 2002 report by the Steering Committee of the Southwestern Pennsylvania Water and Sewer Infrastructure Project (WSIP) of the Western Division of the Pennsylvania Economy League, and largely concurred with by this National Research Council (NRC) committee.

The WSIP's "best estimate" of the total investment required to fix regional water and sewer infrastructure throughout southwestern Pennsylvania without adjusting the institutional structures currently delivering these services is approximately $10 billion (WSIP, 2002). Although high, the costs of these improvements must be reconsidered in the context of the potential costs of inaction, which would include adverse impacts on public health, the environment, and economic growth and possible further federal and state regulatory action or private lawsuits by concerned parties. Compounding the problem, many individual communities in the region often lack the requisite expertise or resources needed to identify and implement the best solutions. The region's extensive governmental fragmentation and the lack of congruence of watershed and political boundaries suggest that the most cost-effective solutions may be obtainable only through a region-wide cooperative approach. Such a cooperative approach could be facilitated by existing public, non-profit, and private organizations as well as by the creation of new organizations. The precedent for such action in response to past crises exists in actions taken during the so-called Pittsburgh Renaissance (see Preface), the formation of ALCOSAN and the Port Authority Transit of Allegheny County, and the Allegheny Regional Assets District.

To help address these issues, the WSIP Steering Committee requested that the NRC undertake a study of the regional wastewater and water quality problems of southwestern Pennsylvania (see Box 1-2).

COMMITTEE AND REPORT

As noted in the Preface, this report was written by the NRC Committee on Water Quality Improvement for the Pittsburgh region organized by the NRC's Water Science and Technology Board. The committee was formed in early 2002 at the request of the Allegheny Conference on Community Development[2] (ACCD) to conduct an independent assessment of the wastewater and water quality problems of the Pittsburgh area and to make recommendations on how these issues and needs of the region can be best addressed by multiple jurisdictions on a cooperative basis (see Box ES-1 for the full statement of task).

Following this introduction, Chapter 2 provides a description of the region's physical setting; a history of the Pittsburgh region's water quality problems; and an overview of the region's demographics, economy, and land use changes. Chapter 3 provides an overview of current water quality conditions in the region, while Chapter 4 details causes of water quality

[2] The ACCD was founded in 1944 as a private, nonprofit organization to unify and coordinate regional transportation and environmental improvement efforts in Greater Pittsburgh. Since its founding, the ACCD has served southwestern Pennsylvania as a prominent private sector leader group dedicated to coordinating civic action by bringing corporate, government, and community leaders together to frame, discuss, and implement civic initiatives. As such, the ACCD in conjunction with the Western Division of the Pennsylvania Economy League initiated the WSIP, culminating in the release of *Investing in Clean Water: A Report by the Southwestern Pennsylvania Water and Sewer Infrastructure Project* (WSIP, 2002), which is discussed throughout this report (see also Appendix B).

BOX 1-2
Study Area

There are multiple ways to define southwestern Pennsylvania as a region for a variety of purposes, such as regulation as part of Region 5 (Southwest Region) of the Pennsylvania Department of Environmental Protection (PADEP); economic and community planning and development by the ACCD and Southwestern Pennsylvania Commission (SPC); U.S. census purposes (e.g., designation of a metropolitan statistical area, or MSA) as illustrated in the following figure. However, for the purposes of this report, and unless otherwise noted, use of the term "southwestern Pennsylvania" or "Pittsburgh region" refers to these 11 counties that also correspond to the study area of the 2002 WSIP report.

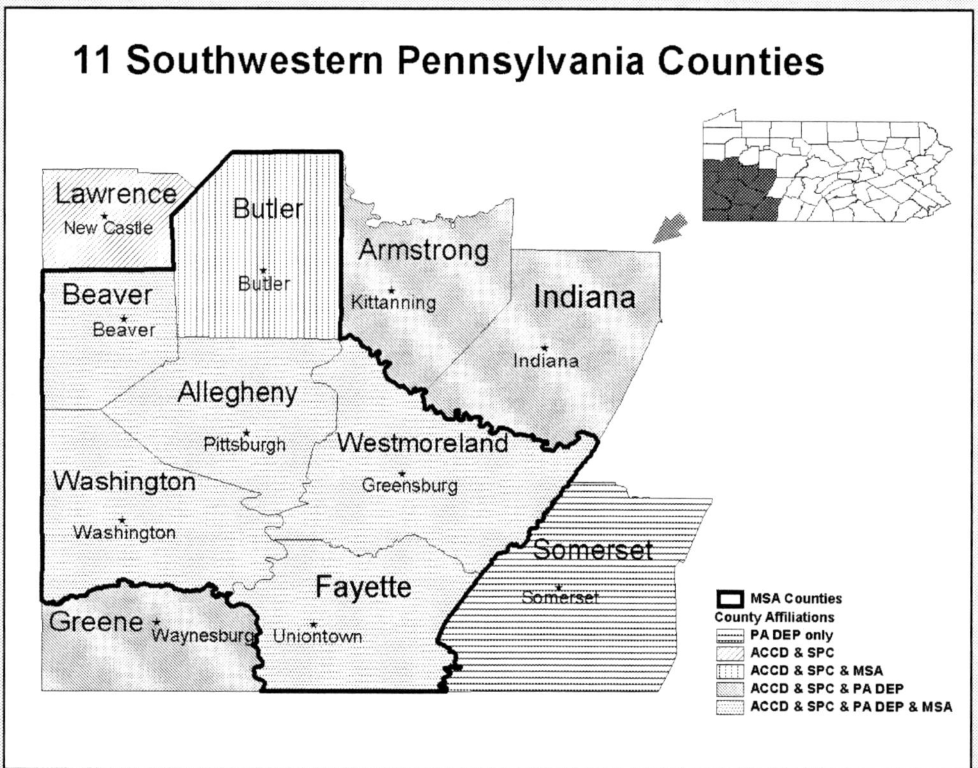

impairment including urban and rural handling of human waste, acid mine drainage, and agricultural and urban runoff. Chapter 5 focuses on decision-making strategies and technical solutions for regional water quality improvement, and Chapter 6 provides and assesses potential institutional and financial solutions for the region's water quality problems. Lastly, during the course of this study of water quality improvement in southwestern Pennsylvania, the committee gained knowledge and insights on several technical, policy, and institutional issues that have broader national implications. These are discussed in Chapter 7. Although detailed conclusions can be found within individual chapters and in the chapter summaries, the report's recommendations pertaining to water quality improvement in southwestern Pennsylvania are contained in Chapters 5 and 6.

REFERENCES

Balbus, J., R. Parkin, and M. Embrey. 2000. Susceptibility in microbial risk assessment: Definitions and research needs. Environmental Health Perspectives 108(9):901-905.

Mershon, S. and J. Tarr. 2003. Strategies for clean air: The Pittsburgh and Allegheny County smoke control movements, 1940-1960. In Devastation and Renewal: An Environmental History of Pittsburgh and Its Region, J. Tarr (ed.). Pittsburgh, PA: University of Pittsburgh Press.

NRC (National Research Council). 2001. Classifying Drinking Water Contaminants for Regulatory Consideration. Washington, DC: National Academy Press.

NRC. 2004. Indicators for Waterborne Pathogens. Washington, DC: National Academies Press.

WSIP (Southwestern Pennsylvania Water and Sewer Infrastructure Project Steering Committee). 2002. Investing in Clean Water: A Report from the Southwestern Pennsylvania Water and Sewer Infrastructure Project Steering Committee. Pittsburgh, PA: Campaign for Clean Water.

2

Regional Water Resources: Physiographical, Historical, and Social Dimensions

As background to understanding the complex water quality problems of southwestern Pennsylvania, this chapter summarizes the region's physical geography; its economic, demographic, and land use trends; and the history of its water supply and wastewater treatment practices. Since the 1970s, the Pittsburgh region has evolved from reliance on heavy industry to an economy based largely on technology, medical research, and higher education. This evolution has been accompanied by the decline of older industrial cities and towns and the spread of low-density development into outlying rural areas. As the region's water quality problems are better understood (see Chapter 3), the need to update the management of the region's surface and groundwater resources commensurately with its new economy and aspirations becomes manifest.

PHYSICAL SETTING

Pittsburgh and southwestern Pennsylvania are intrinsically identified with the "Three Rivers" that drain the region: the Allegheny, the Monongahela, and below their confluence at Pittsburgh, the Ohio River. The Allegheny and Monongahela basins drain more than 19,100 square miles in Pennsylvania, West Virginia, New York, and Maryland (altogether comprising 14 percent of the Ohio River Basin; see Figure 2-1). Historically, these three rivers served as major transportation arteries linking Appalachia with the Midwest. Based on 2001 data from the U.S. Army Corps of Engineers, Pittsburgh is the second busiest inland port in the nation and twelfth busiest U.S. port of any kind. Associated with the three rivers and their tributaries are many sites of historical, ecological, and recreational importance to the region and even the nation.

From its origin in Potter County, Pennsylvania, to its confluence with the Monongahela River in Pittsburgh, the Allegheny River drains 11,805 square miles and flows for 325 miles through 24 counties in Pennsylvania and New York (see Figure 2-2). Its major tributaries include some of the most biologically diverse, scenic, and historic streams in the country. Ecologically, the upper Allegheny River is particularly diverse, providing some of the best habitat for fish and freshwater mussels in northeastern United States. Indeed, about 85 miles of the main stem south of the Allegheny National Forest are part of the National Wild and Scenic

FIGURE 2-1 Ohio River basin showing locks and dams along the Ohio River; boxed area includes the Allegheny and Monongahela River basins and the headwaters of the Ohio River in Pennsylvania. SOURCE: Adapted from ORSANCO; available on-line at *http://www.orsanco.org/rivinfo/basin/basin.htm.*

Rivers System of the United States National Park Service.[1] One of its tributaries, French Creek, is widely known for its diverse natural history, including aquatic species found in this region when European settlement arrived in the mid-eighteenth century.

The Monongahela River drains 7,340 square miles of Maryland, Pennsylvania, and West Virginia (see Figure 2-2). The river rises in the Allegheny Mountains and flows generally northward from Fairmont, West Virginia, through mountainous terrain, farming communities, urban and industrial areas, and coal fields to its confluence with the Allegheny River in Pittsburgh. Over its length of approximately 130 miles, the river is spanned by several locks and dams to facilitate barge traffic. Upstream, a network of dams constructed and operated by the Corps of Engineers since the late 1930s help enhance low flows.

[1] See *http://www.nps.gov/rivers/* and *http://www.nps.gov/rivers/wsr-allegheny.html* for further information about the program and the Allegheny River's designation, respectively.

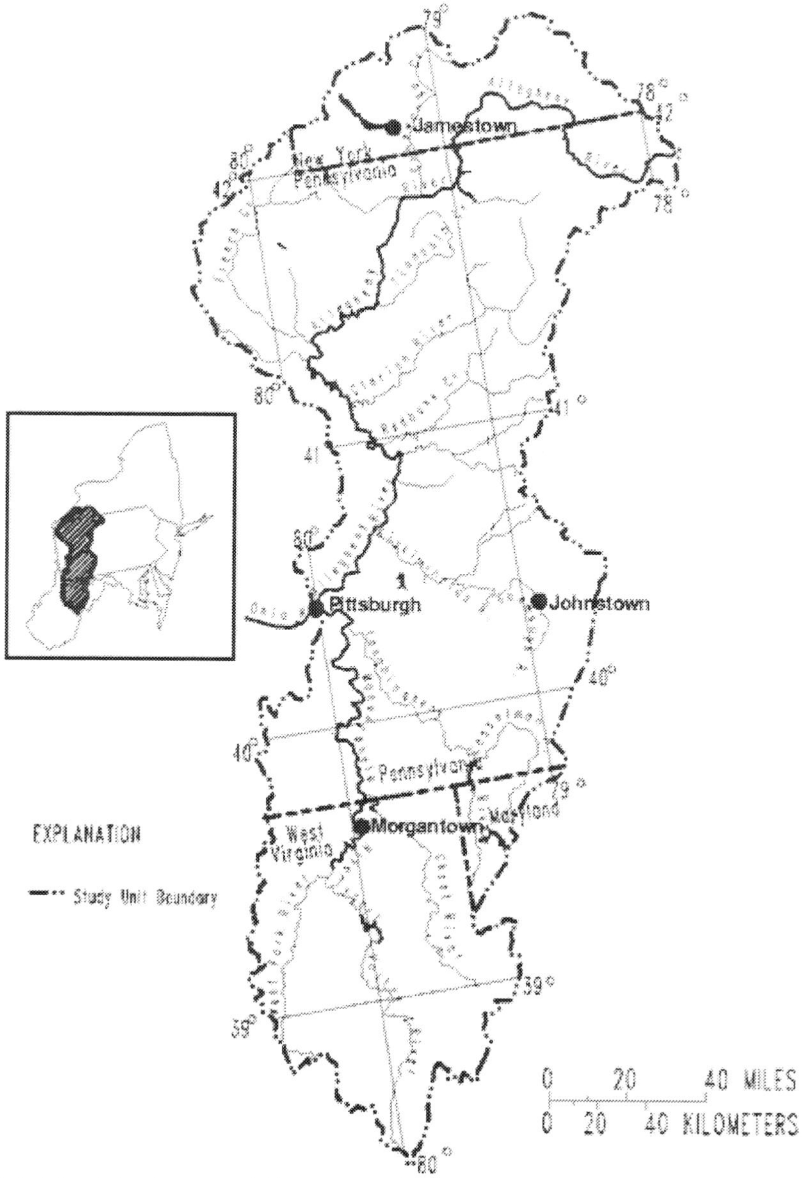

FIGURE 2-2 The Allegheny and Monongahela watershed. SOURCE: USGS, 1995.

Regional Geology, Soils, and Climate

The Pittsburgh region lies primarily in the Appalachian Plateau, which extends southward from New York to Alabama (see Figure 2-3) and is shaped by a geologic history that reflects the bituminous coal fields that have served as an important economic driver for the region for more than a century. However, historical and ongoing extraction of fossil fuels has left the region with a legacy of coal refuse piles, stripped landscapes, and acid mine drainage.

The physical landscape of the region was shaped in part by glaciation. As glaciers moved south during a series of ice ages, they reversed the course of the ancient Monongahela and Allegheny Rivers, which at one time flowed northward into the ancestral Lake Erie basin. As noted by John Harper (1997) of the Pennsylvania Bureau of Topographic and Geologic Survey:

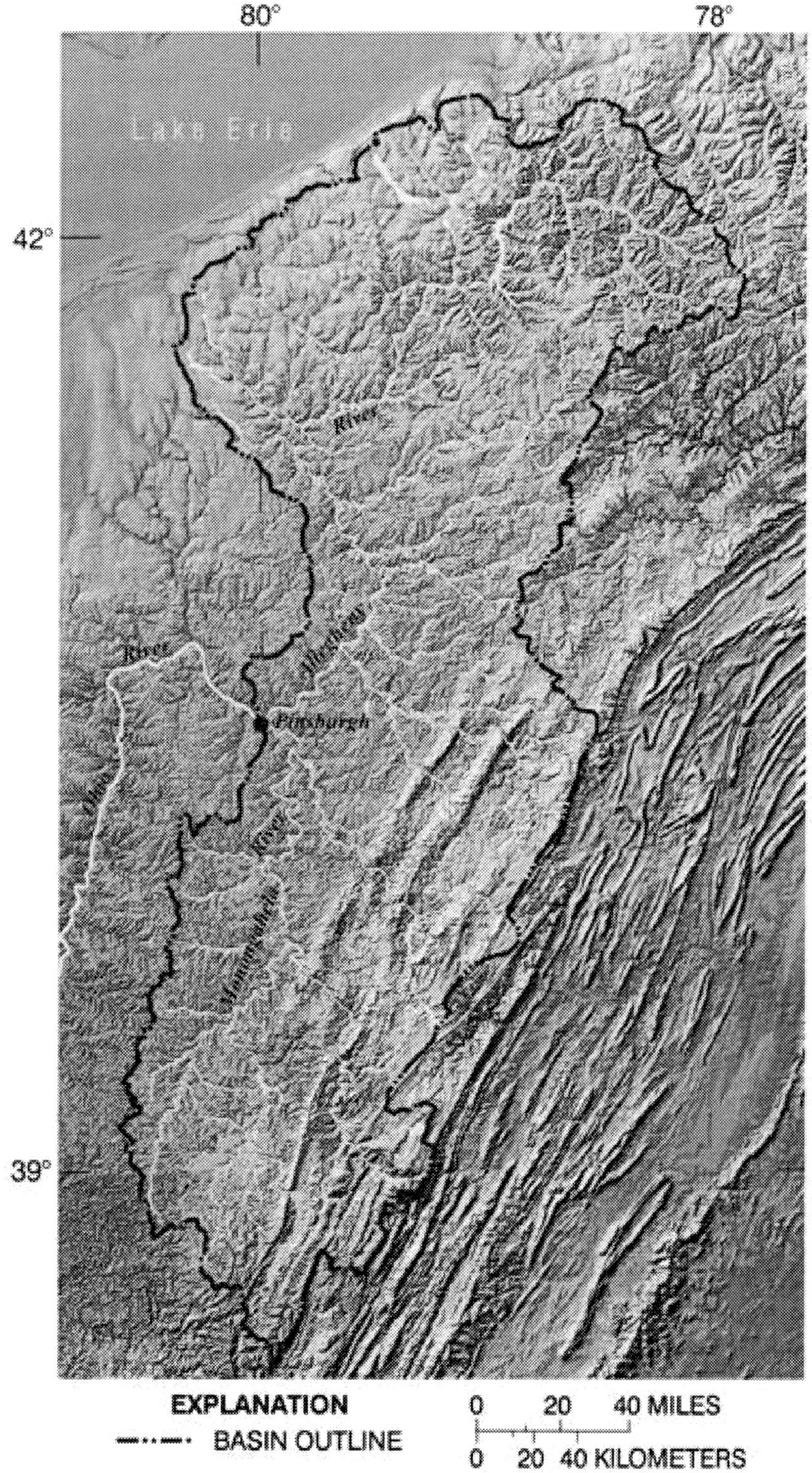

FIGURE 2-3 Topography of the Pittsburgh region. SOURCE: Anderson et al., 2000.

> As the earliest glacier moved into northwestern Pennsylvania about 770,000 years ago, the south-flowing ice blocked the northwest-flowing streams and caused lakes to form along the leading edge of the glacier. Eventually, the lakes became so deep that the water flowed over the divides (hilltops and ridges separating streams), reversing the ancient drainage of the Monongahela, Middle Allegheny, and the Upper Allegheny Rivers . . . The water had to go somewhere. Since it could not flow northward through the ice, it took a southerly course in all of these rivers, carving new water gaps through ridges, and taking over channels from established streams—Nature's version of eminent domain.

The bedrock of the region consists predominantly of Pennsylvanian Age (290-330 million years ago; MYA) cyclic sequences of sandstone, red and grey shale, conglomerate, clay, coal, and limestone and of Permian Age (250-290 MYA) cyclic sequences of shale, sandstone, limestone, and coal. The bedrock is extensively fractured, providing potential avenues for contaminants to enter groundwater. With thin soil cover in many areas and low-permeability, clay-rich soils in some areas, throughout the region it is difficult to locate sites suitable for wastewater treatment and disposal by conventional septic systems.

The climate of the region is generally humid with temperatures that range widely throughout the year. In January, the mean minimum temperature ranges from approximately 12°F at the source of the Allegheny River in Potter County to 20°F near Pittsburgh, while July maximum mean temperatures range from 86°F just south of Pittsburgh to the mid-70s in Potter County. The mean annual precipitation for the region approximates 40 inches and is distributed rather uniformly throughout the year.[2] Annual snowfall totals also average approximately 40 inches. Flooding has occurred as a result of intense precipitation in the region's steep valleys, sometimes exacerbated by ice jams or inadequate urban drainage. Indeed, periodic floods plagued Pittsburgh for much of the twentieth century. The flood of record occurred on March 18, 1936—an historic event throughout the eastern United States—when the Ohio River crested at more than 25 feet above flood stage. In the past 50 years, owing to the presence of several upstream flood control dams constructed by the Corps of Engineers after the 1936 flood, streets in the downtown Pittsburgh business district have been flooded only a few times. The most recent flood in Pittsburgh occurred in January 1996 when the Point State Park and parking lots of the (now demolished and removed) Three Rivers Stadium were inundated.[3]

Southwestern Pennsylvania has a long history of flash floods that ravage farm communities and old industrial towns situated in narrow valleys along local streams and rivers. The infamous Johnstown Flood of 1888, about 100 miles east of Pittsburgh, resulted from collapse of a recreational dam during a heavy rainstorm, killing thousands in the industrial communities downstream.[4]

---

[2] See *http://www.erh.noaa.gov/er/pit/climate.htm#NORMALSN* for further information.

[3] Prior to completion of this report, on September 17-18, 2004, heavy rains associated with the remnants of Hurricane Ivan caused widespread flooding throughout southwestern Pennsylvania. Whereas flooding on the Three Rivers at their confluence (i.e., Point State Park or the "Point") may not have been as great as in 1996, the recent flooding and resultant damages on streams throughout the region were extensive. Thousands of buildings were damaged. Water supply systems were potentially vulnerable to sewer overflows into source waters, but they handled the flood threat well, with no microbial contamination having been detected in drinking water (*http://www.pittsburghlive.com/x/search/s_255848.html*).

[4] Historical (1889) account of the Johnstown flood is available on-line at *http://prr.railfan.net/documents/JohnstownFlood/*.

Regional Biodiversity

The headwaters of the Ohio River in Pennsylvania are home to approximately 300 species of birds, 55 species of mammals, 35 species of reptiles, and 35 species of amphibians (WPC, 2004). Similarly, the diversity of aquatic organisms is exceptional, with 130 native species of fish and about 52 species of freshwater mussels (PABS, 1998). However, a U.S. Geological Survey (USGS) report on water quality in the Allegheny and Monongahela River basins (Anderson et al., 2000; see Chapter 3 and Appendix B) found that the water quality and aquatic life of much of the region had been affected significantly by land development and coal mining activities. That report further states that industrial activity in both large and small towns in the region has resulted in contaminated streambed sediments and contaminated fish tissue. The USGS report also highlights water quality successes, such as the treatment of drainage from active and abandoned mines that has generally resulted in improved water quality. "The general improvement in water quality . . . in sections of the Allegheny and Monongahela Rivers has been accompanied by an increase in the number and species diversity of fish . . . the recovery of rare species is a further indication of the degree of improvement" (Anderson et al., 2000). Of 52 fish species found in the Ohio River in 1818, only one today is no longer found in the river (Pearson and Pearson, 1989). In June 2004, Pittsburgh was chosen to host the 2005 Bassmaster Classic, a contest billed as the world championship of bass fishing (Belko, 2004). Improvements can be attributed to federal and state programs to improve water quality and the efforts of a growing number of grassroot and community watershed organizations.

JOBS, PEOPLE, LAND USE, AND GOVERNANCE

The Allegheny, Monongahela, and Ohio River system has been central to the development, history, and identity of southwestern Pennsylvania. The Three Rivers, as deepened for commercial navigation by locks and dams, connected Pittsburgh with the Ohio-Mississippi Valley waterway system, and the interior hinterlands of western Pennsylvania, West Virginia, and southwestern New York State (see Figures 2-1 and 2-2). Abundant coal and other natural resources, and the availability of convenient water and rail access within and beyond the immediate region, facilitated economic growth in Pittsburgh and surrounding communities from the mid-nineteenth century until the 1950s. The collapse of the steel industry around 1980 dealt the region a serious economic blow. Since the 1970s, the region's economic base has shifted to other sectors, including technology, medical research, higher education, finance, tourism, and other services. However, southwestern Pennsylvania has continued to experience economic weakness, as reflected in population decline, unemployment rates, and other indicators such as poverty level and income.

Indicators of Change

Demographic Trends

The Pittsburgh region has seen population declines in both the central city and the surrounding metropolitan area. Between 1970 and 2000, the population of the City of Pittsburgh fell by 35.7 percent from about 520,000 to about 335,000 (see Table 2-1). Meanwhile, its metropolitan statistical area (MSA; see Box 1-2) population, which stood at 2.5 million in 1970, declined to 2.3 million by 2000, a loss of 7.7 percent (see Table 2-2). During the 1990s, the City of Pittsburgh lost 35,000 residents—a decline of 9.5 percent. By comparison, other northeastern central cities of more than 100,000 in 1990 lost an average of just 1 percent over the 1990s.[5] Furthermore, the Pittsburgh MSA was one of only a handful of metropolitan areas nationally to experience a net population decline in the 1990s (amounting to a loss of 36,000 inhabitants, or 1.5 percent; see Table 2-2). Pittsburgh is not alone in losing population from its central city and inner suburbs. Most older industrial ("rustbelt") cities in the Northeast have experienced loss of population although, unlike Pittsburgh, suburban population growth has generally exceeded inner city losses. This process is commonly referred to as the "hollowing out" of the older urban core, with a "spreading out" or "sprawl" at the metropolitan fringe. According to Sustainable Pittsburgh (2003; see also Appendix B), "seventy percent of [the Pittsburgh Region's] municipalities have fewer residents today than they did 60 years ago." Meanwhile, the Pittsburgh MSA reflected a statewide trend of urbanizing nearly 4 acres for every new resident between 1982 and 1997 (Brookings Institution, 2003).

Today, Pittsburgh's residents are, so to speak, a smaller "family" inhabiting an aging but still elegant "mansion" constructed for a larger household in more prosperous times. The city retains the infrastructure and amenities built for a city of nearly twice its present population, including parks, schools, universities, health care, museums, theaters, water and sewer systems, and (of recent vintage) a light rail public transit system. It also retains a strikingly urbane downtown, excellent civic and commercial architecture, three major professional sports teams(two with relatively new sports stadiums), and a recently completed "green" convention center that is gaining national and international attention.

TABLE 2-1 City of Pittsburgh Population Change: 1970-2000

Year	Pop. (000s)	Decadal Change %	Overall Change %
1970	520	—	—
1980	424	−18.4	—
1990	370	−12.7	—
2000	334	−9.5	1970-2000: −35.7

SOURCE: Adapted from U.S. Census of Population, various tables.

TABLE 2-2 Pittsburgh MSA Population Change: 1970-2000

Year	Pop. (000s)	Decadal Change %	Overall Change %
1970	2,556	—	—
1980	2,571	+0.5	—
1990	2,395	−6.8	—
2000	2,359	−1.5	1970-2000: −7.7

SOURCE: Adapted from U.S. Census of Population, various tables.

[5] For further information see the 2000 Census of Population, available on-line at *http://quickfacts.census.gov/qfd/index.html.*

Despite its metropolitan population decline, the Pittsburgh MSA has been sprawling further onto rural land. According to a Brookings Institution study of urban sprawl from 1982 to 1997, the Pittsburgh MSA lost 8 percent in population but grew by 42.6 percent in urbanized area during that time (Fulton et al., 2001). This resulted in a loss of average density of 35.5 percent, the fourth greatest decline in density among northeastern MSAs during the study period (those exceeding Pittsburgh were Johnstown, Pennsylvania; Portland, Maine; and Utica, New York). Another recent study placed Pittsburgh among the top 20 "land consuming" metropolitan areas nationally, based on its estimated growth in developed land area of 201,000 acres, or 43 percent, between 1982 and 1997 (American Rivers-NRDC-Smart Growth America, 2002). The Pittsburgh MSA was the only metropolitan area of the top 20 to have lost population over the study period.

At the county scale, a picture of wide contrasts in population change and economic activity is presented, with some counties growing, some remaining relatively stable, and some declining (see Table 2-3). Of the total MSA population loss of 36,000 during the 1990s, 35,000 were lost to the City of Pittsburgh; the rest of the MSA outside Pittsburgh thus experienced only a net loss of 1,000 residents. Allegheny and Beaver Counties lost population (–4.1 percent and –2.6 percent, respectively). Butler and Fayette Counties gained population (+14 percent and +2.3 percent, respectively), and Westmoreland and Washington Counties were relatively stable. The five nonmetropolitan (non-MSA; see Box 1-2) counties in the study region were also divided between population losers (Lawrence –1.7 percent; Armstrong –1.5 percent); gainers (Greene +2.8 percent; Somerset +2.3 percent); and Indiana County, which, was stable. In the 11-county region, Allegheny County and 5 out of 6 counties adjoining it either lost population or were stable. Gainers in the region included most notably Butler County directly north of Allegheny (within commuting distance to Pittsburgh) and three rural counties south of Allegheny County, two of which (Greene and Fayette Counties) are within commuting distance of Morgantown, West Virginia.

Economic Trends

In addition to demographics, another important economic indicator is job growth. With the exception of a mild recession in 1990-1991, the last decade of the twentieth century was a period of exceptional job growth in the United States. Total employment in the United States increased by 20.6 percent from June 1990 to 2001 (Fuller et al., 2002). Employment growth in the Commonwealth of Pennsylvania, in comparison, lagged the nation with an overall growth rate of 10.3 percent. Notable exceptions to this trend in the Pittsburgh region were Butler County with employment growth of 32.5 percent and Fayette County with employment growth of 20.2 percent. Employment growth rates in Beaver, Greene, Fayette, Washington, and Westmoreland Counties lagged the nation but exceeded that of the state; employment growth in Allegheny, Armstrong, and Lawrence Counties lagged the state.

Unemployment rates for the region, state, and the United States are summarized in Table 2-4 for June 1990 and June 2001. With the exception of Allegheny County, southwestern Pennsylvania counties had higher unemployment rates than the state and nation in both periods. Particularly noteworthy are the very high unemployment rates in Armstrong, Fayette, Greene, and Indiana Counties. The high unemployment rates in these counties reflect a general tendency

TABLE 2-3 2000 Population Change in the Pittsburgh Region, By County

County	2000 Population	Change from 1990 %
Allegheny	1,281,666	–4.1
Armstrong	72,392	–1.5
Beaver	181,412	–2.5
Butler	174,083	14.5
Fayette	148,644	2.3
Greene	40,672	2.8
Indiana	89,605	–0.4
Lawrence	94,643	–1.6
Somerset	80,023	2.3
Washington	202,897	–0.8
Westmoreland	369,993	–0.1
Pennsylvania	12,281,054	3.4
United States	281,241,906	13.2

SOURCE: Adapted from U.S. Census Bureau and Pennsylvania State University's Cooperative Extension as cited in Pennsylvania State University's Center for Economic and Community Development Pennsylvania Census 2000, available on-line at *http://cecd.aers.psu.edu/census2000/PAcountypop.PDF*.

TABLE 2-4 Unemployment Rates in the Pittsburgh Region (percent)

County	June 1990	June 2001
Allegheny	4.3	3.9
Armstrong	7.1	8.0
Beaver	5.9	5.0
Butler	5.6	4.9
Fayette	8.1	7.1
Greene	9.3	6.8
Indiana	7.6	6.4
Lawrence	6.2	5.8
Somerset	6.7	5.7
Washington	5.6	4.9
Westmoreland	5.7	5.4
Pennsylvania	5.0	4.8
United States	5.3	4.7

SOURCE: Fuller, et al., 2003.

for rural Pennsylvania counties to have greater unemployment than metropolitan counties (Shields, 2002).

Three additional economic indicators are presented in Table 2-5: (1) median income; (2) median value of owner-occupied housing; and (3) poverty rate (i.e., percentage of households with incomes at or below the poverty level). With the exception of Butler County, median incomes in Pittsburgh region counties are below the Pennsylvania state average. Similarly, the median value of owner-occupied homes is below the state average in all counties except Butler. Seven of the eleven counties in the study region have poverty rates in excess of the state rate of 11 percent, and three exceed the U.S. rate of 12.4 percent.

TABLE 2-5 Median Income, Median Value of Owner-Occupied Housing, and Poverty Rate for the Pittsburgh Region

County	Median Household Income (2000)	Median Value of Owner Occupied Housing (2000)	Poverty Rate (1999, %)
Allegheny	$38,329	$83,500	11.2
Armstrong	$31,557	$63,800	11.7
Beaver	$36,995	$83,200	9.4
Butler	$42,308	$105,300	9.1
Fayette	$27,451	$60,600	18.0
Greene	$30,352	$55,800	15.9
Indiana	$30,233	$68,300	17.3
Lawrence	$33,152	$71,100	12.1
Somerset	$30,911	$66,900	11.8
Washington	$37,607	$85,400	9.8
Westmoreland	$37,106	$87,600	8.6
Pennsylvania	$40,106	$94,800	11.0
United States	$41,994	$111,800	12.4

SOURCE: Adapted from U.S. Census Bureau Census 2000 Demographic Profile Highlights, available on-line at *http://factfinder.census.gov/home/saff/main.html?_lang=en.*

Land Use Change and Water Quality

Development and demographic changes in the Pittsburgh region are fraught with significance for water and sewer investment decisions and water quality. There is a substantial literature demonstrating that water quality is impacted by land cover (e.g., Allan et al., 1997; Herlihy et al., 1998; Jones et al., 2001; Roth et al., 1996). In urbanizing watersheds, such as those in the Pittsburgh region, the impact of increasing impervious surface areas is becoming of increasing concern. Indeed, impervious surface is emerging as an important indicator of effects on water quality and biotic quality in streams. Numerous studies have attempted to identify thresholds in the relationship between impervious cover and various types of environmental impacts. A variety of studies indicate that 10 percent impervious cover represents an important threshold for many environmental impacts (e.g., CWP, 2003; Scheuler, 1994).

Impervious surfaces include roads, roofs, parking lots, sidewalks, and other constructed surfaces that are impenetrable to water. Increasing the amount of impervious surface within a watershed affects the hydrologic regime by altering the volume, pattern, and timing of hydrologic flows (CWP, 2003). With an increase in impervious surface area, less precipitation infiltrates the soil and more runs off directly into surface waters. Changes in water flows can lead to physical changes as increased stormwater runoff results in higher periodic stream flow, stream channel enlargement and incision, greater stream bank erosion, and increased sedimentation in the stream channel (CWP, 2003; EPA, 2001; Scheuler, 1994). The water quality impacts of increased stormwater runoff can also be significant. Reduced infiltration means that urban contaminants are carried more directly into surface waterbodies (EPA, 2001). Stream temperatures can be affected by the imperviousness of the watershed (Galli, 1990; Scheuler, 1994). These and other water quality impacts and their implications for humans and aquatic ecosystems are discussed in Chapters 3 and 4.

Apart from the physical impacts of development on water quality through the effects of increased impervious areas, development patterns and population shifts also affect the utilization of existing sewer and water infrastructure and the location of investment in new infrastructure.

As population declines in the central city and inner MSA counties, there is likely to be surplus water and sewer capacity in many of those communities, along with roads, schools, parks, and other infrastructure in place. Much of this infrastructure is old and in need of maintenance and upgrading which has led to sewer "tap-in" restrictions in a number of communities in the Pittsburgh region communities (described later). Meanwhile, new water and sewer infrastructure is under construction or under consideration to serve rapidly growing areas in the outer fringe counties and to remedy existing deficiencies in on-site water and septic facilities.

A balance is needed between the updating and maintenance of water and sewer services already in place versus recreating such capacity to serve outlying development at considerably lower density and therefore higher cost per household. According to a report by the Environmental Law Institute (ELI, 1999; see also Appendix B):

> In the counties surrounding Allegheny, the perceived environmental problem is still the release and potential release of untreated or inadequately treated sanitary sewage—rather than the patterns of development which are producing these problems. Solutions are, in turn, driven by current financing realities and institutional preferences for new construction. These factors promote the extension of existing sewer collection systems to a larger ratebase by sewering larger areas of the region, and encourage the replacement of on-lot systems with sewers and wastewater treatment. While this promotes near term environmental improvement, the effect on development and future growth is significant . . . Common interests of the counties include the need to revitalize the older urban centers and not simply attract greenfields development at the margins of the respective counties.

Similarly, the 2003 report by Sustainable Pittsburgh, Inc.,[6] states, "Fiscal expediency alone rationalizes steering development first to existing communities to simultaneously fix, upgrade, and use in-place surplus capacity as opposed to building new elsewhere." In that report, Carnegie Mellon University President Jared Cohon maintains that, "Southwestern Pennsylvania's waters are a priceless asset for residents, recreation, industry, and agriculture. To adequately protect that resource, we must make greater investments in infrastructure, and spend that money wisely."

In a June 9, 2002, speech to a Pittsburgh Smart Growth Conference, Brookings Institution demographer Bruce Katz warned that Pittsburgh's combination of declining population accompanied by increasing sprawl puts it and similar older industrial cities of the Northeast and Midwest at a competitive disadvantage in comparison with Sunbelt urban areas of the West and South:

> I think from a fiscal perspective, from a social perspective, from an environmental perspective. . . from the perspective of how you compete over time, for new economy firms, for knowledge-based firms, that are seeking quality of life, seeking affordable housing, seeking the natural environment that has been preserved. You are undermining perhaps your competitiveness over time as a region.

---

[6] See Appendix B for a summary of that report, available on-line at *http://www.sustainablepittsburgh.org/citizensvision/CitizensVision.pdf.*

Among several factors that influence the rate of urban sprawl, Katz also listed infrastructure:

> How do we spend the billions of dollars of state money on roads, on water, on sewer, on school facilities, on government facilities, on higher educational institutions? Do we spend this to reinforce, to support areas that we've already invested in—cities, older suburbs, townships—or do we continue to basically use them to facilitate more sprawl?

At the opposite end of the development spectrum, many of the same arguments can be made for providing appropriate infrastructure (centralized or decentralized) to certain outlying communities to assist them in preserving rural character, farmland, and open space vital to tourism and outdoor recreation. Because funds for rural water and sewage projects are limited, many of these areas do not obtain funding until commuter "bedroom community" growth (e.g., Burgettstown, Pennsylvania) or extension of roads and their accompanying development overwhelm or threaten to overwhelm the local capacity. Examination of Pennsylvania Act 537 Official Plan Aging database[7] reveals that many of the more recent plan updates follow this pattern. Lack of comprehensive rural facilities planning to prioritize infrastructure development in areas of significant rural and recreational potential risks being a partner to abandonment of the urban center in promoting sprawl. The goal of making wise investments in water facilities and other infrastructure, as per President's Cohon's challenge above, is greatly hampered by the political fragmentation of metropolitan Pittsburgh and adjoining areas. In addition to the state and federal governments, the region is governed by 11 counties, 595 municipalities, and 492 other water and sewer providers (see Chapter 6 for further information). According to a recent op-ed column in the *Pittsburgh Post-Gazette*, the region "maintains an astonishing 17.7 general purpose governments per 100,000 residents. This overload of government is more than triple the nation's metropolitan average of 6.1 governments per 100,000" (Katz, 2004) for the nation's metropolitan areas.

Stormwater management under the Clean Water Act (CWA) is a critical water-related management function that is complicated by political fragmentation. In particular, efforts by certain jurisdictions to address water quality impairments through "best management practices" (BMPs) may be partially offset or limited by rising stormwater discharges from upstream sources in other communities within the same local watershed. For example, stormwater runoff to Turtle Creek from communities in Westmoreland County adversely affected Allegheny County's ability to maintain water quality and prevent flooding downstream on that creek (McElfish and Casey-Lefkowitz, 2001). The prevalence of such externalities among jurisdictions sharing an urbanizing watershed highlights the need for intergovernmental cooperation through compacts, joint powers agreements, special districts, or other watershed-based institutional arrangements. Such cooperative efforts must address the type, design, and location of new development within watersheds: "Without strict land use planning, aggressive land acquisition programs, and integrated watershed management . . . erosion, sediment, and stormwater control programs cannot do the job" (Watershed Management Institute, 1997). The institutional aspects of regional management of water resources are discussed in detail in Chapter 6. In particular, the

---

[7] For further information, see *http://www.dep.state.pa.us/dep/deputate/watermgt/wqp/WQP_WM/537Map/537Plan.htm*.

concept of a "regional water forum" representing all sectors of stakeholders is recommended for consideration by southwestern Pennsylvania area leaders and citizens.

ECOLOGICAL RESTORATION IN THE REGION

Despite its economic and demographic challenges, key stakeholders in the Pittsburgh region are striving to improve the ecological condition and public enjoyment of its Three Rivers and their many local tributary watersheds. A recent paper by Collins et al. (2003) refers to

> . . . recent struggles to reclaim the natural aspects of the rivers as part of a new post-industrial environmental infrastructure with natural amenities essential to the new economy. This evolving post-industrial reconceptualization frequently involves conflicting efforts of managing the rivers for their scenic and recreational as well as transport opportunities, and restoring and preserving the riverine ecological systems as a public commons.

With the support of area foundations, the Pittsburgh region's main stem rivers have been rediscovered as amenities supporting water-based recreation and ecological habitat reaching virtually into the city center. Local and regional environmental organizations such as the Pennsylvania Environmental Council (PEC, 2003), 3 Rivers 2nd Nature (3R2N) at the Studio for Creative Inquiry at Carnegie Mellon University, and the Nine Mile Run Watershed Association are collaborating in the analysis and restoration of riparian habitat in selected sites (see Box 2-1). They also are promoting environmental education and public interest in the region's abundant aquatic resources. As a result, the level of water quality in local streams, lakes, and rivers, as well as groundwater, is of rising public concern.

HISTORICAL SETTING

Pittsburgh Water Supply and Wastewater Treatment

Throughout its history, one of the most serious environmental problems facing the City of Pittsburgh and its outlying areas has been pollution of its streams and rivers. This pollution, from both domestic and industrial sources, accelerated during the nineteenth and twentieth centuries and severely impacted the quality of the water drawn from the rivers for domestic purposes and for various industries. It also sharply curtailed the use of the rivers for recreation.[8] Current water pollution problems, especially from combined sewer overflows (CSOs), are the result of a series of decisions related to both water supply and wastewater collection and disposal made over the past century-and-a-half to:

1. draw Pittsburgh's water supply from the neighboring rivers;

---

[8] For a comprehensive study of these issues on a national level, see Melosi (2000); for historical information on Pittsburgh infrastructure development, see Tarr (1989).

2. construct a water carriage system and build it following a combined rather than a separate sewer design;
3. use the rivers as the place of sewage disposal based on the concept that running water purified itself or provided adequate dilution to disperse the sewage;
4. filter and also chlorinate drinking water in order to deal with public health concerns rather than to use a protected upcountry source;
5. resist state orders to treat the city's sewage and change the design of the city's sewerage system (Tarr, 1996a); and
6. subsequent to World War II, treat the sewage of Pittsburgh and neighboring communities by creating an authority with a centralized wastewater treatment plant, but allow communities to own and maintain their own sewage collection systems.

Water Supply

For its first quarter century, Pittsburgh citizens drew their water supplies primarily from local ponds and wells. However, these sources became increasingly inadequate to meet domestic, industrial, fire protection, and public health needs. In 1826, after extensive debate over the issue of public or private ownership, the Pittsburgh Select and Common Councils approved the construction of a waterworks. The waterworks was subsequently completed in 1828 at an initial cost of $40,000 and utilized a steam pump to draw water from the Allegheny River, raising it to a million-gallon reservoir located on Grant Hill for gravity distribution throughout the city (Tarr, 1989).

Throughout the nineteenth century, the waterworks expanded in response to population and territorial growth. By 1915, the system had 743 miles of distribution pipes (Lanpher and Drake, 1930). Because of extensive waste and leaky pipes, the city had one of the highest rates of per capita water consumption of any city in the nation (Tarr and Yosie, 2004).

Waterworks funding was the single largest expenditure made by the city during its first 50 years, which is not surprising given the extent to which waterworks costs constituted a substantial part of the total municipal budget of nineteenth century American cities (Tarr, 1989). Pittsburgh's willingness to make such a large expenditure for a public good can be explained by the coming together of a variety of interest groups—including merchants and industrialists, homeowners, fire insurance companies, and physicians and others concerned with the public health—to demand an adequate supply of water. Notably, these investments also occurred during a period of prosperity when many of the city's great buildings and other facilities were being constructed.

Sewerage System

A supply of potable water was only one part of the city's metabolic system; wastewaters from households and industries and stormwater had to be disposed of as well. During most of the nineteenth century, Pittsburghers placed household wastes and wastewater in nearby cesspools and privy vaults, not in sewers. The 1804 borough charter gave the city the right to regulate such receptacles, but the city councils did not enact an ordinance until 1816, when

BOX 2-1
SUMMARY OF COMMUNITY-BASED STREAM RESTORATION PROJECTS

Nine Mile Run

Nine Mile Run is a stream whose watershed drains five Pittsburgh area municipalities, flowing through wooded Frick Park and an urban brownfield before it joins the Monongahela River. The brownfield features a mountain of steel mill slag (15-20 stories high), dumped by Duquesne Slag on the Nine Mile Run Floodplain between 1922 and 1972. The Nine Mile Run Greenway Project (NMR-GP) was established in response to a Pittsburgh housing redevelopment plan for the site that would have buried the stream. The NMR-GP team consisted of diverse academic and outside experts, under the leadership of the Studio for Creative Inquiry at Carnegie Mellon University (CMU). The primary purpose of the project was to explore the cultural and aesthetic issues of postindustrial public space and ecology. The team conducted studies that revealed the potential for aquatic and terrestrial approaches to ecological restoration on the urban brownfield and Nine Mile Run. In addition, the team conducted a series of "community dialogues" that helped develop a set of concept plans used as the framework for a subsequent U.S. Army Corps of Engineers restoration study and plan and a $7 million stream restoration project for the City of Pittsburgh. For further information, see McElwaine (2003) and *http://slaggarden.cfa.cmu.edu.*

A partnership of the Pittsburgh Redevelopment Authority and four developers has recently undertaken a brownfields restoration of an unsightly, contaminated 238-acre slag mountain towering above Nine Mile Run. Following extensive environmental remediation and safety measures (Trout et al., 2001), the team is creating a multiphase "new traditional" urban development with walkable neighborhoods, greenspace commons, trail access to Frick Park and downtown, and housing styles and construction and lot layouts that recreate the look and feel of Pittsburgh's close-knit, old-fashioned "streetcar neighborhoods." Demand for the 65 residences in Phase I, completed in 2002, was so strong that a lottery was held to select buyers from a waiting list of 500 (McKay, 2002).

Nine Mile Run and Stream Daylighting

The aquatic restoration techniques used on Nine Mile Run emerged from water quality studies and water quantity studies based on fluvial geomorphology and wetland science. Ultimately, the

citizen complaints about overflowing filth and smells from privy vaults caused the councils to approve fines in the case of nuisances. Privy vaults and garbage continued to foul the streets and pollute the rivers with waste, despite the cleaning efforts of private scavengers under city contract. In 1875, the Pittsburgh Board of Health observed that privy vault nuisances were the major health issue facing the city (Tarr, 1989).

The provision of running water to households and the adoption of water-using appliances such as sinks, showers, and water closets, exacerbated the problem of overflowing cesspools and privy vaults. On the one hand, improving the availability and volume of water supply was a benefit, but on the other hand it had the potential to deliver contaminated water with resulting impacts on public health.[9] In many cases, in order to dispose of the wastewater, households connected these appliances to existing wastewater disposal sinks (i.e., cesspools and

[9] Before the availability of piped-in water, Americans had found various ways—such as the use of cisterns, wells and local ponds—to provide a water supply for household plumbing (see Ogle, 1996).

restoration techniques included "stream daylighting," the act of unburying, and renaturalizing or restoring a stream and its channel that flowed through Pittsburgh's Frick Park. The project team also developed concept plans for the redesign and reconstruction of the main stream channel to address a century of erosion caused by stormwater from nearby sewers and roads. By 2005, Nine Mile Run is slated to follow a new streambed, its natural sinuosity restored and wetlands preserved and extended, reducing erosion and allowing for more storage of stormwater during a flood event. Terrestrial restoration was based on a botanical assessment of existing steep slope slag vegetation and greenhouse studies of soil nutrients and native seed sources that held the promise of an ecologically viable habitat (in slag soils) linking the interior forests of Frick Park to the Monongahela River corridor. The plan was for increased landscape diversity, with native plants and a stream that was redesigned for current flow conditions, restoring the potential for a warm water fishery.

3 Rivers 2nd Nature

3 Rivers 2nd Nature (3R2N) is a five-year project funded by the Heinz Endowments that addresses the meaning, form and function of the Monongahela, Allegheny, and Ohio Rivers and riverbanks and 53 tributary streams within Allegheny County. This project builds on the successful linking of ecosystem analysis and community dialogue developed by CMU's Studio for Creative Inquiry in the Nine Mile Run watershed project discussed above. The project team works with scientists, engineers, planners, and policy experts from the Three Rivers Wet Weather Demonstration Program (3RWW), Allegheny County Sanitation Authority, Allegheny County Health Department, University of Pittsburgh, and private practice. The project has developed partnerships to conduct fieldwork and establish an ecological baseline for Allegheny County and undertakes terrestrial studies of riverbanks and forests and aquatic studies of stream biology and fisheries. It also conducts dry weather physical chemistry and bacterial indicators for pathogen studies in rivers and streams. Like the Nine Mile Run project, it has established a series of river dialogues with local citizens. The project works with communities to address the cultural understanding of living rivers and water quality, as well as the potential to preserve and restore natural ecosystems in a postindustrial urban setting. The 3R2N project will develop a waterfront plan and a series of waterfront node designs with artists, landscape architects, and ecological designers as the project comes to a close in late 2004. For more information, see *http://3r2n.cfa.cmu.edu*.

privyvaults). In 1881, for instance, about 4,000 of the 6,500 water closets in Pittsburgh were connected to privy vaults and cesspools but only about 1,500 to street sewers (Yosie, 1981).

Such conditions raised the possibility of waterborne disease outbreaks and highlighted the need for improved sanitation and construction of a sewer system. The city first constructed underground sewers in the commercial district, mainly to prevent flooding by draining urban runoff. By 1866, the commercial district had a "fairly adequate" system of main sewers (Thompson, 1948). Other areas of the city, however, had much more uneven and haphazard service. By 1875, the city had about 25 miles of sewers (brick and pipe), mostly for stormwater drainage. These early sewers suffered from design faults and were often either undersized or oversized and subject to constant clogging. Sewers did not conform to topography (the city lacked topographical maps until the 1870s) or follow an overall engineering plan since the city commonly built sewers in response to council members' attempts to meet specific constituent demands. In addition, many households constructed their own sewers, which were often unrecorded and not reliable (Tarr, 1989).

Debate raged between engineers and physicians about the design of a new centralized sewer system. Should it be a separate, small pipe system that carried only domestic and industrial wastes (the technology advocated by the famous sanitarian George E. Waring, Jr.), or should it be a larger, combined system that could accommodate both wastewater and stormwater in one pipe (a design then favored by many sanitary engineers)? Physicians argued that the separate system was preferable because it would protect health by removing wastes from households before they had begun to decompose, thus generating disease-causing sewer gas. They considered stormwater a secondary matter, best handled by surface conduits. Engineers took a different position and maintained that sanitary wastes and stormwater were equally important; therefore, a large pipe system that would accommodate both was more economical. After years of debate, the superior stormwater removal virtues of the combined system convinced city officials, and by the late 1880s, Pittsburgh had begun to build a system of large combined sewers (Tarr, 1996b; Tarr and Yosie, 2004).

Between 1889 and 1912, civil engineers from the new Bureau of Engineering of the Public Works Department constructed more than 412 miles of sewers, almost all of the combined type. The combined sewer carried both stormwater and domestic sewage in one pipe, but because of the fluctuations in the size of rainstorms, the system included overflow valves at critical points throughout the system. The construction of the planned centralized sewerage system signified a movement away from the "piecemeal, decentralized approach to city-building characteristic of the 19th century" (Peterson, 1979). In constructing a large centralized combined sewer network, Pittsburgh was following the lead of other large American cities such as Boston, Chicago, and New York. Like other cities, Pittsburgh discharged its untreated sewage directly into neighboring waterways, with 47 public sewer outlets into the Monongahela River and another 98 into the Allegheny and Ohio Rivers (Tarr and Yosie, 2004).

Typhoid Fever and Water Filtration

While Pittsburgh was discharging its untreated sewage into its neighboring rivers, upstream communities were also building sewers and discharging their wastes into the same rivers. By 1900, more than 350,000 inhabitants in 75 upstream municipalities were discharging their untreated sewage into the Allegheny River—the river that provided drinking water for most of Pittsburgh's population. Some of Pittsburgh's own sewers discharged into the river at locations above the city's water supply system intakes. The resulting pollution gave Pittsburgh the highest death rate from typhoid fever of the nation's large cities, more than 100 deaths per 100,000 people from 1873 to 1907.[10] In contrast, in 1905 the average death rate for northern cities was 35 per 100,000 persons (Koppes and Norris, 1985). At least one reason why the poorer, immigrant neighborhoods may have had higher death rates is that although even these neighborhoods were required to be sewered, the communal toilet facilities themselves were often not water-carriage and did not convey the waste into the sewers (Byington, 1910).

Concerned over the growing typhoid mortality and morbidity rates, in the early 1890s various professional groups formed a joint commission to study the question of water pollution

[10] Typhoid fever death rates were highest in working class immigrant and African-American living areas. Thus, of the nine wards with the highest typhoid death rates in the 1890s, seven were working class. According to the municipal Health Department, Pittsburgh appeared "as two cities, one old and congested with a high mortality, and the other new and spacious with a very low death rate" (PADOH, 1911).

and water quality. The commission's report, issued early in 1894 and based on chemical, bacteriological, and statistical methods, found Pittsburgh water supply "not only not up to a proper standard of potable water but . . . actually pernicious" (Tarr and Yosie, 2004). That commission recommended that the water of Pittsburgh be treated using a slow-sand filtration system. This recommendation was followed in 1896 by the mayor's appointment of a Pittsburgh Filtration Commission. In 1899, after detailed investigation by outside consultants, the commission recommended that the city construct a slow-sand filtration plant drawing water from the Allegheny River. Thus, Pittsburgh joined cities such as Philadelphia, Cincinnati, and Cleveland that took their water from neighboring rivers and filtered it, whereas other cities, such as Boston, Newark, and New York, chose to rely on protected watersheds—an option rejected by the filtration commission because of high costs. In December 1907, the water department delivered the first filtered water and the city's typhoid rates began to drop. In 1912, the city began chlorinating its water supply and Pittsburgh's death rate from typhoid fever dropped to the national average for large cities (Lanpher and Drake, 1930; Tierno, 1977).

Sewage Disposal: Retaining the Rivers as Sewers

Water filtration provided one safety net with regard to sewage-contaminated water, but many sanitarians and public health physicians during the first decades of the twentieth century believed it also necessary to treat the city's sewage prior to release in nearby streams and rivers. Many professional and business groups had begun to protest sewage disposal by dilution in streams. They demanded that municipalities treat their sewage and called for state laws against stream pollution (Tarr, 1996a). In the early twentieth century, states such as Connecticut, Massachusetts, New Jersey, New York, Ohio, and Pennsylvania, responding to a series of unusually severe typhoid epidemics, gave their health boards increased power to control sewage disposal in streams (Tarr, 1996a).

In 1905, a law "to preserve the purity of the waters of the state for the protection of the public health" (the Purity of Waters Act[11]) was passed by the Pennsylvania state legislature in response to a typhoid epidemic in 1903. The act forbade the discharge of any untreated sewage into state waterways by new municipal systems. Although it permitted cities already discharging their untreated sewage to continue the practice, it required them to secure a permit from the state health commission if they wished to extend their systems (Snow, 1907).

Although Pittsburgh was filtering its drinking water after 1907, the city continued to dispose of its untreated sewage in neighboring rivers, endangering the water supply of downstream communities. In 1910, the city requested that the Pennsylvania Department of Health (PADOH) grant it a permit allowing it to extend its sewerage system. The department, then headed by Dr. Samuel G. Dixon, responded by requesting a "comprehensive sewerage plan for the collection and disposal of all of the sewage of the municipality" before it would grant the permit (Tarr, 1996a). In addition, PADOH argued that in order to attain efficient treatment, the city should consider changing from a combined sewer system to a separate sewer system. The department's chief engineer, F. Herbert Snow, maintained that the plan was needed to protect the

[11] The Purity of Waters Act was the first Pennsylvania state legislation specifically related to water quality. The act sought to preserve the purity of the waters of the state for the protection of the public health.

public health of communities that drew their water supplies from rivers downstream from Pittsburgh (Gregory, 1974; Tarr, 1996a).

The City of Pittsburgh responded to the PADOH order by hiring the engineering firm of Allen Hazen and George C. Whipple to act as consultants for the required study. The engineering firm argued in its report that a Pittsburgh sewage treatment plant would not free the downstream towns from threats to their water supplies from the need to filter them, since other upstream communities would continue to discharge raw sewage into the rivers. The method of disposal by dilution in rivers, they maintained, was sufficient to prevent nuisances, particularly if storage reservoirs were constructed upstream from Pittsburgh to augment flow during periods of low stream discharge. Hazen and Whipple argued that there was no case "where a great city has purified its sewage to protect public water supplies from the stream below" (Tarr, 1996a). "Rivers are the natural and logical drains and are formed for the purpose of carrying the wastes to the sea," argued N.S. Sprague, superintendent of the Pittsburgh Bureau of Construction in forwarding the Hazen-Whipple report to the director of public works (Tarr and Yosie, 2004).

Hazen and Whipple's most powerful argument addressed the lack of economic feasibility of converting Pittsburgh's combined sewerage system to separate sewers and building a sewage treatment plant. There was no precedent, they claimed, for a city replacing the combined system with the separate system "for the purpose of protecting water supplies of other cities taken from the water course below" (Tarr, 1996a). The report calculated that financing such a project would cause the city to exceed its municipal indebtedness level and thus violate state law. Moreover, because the sewage treatment plant was intended for the protection of downstream communities, it would not give Pittsburgh any direct benefits, while downstream cities would still have to filter their water to protect against waterborne disease. Hence, they concluded that "no radical change in the method of sewerage or of sewage disposal as now practiced by the City of Pittsburgh is necessary or desirable."

Although the engineering press received the Hazen-Whipple report with enthusiasm, PADOH health commissioner Dixon deemed it an insufficient response to his original request for a long-range plan for a comprehensive regional sewerage system. Dixon argued that water pollution had to be viewed from a health rather than a nuisance perspective and that the immediate costs of sewage treatment would be outweighed by the long-range health benefits. The time had come, Dixon stated, "to start a campaign in order that the streams shall not become stinking sewers and culture beds for pathogenic organisms . . ." (Tarr, 1996a). Given the political climate and financial limitations of the city, however, Dixon had no realistic means by which to enforce his order. In 1913 he capitulated and issued Pittsburgh a temporary discharge permit.

Pittsburgh Treats its Sewage

For Pittsburgh to be compelled to treat its sewage would require major policy and value changes on the part of both government officials and the public. In 1923, the Pennsylvania General Assembly enacted legislation proposed by then Governor Gifford Pinchot that created, within the Department of Health, a Sanitary Water Board whose function was to balance economic growth and improved water quality (Saville, 1931; Stevenson, 1923). The board created a stream classification system that designated streams into three categories for municipal and industrial users: (1) steams that were relatively clean and pure; (2) streams in which pollution existed but could be controlled; and (3) streams that were so polluted they could not be used as public water supplies or for fishing and recreational purposes without treatment and, therefore, could continue to be used for the discharge of untreated wastes. Pittsburgh rivers fell in the third category; that is, they were to continue to be used as open sewers.

Although some local advances in sewage treatment were made in the 1920s and 1930s, most of the sewage discharged by municipalities into streams in the Ohio basin remained untreated. The sewage and industrial wastes from Pittsburgh and surrounding municipalities formed the basin's largest pollution load and created a notable oxygen depletion zone in the Ohio River below the city. By the late 1930s, offensive sights and smells from the rivers had increased, causing "unsightly and malodorous conditions along all water fronts," as sewage from the city as well as from upstream communities overwhelmed stream oxidation capacity (USPHS, 1944).[12] Many water supplies suffered from problems of taste and odor, and sewage contamination threatened to adversely impact public health. Although filtration and chlorination had sharply reduced typhoid death rates, death rates from diarrhea and enteritis remained elevated. In addition, ongoing leakage increased the costs of water filtration for Pittsburgh residents. These obvious water quality problems caused a gradual increase in citizen awareness that water pollution had adverse economic consequences, reduced environmental quality, and threatened the public health.

In 1937, in response to the demands of various stakeholders including conservation groups and public health authorities for improved stream quality, the Pennsylvania General Assembly passed the Clean Streams Act (Casner, 1999). This act gave the Sanitary Water Board the power to issue and enforce waste treatment orders to all municipalities and most industries, notably excluding acid drainage from coal mines.[13]

Faced by a state law that provided the state authority with enforcement powers, the city and other western Pennsylvania towns were forced to seriously consider means to reduce their river pollution. Movement toward reform began in the late 1930s, only to be interrupted by the outbreak of World War II. In 1944, however, with the war coming to an end, the Sanitary Water Board announced comprehensive plans to reduce pollution of Pennsylvania streams. In a major step, the board required that all municipalities treat their sewage "to a primary degree." In June 1945, it issued orders to the City of Pittsburgh and 101 other Pennsylvania municipalities, including more than 90 Allegheny County industries, to cease the discharge of untreated wastes into state waterways. State officials decreed that these communities comply with the treatment orders by May of 1947 (Yosie, 1981).

Municipal and county officials reacted with consternation to the 1945 state waste treatment orders as they confronted difficult administrative and technological issues. The previous argument that state spending limits prevented spending for sewage treatment plants could no longer be used, since new legislation and court decisions in 1934 and 1935 permitted local governments to create authorities and exercise user charge financing to provide public services such as sewage treatment (Reader, 1954). Among the most pressing issues were

[12] Field tests conducted by PADOH, for instance, showed that the dissolved oxygen level at many Ohio River locations was below 4 parts per million—the minimum level necessary for maintenance of aquatic life. In 1944, a U.S. Public Health Service (USPHS, 1944) study of the Ohio River reported that the organic waste load reaching the Ohio River from Pittsburgh and its suburbs had a population equivalent of 1,334,300.

[13] Some authorities had argued that the acid mine drainage neutralized bacterial wastes (see Casner, 1999).

questions as to whether sanitary policy should be determined by each municipality or by a regional agency, how sewage treatment should be financed, whether a central treatment facility or multiple disposal plants would be most efficient, and what form of treatment technology made the most sense given the existing system of sewage collection and drainage and local conditions of population density and topography (Yosie, 1981).

The county commissioners proceeded to survey affected Allegheny County municipalities to determine if they would support the incorporation of a special district governmental authority to plan and implement a waste disposal program. Commissioner John J. Kane argued that a countywide waste treatment agency financed by revenue bonds could service participating localities without increasing their level of bonded indebtedness. User charges collected from households and industries connected to the treatment system could provide a self-financing mechanism for efficient and professional waste disposal service. Finally, these officials pointed out that the construction of an authority-managed treatment system would benefit the Pittsburgh metropolitan economy because of the demands it would create for building materials, labor, and technicians.

As a result of a set of meetings sponsored by the county commissioners with Allegheny County municipalities, a pro-authority consensus developed. Seventy-four Allegheny County communities in receipt of state treatment orders resolved that their individual waste disposal problems could be best solved "on a county-wide basis." This provided the support that the county commissioners needed, and on December 19, 1945, acting under the recently passed Pennsylvania Municipal Authorities Act, the commissioners adopted a resolution to create a sanitary authority. Three months later, in March, 1946, the Secretary of the Commonwealth of Pennsylvania officially approved the formation of the Allegheny County Sanitary Authority (ALCOSAN) (Laboon, 1973). The county commissioners appointed John P. Laboon, who had previously served as Allegheny County Public Works Director, as chief engineer. The ALCOSAN board subsequently elected Laboon as chairman.

The ALCOSAN research staff conducted a number of studies and made several recommendations in 1945 and 1946 regarding the creation of an integrated sewage treatment system utilizing activated sludge technology. An early investigation by ALCOSAN showed that county municipalities with a population of 678,000 and industries collectively discharged 65 million gallons of wastewater per day. The authority decided to remove 50 percent of the biochemical oxygen demand generated by wastes entering its proposed activated sludge treatment plant, thereby complying with or exceeding discharge standards established by the Sanitary Water Board (Laboon, 1973). The research staff recommended that a single plant (located on a site situated on the north bank of the Ohio River north of Pittsburgh) would be more cost-effective than multiple plants and lead to fewer siting and odor objections from local populations (Yosie, 1981).

The ALCOSAN staff was also responsible for construction of an extensive intercepting sewer system, connecting various outfalls throughout the service district, and transporting wastes to the treatment works. Because the building of the interceptors promised extensive disruptions of transportation and industrial activity, ALCOSAN officials recommended subaqueous sewer construction (Laboon, 1973). As noted previously, individual municipalities would be responsible for owning and maintaining their own collection systems.

Although the financial and construction details of ALCOSAN were essential, the authority would become feasible only through delicate political negotiations with the City of Pittsburgh and with Allegheny County municipalities protective of their independence. Intensive negotiations brought about agreements between the city and ALCOSAN, the essence of which was to enlarge the number of communities permitted into the service district and to increase city control over the board (Yosie, 1981). Initially, 67 surrounding municipalities joined ALCOSAN, but the refusal of others (see Box 2-2) to join reflected the governmental fragmentation of Allegheny County and the region and the difficulties of securing consensus on governmental issues.

To secure financing (estimated at $82 million in 1948), the city and ALCOSAN decided to utilize non-debt revenue bonds financed by the imposition of waste treatment charges on those households and industrial plants located within the service district (ALCOSAN, 1948). To determine service charges, sanitary officials metered water consumption levels at the household tap or plant site. State legislation required water utilities to furnish water meter readings to ALCOSAN and empowered the agency to terminate water service to individual customers delinquent in their payment of sewage treatment bills (Laboon, 1973). On October 1, 1958, after a four-year loan of $100 million had been secured, official dedication of the plant occurred. Wastewater treatment in and around the City of Pittsburgh became a long awaited reality (Yosie, 1981; Tarr and Yosie, 2004). However, the basic combined sewer collection system for the City of Pittsburgh remained, as it did for other neighboring municipalities, and continued to discharge untreated raw sewage through overflow relief valves into streams and rivers under wet weather conditions.

Sewer Tap-in Restrictions

The presence of aging and inadequate sewer systems in some of the region's municipalities has slowed development and caused serious problems for towns seeking to increase their tax bases. During the 1990s, the Pennsylvania Department of Environmental Protection (PADEP) imposed tap-in restrictions on a number of communities in the Pittsburgh region because of problems such as wet weather overflow, overflows at treatment plants, and leaking sewers. Among those affected, for instance, were Rostraver Township, Pleasant Hills, Plum Boro, Cranberry, and East Huntingdon Township. In 2002, PADEP imposed tap-in restrictions on 31 Allegheny County municipalities including Bethel Park, Ross, Shaler, and Upper St. Clair. In Upper St. Clair, construction of new homes was put on hold because of such restrictions. Retail business was affected also, and in Shaler Township a new shopping mall was delayed and one retailer left the town because it was unable to tap into Shaler's sewers (Fitzpatrick, 2002).

ALCOSAN and Sanitary Sewer Overflow

Several communities that were members of ALCOSAN had sanitary sewers rather than combined sewers (see also Figure 1-2). Stormwaters were intended to be handled by storm sewers, surface channels, or surface runoff into neighboring streams. Sanitary sewer flows, however, often included runoff from connected roof drains. Such connections were, according to one authority, the "construction standard" in Allegheny County and were common in other cities as well. These connections resulted in the sanitary sewer flows often being diluted. Under the system design planned by the authority, all flows into the treatment plant were to be regulated,

BOX 2-2
PENN HILLS: A CASE OF WASTEWATER MISMANAGEMENT[1]

The Borough of Penn Hills is a suburban community located directly east of the City of Pittsburgh on roughly 19 square miles of hilly land (see Figure 1-1). It occupies parts of three watersheds: the Allegheny River on the northwest; Turtle Creek on the south, which runs into the Monongahela River; and Sandy Creek, a minor watershed in the central west region that drains to the Allegheny River. It has a total of 13 drainage areas. Originally known as Penn Township, it had a population of 15,000 in 1940 but experienced explosive growth in the post-World War II period, reaching a population peak of 62,000 in 1970. The federal Works Projects Administration's real property inventory estimated that approximately 40 percent of the homes in the township had no indoor plumbing in 1934, depending on wells for water and probably on-lot septic tanks or privy vaults for waste disposal. In the postwar growth period these conditions changed rapidly. Water was provided by two quasi-public utilities, and developers supplied most new housing developments with sanitary sewers since Penn Hills subsurface conditions did not lend themselves to the use of septic systems. Some septic systems remained, however, as in the oldest area of the township bordering Pittsburgh, where some homes had septic tanks while others were connected to a combined sewer. In addition, Penn Hills was an old mining area, and as in many such communities, some homes still had straight pipes run into the old mines for sewage disposal. Few provisions were made for stormwater, with the expectation that it would drain on the surface or be channeled via short storm sewers into local creeks.

Penn Hills began treating part of its sewage before World War II, and by 1955 the township possessed six treatment plants of various sizes that provided primary and secondary treatment. The reliance on multiple plants rather than a single centralized plant was necessitated by the hilly topography. By 1988, there were five such treatment plants, four of which had been constructed in the previous 10 years or so. By this date, most homes utilizing septic tanks had been forced to connect to sewers, although some newer developments on the fringe of the municipality still utilized such subsurface disposal methods.

Penn Hills began to experience problems with its wastewater collection and treatment system during the period of its most rapid growth in the 1970s and 1980s. These problems related largely to a failure to provide funding for proper system maintenance, inefficient administration, and poor land development practices. Penn Hills also possessed a political culture that strongly resisted raising taxes in order to pay for improved services or to maintain those that existed. Even when sewer fees were levied, they were frequently not collected. Inadequate finances meant that the sewerage system was poorly maintained, with extensive inflow and infiltration problems, frequent incidences of sewage backed up in basements, and a failure to effectively provide for stormwater flows. In their haste to build, developers had exacerbated the system difficulties by building on steep slopes and hooking up roof gutters and driveway drains into sanitary sewers, even though the township began prohibiting such procedures in the late 1950s.

The diversion of stormwater to sanitary sewers resulted in a large volume of stormwater entering the sanitary system. On wet weather days the normal flow was increased 5 to 30 times in volume. This led to hydraulic overloading of the existing wastewater treatment plants and state

and they required that sanitary sewer overflow relief be provided. Sanitary sewer overflows (SSOs) were prohibited under the 1987 amendments to the Clean Water Act but action was not taken against ALCOSAN until 1996.

In the years following its founding, ALCOSAN built additional facilities, improved its treatment processes to meet higher levels set first by the Pennsylvania Sanitary Water Board and later by the U.S. Environmental Protection Agency (EPA) under the CWA. In fall 2003, 26 and prohibition of any new tap-ins, slowing development in the borough. All the Penn Hills wastewater treatment plants utilized secondary treatment with activated sludge as the secondary process. Waste biosolids (sludge) from the activated sludge process were sometimes allowed to discharge into streams via a bypass mechanism during increases in the incoming flows to the plant. In addition, untreated wastewater was permitted to bypass the treatment plant during periods of wet weather because of the existence of sanitary sewer overflows.

When the City of Pittsburgh and other Allegheny County municipalities created ALCOSAN in the 1950s, Penn Hills refused to join because it already had its own treatment plants and was unwilling to forgo its capital investments or the fees and patronage that went with their operation. In addition, it disliked the fee system that ALCOSAN had established. Two peripheral parts of the township, however, one on the northwest and one on the southwest, were connected with sewer systems in the neighboring township of Verona and with that of the City of Pittsburgh because of watershed boundaries. Both of these systems became part of ALCOSAN and Penn Hills was required to pay monthly fees for using their facilities. This fee situation caused further political discontent within the municipality. Throughout the 1980s and early 1990s, as state and federal water quality programs and contaminant standards were strengthened, Penn Hills experienced increasing problems with its sewerage system. These were compounded because of an inadequate funding base to upgrade the collection system and the treatment plants.

One of the major difficulties was created by the decision of Penn Hills to cease dewatering its sludge and end its contracts for landfill disposal, a decision based on a financial shortfall. Sludge was permitted to build up at the treatment plants and then flow into the river on wet weather days along with untreated sewage. The result was a rise in coliform counts in the Monongahela River and a clear violation of the Clean Water Act. In 1989 and 1990, the Pennsylvania Department of Environmental Resources (PADER) ordered Penn Hills to improve conditions at its sewage treatment plants but the municipality failed to act and failed to provide legally required sewage discharge reports to the U.S. Environmental Protection Agency (EPA).

In August 1991, the U.S. Attorney's Office, acting on behalf of the EPA and PADER, asked for a preliminary injunction to force Penn Hills to stop dumping raw sewage into the Monongahela. A court granted the injunction and, in September 1993, followed with a permanent injunction obliging Penn Hills to comply with the Clean Water Act by January 1996. The joint EPA-Pennsylvania lawsuit accused Penn Hills of discharging pollutants in excess of its National Pollutant Discharge Elimination System (NPDES) permit limits; discharging raw sewage through unlawful bypasses within its sewerage system; failing to properly dispose of sludge; failing to properly maintain and operate its facilities; and failing to accurately monitor and report as required under its NPDES permits. In addition, charges were brought against two sewage plant employees for falsifying discharge records, with both being found guilty and one going to prison. During the next five years, Penn Hills spent approximately $50 million to upgrade its sewage system in order to prevent new violations of the Clean Water Act. A settlement of the lawsuit was reached in July 1998 in which Penn Hills agreed to pay a $525,000 penalty, improve its wastewater treatment operations, eliminate unauthorized overflows, and improve monitoring and reporting. In addition, it agreed to become a member of ALCOSAN (EPA, 1998).

---

[1] Unless otherwise stated, information included in this box is derived from Tarr et al. (2002).

55 Allegheny County communities served by ALCOSAN had negotiated consent orders with the EPA, PADEP, and the Allegheny County Health Department for CSOs and SSOs, respectively (see also Chapter 1, footnote 1).

Water Supply and Wastewater Disposal Practices in Rural Southwestern Pennsylvania

Sanitary conditions in rural southwestern Pennsylvania and rural areas throughout the nation were generally considered primitive before the post-World War II period. As late as 1943, a text *Municipal and Rural Sanitation* noted that "At the present time, and in all probability for many years to come, excreta will be disposed of without water carriage at the vast majority of farmhouses, at residences in the smaller towns and villages . . ." (Ehlers and Steel, 1943). Studies by PADOH confirmed that rural farm areas and small towns often were poorly served by sanitary facilities and often relied on wells for drinking water located close to privies (PADOH, 1915).

The most common means for disposal of human waste was the privy vault. Privies differed by type, including surface privies where excreta accumulated on top of the ground and liquids were allowed to leach away; pit privies in which excreta fell into a pit in the ground; and drop privies that overhung brooks or rivers where the excreta dropped into the water to be carried away by the stream (Rosenau, 1927). Pit privies were considered threats to neighboring wells, while drop privies threatened downstream water supplies. Cesspools, usually of the percolating type, were also frequently used. In the smaller towns, some public sewers existed, although private sewers (straight pipes) from water closets or privies frequently drained into neighboring waterbodies. All of these types of privies and conditions existed in southwestern Pennsylvania during the early decades of the twentieth century (PADOH, 1915).

Some small towns had especially bad conditions. A 1915 survey by PADOH of the Allegheny River basin, for instance, noted that in the small town of Derry in Westmoreland County (about 2,000 population), about 200 small private sewers discharged into tributaries of the Allegheny River. In addition, many Derry residents used privies "of a shallow type, almost universally overflowing." According to that report, all of the streams in the community were "badly polluted" (PADOH, 1915).[14]

An unusual feature of southwestern Pennsylvania was the number of small towns or "patch towns" established by coal mining companies. Coal companies constructed these towns in order to provide housing for miners near the mines. Homes in these patch towns were built according to a rather common model, with four "holer" privy vaults located in the back. Water was often provided through a spigot on the side of the house or, in more primitive communities, by wells. Many of these patch communities still exist today, and the company housing has long since been sold to private individuals. Although a substantial number of these homes have been modernized in recent years, many did not receive modern indoor plumbing until after the 1950s and some still rely on "straight pipes" to dispose of their untreated sewage into neighboring streams.[15]

In the 11-county Pittsburgh region today (see Box 1-2), Allegheny County is the only county with a health department. Throughout most of the twentieth century, PADOH had responsibility for all public health activities outside of incorporated municipalities and provided health services for the rural areas of the state. The state appointed and paid for a part time

---

[14] In response to an order of the PADOH, Derry constructed a system of sanitary sewers and sedimentation tanks circa 1915.

[15] See the Virtual Museum of Coal Mining in Western Pennsylvania at *http://patheoldminer.rootsweb.com/* for further information.

medical director for each county, providing public health nurses, school medical inspectors, and sanitary inspectors (Smillie, 1939). With regard to water supply and sewerage, the state conducted sanitary surveys of various rivers basins, while after 1923 the Sanitary Water Board had the authority to grant permits for sewage treatment and construction of new sewers.

New residential development in rural areas increasingly distant from Pittsburgh, as well as older rural housing (both individual and community) and older coal mining patch towns, often did not have practical access to large municipal central sewerage systems. Thus, these communities and developments relied on a literal patchwork of community and on-lot sewage disposal methods. The lack of coordination, standardization, and oversight of system siting, design, construction, and maintenance resulted in inadequate and malfunctioning sewer systems in these areas. Substandard wastewater systems include straight piping of sewage from homes to streams or ditches, septic tanks discharging directly into streams or ditches, drywells and cesspits, homes connected to neighborhood or community straight pipes ("wildcat sewers"), sewage discharges into abandoned underground coal mines, and failed or malfunctioning community "package plant" treatment systems. In addition, because many of these substandard systems predate modern sewage regulations and ordinances, no records exist for a majority of them.

According to the Pennsylvania Association of Township Supervisors, the post-World War II housing boom in Pennsylvania resulted in an overloading of urban area central treatment systems, a proliferation of poorly operated small treatment systems, and between 500,000 and 1,000,000 malfunctioning on-lot systems throughout the Commonwealth. Concerns about surface water and groundwater deterioration, and the accompanying public health risks, led to the passage in 1966 of Act 537, the Pennsylvania Sewage Facilities Act.[16] This act was subsequently amended to expand options for local-level cooperation and enforcement and to expand on-site wastewater treatment options (ELI, 1999).

Although Act 537 has addressed many sewage problems, southwestern Pennsylvania's rural areas still face significant sewage disposal challenges. Many sewage problems accompany substandard, deteriorating housing where residents rely on fixed retirement income, low income jobs, or unemployment compensation. Local tax bases are often inadequate to support the required staffing and resources for data collection to adequately define local problems, to devise the sometimes unconventional solutions needed to address the problems, and to provide necessary management and record keeping. In some of these areas there may be added pressures from adjacent new home development or from economically struggling neighboring farms likely to be sold for new or second-home development.

In 1997, PADEP issued a guidance report *Policy Establishing New Program Direction Policy for Act 537 Comprehensive Planning* recognizing the special needs of rural municipalities (e.g., low development density, lack of available funding) and describing the department's role in assisting these municipalities in finding both technical and financial solutions to sewage

---

[16] Act 537 requires each Pennsylvania municipality to prepare and periodically update an official sewage facilities plan that is intended to provide a level of scrutiny to infrastructure decisions that might not otherwise take place if the only review were that provided under local zoning and site plan approval requirements (ELI, 1999). Such plans are intended to identify how sewage will be handled and properly disposed of in each municipality; they also lay out how the necessary sewer conveyance and treatment facilities will be located, constructed, and maintained. When a new development requires the extension of sewer lines or the construction of additional capacity for wastewater treatment, the municipality is required to prepare and approve a plan revision, which is then submitted to PADEP.

problems. That report states that some low-income rural areas may require up to 90 percent grants to afford sewerage projects using conventional methods. It acknowledges that phased implementation and long-term goal setting may be needed to address rural sewage treatment issues, and it encourages the development of local management programs and decentralized ("noncentralized") sewage treatment alternatives (PADEP, 1997a).

Also in 1997, PADEP issued the policies and procedures document *Impact of Use of Subsurface Disposal Systems on Groundwater Nitrate Nitrogen Levels* (PADEP, 1997b). That document details the requirements for technical studies, siting, and system design and technologies to avoid nitrate contamination of groundwater in vulnerable areas.

Acid Mine Drainage

The sulfuric acid discharge from coal mines has been the most pervasive and widespread water pollution problem in southwestern Pennsylvania's industrial history (see also Chapters 3 and 4). The region's bituminous coal has a high sulfur content and produces enormous acid loads. Drainage from extensive networks of abandoned underground mines; thousands of small, abandoned "country bank" mines (used for local domestic purposes); and large numbers of active commercial mining operations contributes to stream degradation throughout the region. Historically, acid mine drainage (AMD) destroyed fish communities and altered the flora along both small streams and major rivers, caused millions of dollars of damages to domestic and industrial water users, and increased the costs of water and sewage treatment (Casner, 1994).

Acid mine drainage also affects human quality of life and public health. AMD diminishes the quality of drinking water sources and impairs water delivery. Its corrosive action has destroyed pipes and pumps, forcing water authorities to build neutralizing plants as a component of their overall water treatment systems.

The coal industry also has incurred damages from AMD. The most immediate problems occurred at the mines themselves, where the corrosive action of the acidic water damaged and destroyed the pipes and pumps installed to remove water. Damaged machinery increased expenses and also hindered timely coal extraction, while leaks caused by corrosion increased the cost of energy production and damaged equipment. In highly acidic environments, bronze, lead, and wooden linings and covers were used to protect mining equipment (Casner, 1994).

As acid mine discharges moved into neighboring streams and rivers, other industries encountered degraded water, making treatment a necessity. The effect on the steel industry was severe, requiring all mills to have water treatment facilities. Railroads, which until the 1950s largely used coal-burning locomotives that depended on frequent water stops, were affected adversely as well. Their costs included larger coal bills, boiler repairs, and cleaning (Bardwell, 1953). American railroads were forced to build water treatment plants in an effort to avoid or reduce such costs. By 1934, rail lines had built 1,200 water treatment plants and were treating 90 billion gallons per year. Regional railroads developed water supply systems by building reservoirs above mining districts and running pipelines along rail networks (Crichton, 1927).

By the 1920s, municipalities, recreational users of the rivers, and industry in southwestern Pennsylvania had begun pushing government for solutions to counter the growing burden of AMD. For government policy makers, acidic water offered a significant engineering and policy dilemma because it hindered effective water treatment and presented an indirect public health threat. However, the economic importance of coal production in southwestern

Pennsylvania inhibited coercive action. Indeed, the coal industry argued emphatically that no "suitable" method existed for the treatment of acid mine water. Even if authorities contemplated action, the industry had achieved legal protection in a precedent set by the seminal 1886 case of *Pennsylvania Coal Company v. Sanderson and Wife*, which concerned destruction of the water supplies for a family farm near Scranton, Pennsylvania. In this case the Pennsylvania Supreme Court maintained that "the right to mine coal is not a nuisance in itself" and that the acidic substances entered the stream via natural forces that were beyond company control. The justices also considered the economic importance of the coal industry and its provision of jobs, arguing that "the trifling inconvenience to particular persons must sometimes give way to the necessities of a great community" (Casner, 2004). In 1905, when the Pennsylvania legislature passed the Purity of Waters Act, it specifically exempted "waters pumped or flowing from coal mines." In 1923, at then Governor Gifford Pinchot's initiative, and as noted previously, the Pennsylvania legislature created the Sanitary Water Board with investigatory and advisory powers. Again, however, state legislators specifically exempted AMD from possible proposed restrictions, upholding its protection under state law (Broughton et al., 1973; Wolman, 1947).

The first significant challenge to the protected status of coal mining came in 1923, in the case of the *Pennsylvania Railroad v. Sagamore Coal, et al.* when the Pennsylvania Supreme Court rendered a decision that undermined the 1886 Sanderson doctrine. In this case, the Pennsylvania Railroad successfully sued several coal mining companies for polluting its reservoir in the Indian Creek watershed (located 65 miles southeast of Pittsburgh) on the grounds that the acid pollution created a public health nuisance since it polluted the water supplied to several regional water companies. The court ruled that coal companies possessed "no right of any kind" to discharge acidic water into streams when the public made use of the water (Casner, 2004).

In the 1920s, pressures to resolve problems resulting from AMD emerged from different stakeholders including domestic and individual water users, industrial users, and sportsmen's groups. Various strategies were suggested, but the one that gained the most favor involved sealing abandoned coal mines. Sealing of mines causes flooding of the mine voids, which substantially reduces the amount of oxygen, and thus acid production, in those voids. In 1924, the Pennsylvania Supreme Court ordered the mines above Indian Creek sealed in order to prevent further contamination of the Pennsylvania Railroad's reservoir. In the 1920s, the U.S. Bureau of Mines studied mine sealing as a pollution abatement method. The technique initially seemed an inexpensive and simple remedy to an expensive and complex problem, but it failed to take into account the number and size of abandoned mines, natural geologic factors, and industry's strong preference for the government to pay for sealing (Casner, 2004).

In the 1930s, mine sealing projects were undertaken under the auspices of the federal Civil Works Administration and the Work Projects Administration (WPA). In the first two years of the WPA program (1935-1937), sealing crews covered more than 47,000 openings at 1,527 sites in four states including Pennsylvania, Ohio, West Virginia, and Kentucky. Pennsylvania secured the most openings, with a reported 30,000 at 317 mines in 22 counties. The sealing temporarily produced the desired effect, and in 1940 the U.S. Public Health Service estimated that the average residual load of acid on the main Ohio River measured 48 percent of what it was prior to the sealing project (USPHS, 1944). On the Monongahela River, sealing reduced acid loads by 51 percent. On the Kiskiminetas River, a tributary of the Allegheny River and at that time the primary source of acidic water affecting the Pittsburgh water supply, reduction efficiency at abandoned mines achieved one of the highest levels with a decline of 78 percent

(Casner, 2004). Most public authorities through the 1950s considered the project an unqualified success. The program had reduced acid concentrations in the waters of the Ohio River basin and served as an excellent example of the benefits of federal cooperation with state efforts in water pollution control. Mine sealing, however, produced only temporary relief from acid mine drainage because seals frequently broke down and allowed air and water to enter or water to escape the mines. By the 1960s, it was clear that AMD pollution of Pennsylvania streams remained a major state environmental problem.

Beginning in the late 1960s, Pennsylvania, West Virginia, and the federal government enacted legislation requiring active mining operations to treat polluted water prior to discharge. The Clean Water Act instituted the National Pollutant Discharge Elimination System (NPDES), which required all "point sources" of pollution to apply for an NPDES permit and meet discharge water quality standards. Both Pennsylvania and West Virginia were granted primacy such that their respective agencies received authority under the Federal Water Pollution Control Act of 1972 to issue and enforce NPDES permits. Permitted discharge limits for mining operations are typically governed by "technology-based limits." The act (which in 1977 became known as the Clean Water Act; see also Box 1-1) was amended by Congress in 1987 to establish the Section 319 Nonpoint Source Management Program, recognizing that regulated point sources in many regions of the country, accounted for only a minor share of the pollutant loadings.

While the CWA initiated federal oversight of pollution, creation of the federal Office of Surface Mining, Reclamation and Enforcement (OSM) and its Surface Mining Control and Reclamation Act (SMCRA) of 1977 brought federal oversight to the permitting of new coal mines. Title IV of SMCRA identified abandoned mine lands (AMLs) as mines that were abandoned or left in an inadequate state of reclamation prior to August 3, 1977. Title IV also created the Abandoned Mine Land Reclamation Fund supported by a tax on coal production. A portion of these funds has been distributed to eligible states and used to reclaim abandoned mines, reduce hazards, and make water quality improvements. As of 2003, the Abandoned Mine Land Reclamation Fund had a balance of about $1.5 billion (OSM, 2003; see Chapter 4 for further information). Section 403 of SMCRA assigns the following priorities to the expenditure of AML funds:

1. protection of public health, safety, general welfare, and property from extreme danger of adverse effects of coal mining practices;
2. protection of public health, safety, and general welfare from adverse effects of coal mining practices;
3. restoration of land and water resources and the environment previously degraded by adverse effects of coal mining practices, including measures for the conservation and development of soil, water (excluding channelization), woodland, fish and wildlife, recreation resources, and agricultural productivity;
4. protection, repair, replacement, construction, or enhancement of pubic facilities such as utilities, roads, recreation, and conservation facilities adversely affected by coal mining practices; and
5. development of publicly owned land adversely affected by coal mining practices including land acquired as provided in this title for recreation and historic purposes, conservation and reclamation purposes, and open space benefits.

The magnitude of the abandoned mine land problem dictated that much of the historic funding was devoted to priorities 1 and 2, while significant efforts to use the fund to remediate water (AMD) issues only began around 1995. Significant programs for AMD cleanup have been developed and are administered by the Commonwealth of Pennsylvania, EPA, OSM, and various watershed organizations.

SMCRA requires that any new mining permit be accompanied by a bond to cover the cost of reclamation in the event that the permittee is financially unable to do so. These bonds have not been assessed in amounts adequate to treat AMD and are based on surface disturbance. Typically, bond rates are less than $5,000 per acre. Bond forfeiture results from a finding by the state regulatory agency that the company is unable to fulfill its environmental requirements under its mining permit. The state then uses the bond amount to reclaim the surface disturbances.

Prior to enactment of the SMCRA, Pennsylvania established Operation Scarlift in the late 1950s specifically to deal with abandoned mines. Operation Scarlift constructed a series of lime neutralization treatment stations to neutralize some AMD discharges in severely affected watersheds. However, it was funded by a revenue bond, which when exhausted caused the program to become inactive in the 1970s. In recent years, Pennsylvania has instituted statewide programs to deal specifically with AMD discharges from bond forfeiture sites.

As noted previously, mine drainage is classified as a point source if it originates from an active, post-August 1977 mine. Discharges from these mines are governed by their respective NPDES permits. Discharges from mines that were abandoned prior to August 3, 1977, are considered nonpoint sources. They are unregulated and in most coal field watersheds are responsible for the overwhelming majority of metal ion and acidity loadings to surface waters. Policies regarding the states' responsibilities in maintaining NPDES permit conditions on bond-forfeited AMD sites are an emerging issue in both Pennsylvania and West Virginia. Recently, in Pennsylvania, the liquidation of LTV Corporation's coal assets placed five large underground coal mines along the Monongahela River under state responsibility. As a result, PADEP is evaluating ways to either operate LTV's AMD treatment plants or find more efficient methods for treating AMD (Hopey, 2003).

Current and Anticipated AMD Loadings in the Pittsburgh Basin

Coal mines in the Pittsburgh basin,[17] which generate about 5,500 tons of dissolved iron annually, also contribute to an acid loading of about 16,000 tons contained in 19 billion gallons of water (see Chapter 4 for a detailed discussion of the characteristics and effects of AMD in southwestern Pennsylvania). About one-third of the basin's discharge from mines is treated by the mining industry. Abandoned mines generate the remainder and currently pollute many of the

[17] The term "Pittsburgh basin" refers to the commonly accepted geological definition of the regional synclinal structure containing the Pittsburgh Coal Seam. It is the primary coal seam influencing the water quality of the Monongahela River and the most heavily mined coal seam in southwestern Pennsylvania. Thus, the data included in this section reflects only AMD discharges from the Pittsburgh Coal Seam. Deeper coal seams, such as the Freeport and Kittanning Seams, only outcrop around the northern and eastern margins of the same synclinal structure (i.e., areas north of Allegheny County and east of Fayette and Westmoreland Counties).

major tributaries to the Monongahela and Ohio Rivers. If all of this mine water were untreated however, it would be sufficient to add substantial metal loadings and acidity to already impaired tributaries with possible localized, severe effects on the Monongahela River. The Monongahela River's alkalinity at the West Virginia-Pennsylvania state line is about 36 mg/L and its low flow is about 6,000 cubic feet per second. This rate of mine drainage would supply about 200,000 tons of alkalinity per year. The majority of flooded mines are currently discharging net alkaline water, with soluble Fe^{2+} concentrations in the range of 25 to 100 mg/L. Given the high volumes of these mine discharges, iron staining and oxygen depletion in the Monongahela River are more likely to be problematic than acidity. Thus, under all but low-flow periods (late summer, early fall); dilution will likely ensure that effects of additional mine pool discharges would be localized, with affected plumes extending along the banks of the river for miles. During low-flow periods, water movement between navigation pools is extremely slow and oxygen deficits in the rivers would be exacerbated by mine drainage. Probably the worst-case scenario would entail a neutral, net alkaline Monongahela River at the Pittsburgh Point (confluence of Allegheny and Monongahela Rivers) with enough suspended ferric hydroxide to color the river orange. Oxidation of ferrous to ferric ion would contribute to the river's oxygen deficit, but a discussion of the effect on fish populations is beyond the scope of this report.

SUMMARY

The Allegheny, Monongahela, and Ohio River system has been central to the development, history, and identity of southwestern Pennsylvania. Abundant coal and other natural resources and the availability of convenient water and rail access within and beyond the immediate region facilitated economic growth and, at the same time, extensive air and water pollution in the City of Pittsburgh and surrounding communities from the mid-nineteenth century through the 1950s. With the decline of the steel industry in the late twentieth century, the region's economic base shifted to other sectors, including medical research, technology, and higher education. While there has been a remarkable transformation and recovery of the region's economy in the last two decades, many communities in southwestern Pennsylvania continue to experience significant economic problems resulting from the decades-long decline in mining and traditional manufacturing sectors. As a result, the population of the City of Pittsburgh declined steadily from about 520,000 in the 1970s to its present level of about 335,000. Despite this net loss of population, the Pittsburgh metropolitan area has been sprawling further onto rural land at rates that exceed other cities in the northeastern United States.

Although the environmental quality of the 11 counties of southwestern Pennsylvania and the City of Pittsburgh has improved dramatically in recent decades, pervasive water quality problems remain a legacy that transcends municipal, county, and even state lines. In this regard, acid mine drainage, effluent from on-lot septic systems, and raw sewage continue to enter local streams, the region's three major rivers, and underlying groundwater in both urban and rural areas. These problems threaten the region's public health, environment, economy, and image.

Many of the region's current urban water quality problems can be traced to historical water supply and wastewater infrastructure decisions. The City of Pittsburgh, ALCOSAN, and its 83 serviced communities are facing extensive and costly regulatory action under the federal Clean Water Act for both combined and sanitary sewer overflows. Furthermore, some sewage-related water quality problems persist even in dry weather because the presence of aging and deteriorating septic systems and sewer pipes that are a major source of sewage contamination to groundwater supplies. This problem is exacerbated by the fact that southwestern Pennsylvania is dominated by poor shallow soils, a high water table, and sloped terrain, making the region one of the most challenging in the country for use of on-site sewage treatment and disposal systems such as septic tanks and leach fields.

REFERENCES

ALCOSAN (Allegheny County Sanitation Authority). 1948. Report on the Proposed Collection and Treatment of Municipal Sewage and Industrial Wastes by the Allegheny County Sanitary Authority. Pittsburgh, PA.

Allan, J., D. Erickson, and J. Fay. 1997. The influence of catchment land use on stream integrity across multiple spatial scales. Freshwater Biology 37:149-161.

American Rivers, NRDC (National Resources Defense Council), Smart Growth America. 2002. Paving Our Way to Water Shortages: How Sprawl Aggravates the Effects of Drought. Washington, DC.

Anderson, R., K. Beer, T. Buckwalter, M. Clark, S. McAuley, J. Sams, and D. Williams. 2000. Water Quality in the Allegheny and Monongahela River Basins: Pennsylvania, West Virginia, New York, and Maryland (1996-98). Denver, CO: U.S. Geological Survey.

Bardwell, R. 1953. Water treatment for railway systems. Water Works and Sewage 80:9.

Belko, M. 2004. Pittsburgh lands Bassmaster Classic for 2005. Pittsburgh Post-Gazette, June 16. Available on-line at *http://www.post-gazette.com/pg/04168/332700.stm*. Accessed June 16, 2004.

Brookings Institution. 2003. Back to Prosperity: A Competitive Agenda for Renewing Pennsylvania. Washington, DC: Brookings Institution.

Broughton, R., T. Koza, and F. Selway. 1973. Acid mine drainage and the Pennsylvania courts. Duquesne Law Review 11:499-555.

Byington, M. 1910. A Homestead Court in Homestead: The Households of a Mill Town, 1907-1908. New York: Charities Publishing Committee.

Casner, N. 1994. Acid water: A history of coal mine pollution in western Pennsylvania, 1880-1950. Ph.D. dissertation, Carnegie Mellon University.

Casner, N. 1999. Polluter versus polluter: The Pennsylvania railroad and the manufacturing of pollution policies in the 1920s. Journal of Policy History 11:179-200.

Casner, N. 2004. Acid mine drainage and Pittsburgh's water supply. In Devastation and Renewal: An Environmental History of Pittsburgh and Its Region, J. Tarr (ed.). Pittsburgh, PA: University of Pittsburgh Press.

Center for Watershed Protection (CWP). 2003. Impacts of impervious cover on aquatic systems. Watershed Protection Research Monograph No. 1. Ellicott City, MD: CWP.

Collins, T., E. Muller, and J. Tarr. 2003. Pittsburgh rivers: From urban industrial infrastructure to environmental infrastructure. Paper delivered at Conference on Rivers in History: Designing and Conceiving Waterways in Europe and North America. Pittsburgh, PA: Carnegie Mellon University.

Crichton, A. 1927. Disposal of drainage from coal mines. Proceedings of American Society of Civil Engineers 53:1656-1666.

Ehlers, V., and W. Steele. 1943. Municipal and Rural Sanitation, 3rd Ed. New York: McGraw-Hill.

ELI (Environmental Law Institute). 1999. Plumbing the Future: Sewerage and Sustainability in Western Pennsylvania. Washington, DC: ELI.

EPA (Environmental Protection Agency). 1998. U.S. and Pennsylvania settle clean water lawsuit against Penn Hills. EPA Environmental News.

EPA. 2001. Our Built and Natural Environments: A Technical Review of the Interactions Between Land Use, Transportation, and Environmental Quality. EPA 231-R-01-002. Washington, DC: EPA.

Fitzpatrick, D. 2002. Pipe dreams: Lack of sewer, water hookups may keep region out of running for projects. Pittsburgh Post-Gazette. Available on-line at *http://www.post-gazette.com/businessnews/20020721sewer0721p3.asp*. Accessed June 18, 2004.

Fuller, T., M. Shields, and S. Smith. 2002. Road to 2003: Update on Pennsylvania: The Economy, Jobs, Forecasts and Telecommunications. University Park, PA: Pennsylvania State University Center for Economic and Community Development.

Fuller, T., M. Shields, and S. Smith. 2003. Road to 2004: Update on Pennsylvania: The Economy, Jobs, Forecasts and Telecommunications. University Park, PA: Pennsylvania State University Center for Economic and Community Development.

Fulton, W., R. Pendall, M. Nguyen, ad A. Harrison. 2001. Who Sprawls Most? How Growth Patterns Differ Across the U.S. Washington, DC: Brookings Institution.

Galli, J. 1990. Thermal impacts associated with urbanization and stormwater management best management practices. Washington, DC: Metropolitan Washington Council of Governments, Maryland Department of the Environment.

Gregory, G. 1974. A study in local decision making: Pittsburgh and sewage treatment. Western Pennsylvania Historical Magazine 57:25-42.

Harper, J. 1997. The formation of Pittsburgh's Three Rivers. Pennsylvania Geology 28(3/4):4.

Herlihy, A., J. Stoddard, and C. Johnson. 1998. The relationship between stream chemistry and watershed land cover in the mid-Atlantic region of the U.S. Water, Air, and Soil Pollution 105:377-386.

Hopey, D. 2003. LTV OKs plan to treat mine drainage: Trust fund set up to pay for cleanup, reclamation. Pittsburgh Post-Gazette. Available on-line at *http://www.post-gazette.com/pg/03247/217887.stm*. Accessed April 19, 2004.

Jones, K., A. Neale, M. Nash, R. van Remortal, J. Wickham, K. Riitters, and R. O'Neill. 2001. Predicting nutrient and sediment loading to streams from landscape metrics: A multiple watershed study from the United States mid-Atlantic Region. Landscape Ecology 16:301-312.

Katz, B. 2002. The new metropolitan agenda: Speech to Southwestern Pennsylvania Smart Growth Conference, June 9.

Katz, B. 2004. Pittsburgh: The road to reform. Op-ed column. Pittsburgh Post-Gazette, January 18. Available on-line at *www.brookings.edu/views/op-ed/katz/2004118.*

Koppes, C., and W. Norris. 1985. Ethnicity, class, and mortality in the industrial city. Journal of Urban History 11:259-279.

Laboon, J. 1973. Chronological highlights of the history of the Allegheny County Sanitary Authority. Manuscript: 1-2. Pittsburgh, PA.

Lanpher, E., and C. Drake. 1930. City of Pittsburgh: Its Water Works and Typhoid Fever Statistics. Pittsburgh, PA: City of Pittsburgh.

McElfish, J., Jr., and S. Casey-Lefkowitz. 2001. Smart Growth and the Clean Water Act. Washington, DC: Northeast-Midwest Institute.

McElwaine, A. 2003. Slag in the park. In Devastation and Renewal: An Environmental History of Pittsburgh and Its Region, J. Tarr (ed.). Pittsburgh, PA: University of Pittsburgh Press.
McKay, G. 2002. Homes in Summerset at Frick Park blend old-time style, new urban amenities. Pittsburgh Post-Gazette, February 16.
Melosi, M. 2000. The Sanitary City: Urban Infrastructure in America from Colonial Times to the Present. Baltimore, MD: Johns Hopkins University Press.
Ogle, M. 1996. All the Necessary Conveniences: American Household Plumbing, 1840-1890. Baltimore, MD: Johns Hopkins University Press.
OSM (Office of Surface Mining). 2003. Abandoned Mine Land Reclamation: Reclamation of Abandoned Mine Land That Took Place Before the Surface Mining Law Was Passed n 1977. Available on-line at *http://www.osmre.gov/annualreports/03aml.pdf.*
PABS (Pennsylvania Biological Survey). 1998. Inventory and Monitoring of Biotic Resources in Pennsylvania: Current Ecological and Landscape Topics, Volume 1. Harrisburg, PA: Department of Conservation and Natural Resources.
PADEP (Pennsylvania Department of Environmental Protection). 1997a. Policy Establishing New Program Direction Policy for Act 537 Comprehensive Planning. 362-2206-007. Harrisburg, PA: Bureau of Water Quality Protection.
PADEP. 1997b. Impact of Use of Subsurface Disposal Systems on Groundwater Nitrate Nitrogen Levels. 362-2207-004. Harrisburg, PA: Bureau of Water Supply and Wastewater Management.
PADOH (Pennsylvania Department of Health). 1911. Fourth Annual Report of the State Department of Health. Harrisburg, PA: PADOH.
PADOH. 1915. Report on the Sanitary Survey of the Allegheny River Basin. Harrisburg, PA: PADOH.
Pearson, W., and B. Pearson. 1989. Fishes of the Ohio River. Ohio Journal of Science 89(5):181-187.
PEC (Pennsylvania Environmental Council). 2003. Three Rivers Conservation Plan: Draft. Pittsburgh, PA: PEC.
Peterson, J. 1979. The impact of sanitary reform upon American urban planning. Journal of Social History 13:84-89.
Reader, F. 1954. Financing municipal sewage treatment facilities in Pennsylvania by use of municipal authorities. Dickinson Law Review 58:335-336.
Rosenau, M. 1927. Preventive Medicine and Hygiene, 5th Ed. New York: D. Appleton and Co.
Roth, N., J. Allan, and D. Erickson. 1996. Landscape influences on stream biotic integrity assessed at multiple scales. Landscape Ecology. 11:141-156.
Saville, T. 1931. Administrative control of water pollution. Transactions, American Institute of Chemical Engineers 27:74-77.
Scheuler, T. 1994. The importance of imperviousness. Watershed Protection Techniques 1:100-111.
Shields, M. 2002. Pennsylvania's Rural Economy: An Analysis of Recent Trends. Pennsylvania State University's College of Agricultural Sciences Agricultural Research and Cooperattive Extension. Available on-line at *http://cecd.aers.psu.edu/pubs/PA_rural_economy_trends.pdf.*
Smillie, W. 1939. Public Health Law, 2nd. Ed. New York: The Commonwealth Fund.
Snow, F. 1907. Administration of Pennsylvania laws respecting stream pollution. Proceedings of the Engineers' Society of Western Pennsylvania 23:266-283.
Stevenson, W. 1923. Pennsylvania sanitary water board. Engineering News Record 91:684-85.
Sustainable Pittsburgh. 2003. Southwestern Pennsylvania Citizens' Vision for Smart Growth: Strengthening Communities and Regional Economy. Pittsburgh, PA: Sustainable Pittsburgh.

Tarr, J. 1989. Infrastructure and city-building in the nineteenth and twentieth centuries. In City at the Point: Essays on the Social History of Pittsburgh, S. Hays (ed.). Pittsburgh, PA: University of Pittsburgh Press.

Tarr, J. 1996a. Disputes over water-quality policy: Professional cultures in conflict, 1900-1917. In The Search for the Ultimate Sink: Urban Pollution in Historical Perspective, J. Tarr (ed.). Akron, OH: University of Akron Press.

Tarr, J. 1996b. The separate vs. combined sewer problem: A case study in urban technology design choice. In The Search for the Ultimate Sink: Urban Pollution in Historical Perspective, J. Tarr (ed.). Akron, OH: University of Akron Press.

Tarr, J., and T. Yosie. 2004. Critical decisions in Pittsburgh water and wastewater treatment. In Devastation and Renewal: An Environmental History of Pittsburgh and Its Region, J. Tarr (ed.). Pittsburgh, PA: University of Pittsburgh Press.

Tarr, J., S. Mershon, et al. 2002. Sewerage problems in Penn Hills, PA, 1930-1997, unpublished. Pittsburgh, PA: Carnegie Mellon University.

Thompson, J. 1948. A financial history of the City of Pittsburgh, 1816-1910. Ph.D. dissertation, University of Pittsburgh.

Tierno, M. 1977. The search for pure water in Pittsburgh: The urban response to water pollution, 1893-1914. Western Pennsylvania Historical Magazine 60:23-36.

Trout, H., S. Pfaff, and J. Matviya. 2001. An Overview of the Summerset at Frick Park Project. In proceedings Brownfields 2001, Chicago.

USGS (United States Geological Survey). 1995. National Water Quality Assessment Program- The Allegheny-Monongahela River Basin. Available on-line at *http://pa.water.usgs.govreports/fs_137-95/report.html.* Accessed April 4, 2004.

USPHS (United States Public Health Service). 1944. Ohio River Pollution Control: Report of the Ohio River Committee. 78th Cong., 1st sess. H. Doc. 266.

Watershed Management Institute. 1997. Institutional Aspects of Runoff Management: A Guide for Program Development and Implementation. Crawfordville, FL: Watershed Management Institute.

Wolman, A. 1947. State responsibility in stream pollution abatement. Industrial and Engineering Chemistry 39:561-565.

WPC (Western Pennsylvania Conservancy). 2004. Conservation Reserve Enhancement Program for Ohio River Basin in Western Pennsylvania. Pittsburgh, PA: WPC.

Yosie, T. 1981. Retrospective analysis of water supply and wastewater policies in Pittsburgh, 1800-1959. Doctor of Arts dissertation, Carnegie Mellon University.

3

Water Quality in the Region

Surface water and groundwater in southwestern Pennsylvania often contain many different pollutants from a variety of sources. This chapter provides an overview of the types of water quality problems in the region. Specifically, it provides an introduction to water quality standards, an overview of aquatic pollutants by broad classes, and a summary of current water quality conditions in the Pittsburgh region. In doing so, it provides the background needed to understand the causes of water quality impairment discussed in Chapter 4.

WATER QUALITY STANDARDS

The health of waterbodies across the United States is determined by comparing certain measured physical, chemical, and biological parameters within those waters to water quality standards. In this regard, water quality standards are currently the foundation of the water quality-based control program mandated by the federal Clean Water Act (CWA).[1] These standards are set individually by states[2] in accordance with the CWA. Each water quality standard consists of two primary and distinct parts: (1) designated beneficial use(s) of the waterbody and (2) narrative and numeric water quality criteria for biological, chemical, and physical parameters that measure attainment of designated use(s). For example, a water quality standard for dissolved oxygen in surface waters would list the various oxygen concentrations required for waterbodies meeting different uses. New or revised water quality standards are subject to review and approval by the U.S. Environmental Protection Agency (EPA). The CWA also authorizes the EPA to promulgate superseding federal water quality standards. Designated uses represent not only scientific understanding but also value judgments about what a waterbody can and should be used for, whereas criteria reflect only scientific information.

Designated Uses

The CWA requires states to designate a use for each waterbody in their jurisdiction. The primary goal of the CWA, and the minimum that should be attained in all states, is that surface

[1] See Box 1-1 and *http://www.epa.gov/waterscience/standards/* for further information.
[2] The term "state" collectively includes territories, American Indian tribes, the District of Columbia, and U.S. interstate commissions.

waters in the United States should be "fishable and swimmable."[3] These two broad uses have been significantly elaborated on by the states, such that in Pennsylvania all surface waters have been designated for uses that include warm-water fish and other aquatic life use, recreational use, and drinking water supply. In addition to these uses, some waters are of exceptional quality (designated as high quality or exceptional value waters), and some of these may be protected for cold-water fish. As described later, water designated for these higher-end uses must meet more stringent water quality criteria. The most common designated uses are described below, with particular attention to drinking water uses of waters in southwestern Pennsylvania.

Drinking Water

Public health depends on provision of adequate quantities of drinking water free of harmful concentrations of human pathogens and chemical pollutants. Provision of clean, safe drinking water depends on the quality of both the source water and the treatment and distribution systems. Thus, assigning the appropriate use designation and then meeting water quality standards in source waters is the first step in providing safe drinking water (EPA, 2002a).

In southwestern Pennsylvania, drinking water is taken from a variety of sources. While the urban core in Allegheny County (see Chapter 6 for further information) is served predominately by public water services utilizing surface water sources, other counties in the area rely more heavily on public and private groundwater sources. Figure 3-1 shows the distribution of sources by population served for each county. Because population density for the region is highest in Allegheny County, which relies heavily on surface water, the majority of people in the region rely on treated surface water for their drinking water (see Figure 3-2). Major surface water sources of drinking water in the region include the Allegheny River, the Monongahela River, the Ohio River, the Youghiogheny River, Beaver Run, and Indian Creek.

Section 1453 of the Safe Drinking Water Act (SDWA) Amendments of 1996 requires states to develop a Source Water Assessment and Protection (SWAP) program to assess the drinking water sources (not "finished" waters already treated to meet various drinking water standards) serving public water systems for their susceptibility to pollution.[4] A state's SWAP is required to (1) delineate the boundaries of the areas providing source waters for all public water systems, and (2) identify (to the extent practicable) the origins of regulated and certain unregulated contaminants in the delineated area to determine the susceptibility of public water systems to such contaminants. The key objective for conducting source water assessments is to support the development of local, voluntary source water protection programs. In conducting such assessments, each state must use all reasonably available hydrologic information (such as water flow, recharge, discharge) and any other information deemed necessary to accurately delineate the source water assessment areas.

In order to protect public health, treatment of surface waters used for drinking water is mandated. Large water service suppliers in the region that utilize surface water are listed in Table 3-1. While these large systems provide significant populations with water, there are also many smaller water service providers in the region, many of which rely heavily on groundwater

---

[3] It should be noted that exceptions to the fishable, swimmable use exist. For example, in Pennsylvania a portion of the Delaware Estuary and water in the vicinity of the harbor at Erie do not fully support and are not expected to support the "fishable and swimmable" goal of the CWA.

[4] Further information on SWAP can be found at *http://www.epa.gov/safewater/protect/swap.html*.

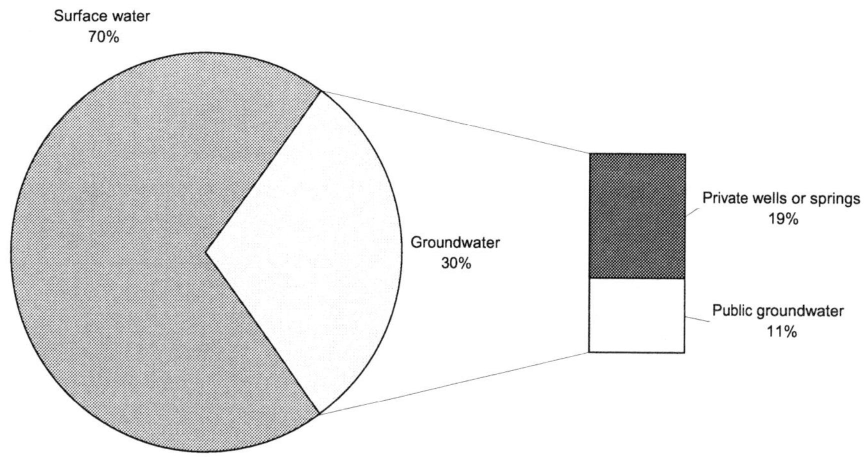

FIGURE 3-1 Percentage of southwestern Pennsylvania county populations served by groundwater and surface water. SOURCE: USGS, 1995.

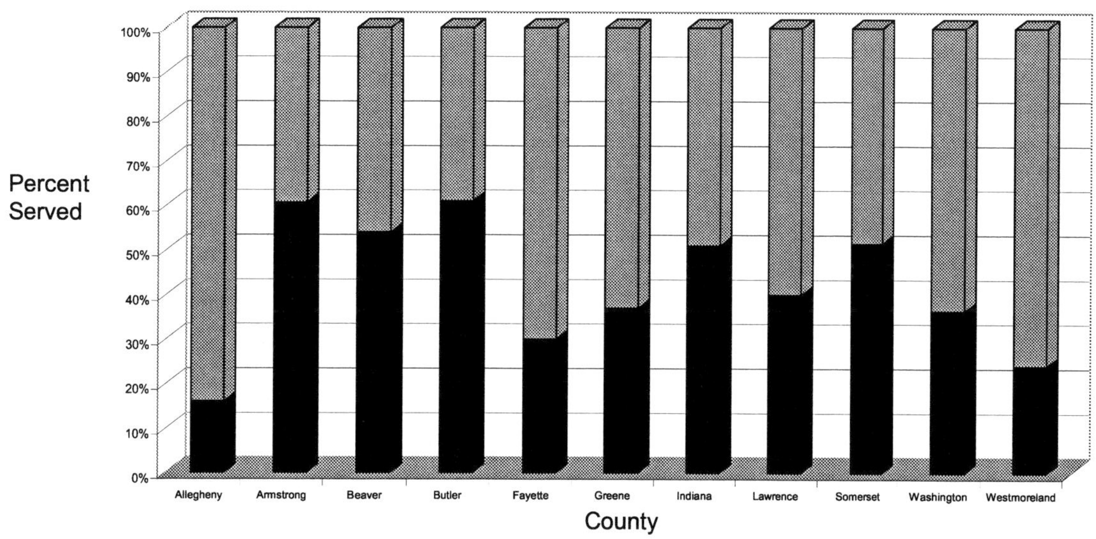

FIGURE 3-2 Source waters for drinking water in southwestern Pennsylvania by percentage served. NOTE: Black represents groundwater; white represents surface water. SOURCE: USGS, 1995.

as sources. Table 3-2 indicates that nearly 90 percent of the community water systems in Pennsylvania serve fewer than 10,000 persons. Nationally, about 94 percent of community water systems in the United States (more than 54,000 systems nationwide) served populations of 10,000 or fewer in 1993, but only 21 percent of the U.S. population was served by systems providing water to 10,000 or fewer people (NRC, 1997). About two-thirds of the small systems in southwestern Pennsylvania serve 500 or fewer persons, which has contributed to a proliferation of management and operational organizations across the region (as discussed in Chapter 6). The smallest systems often lack the financial resources and technical skills

TABLE 3-1 Public Water Systems Serving Populations of 100,000 or More in Southwestern Pennsylvania

Water Supplier	Principal County Served	Population Served
Pennsylvania-American Water Company-Pittsburgh	Allegheny	569,300
Pittsburgh Water & Sewer Authority	Allegheny	250,000
Westview Borough Municipal Authority	Allegheny	200,000
Wilkinsburg-Penn Joint Water Authority	Allegheny	120,000
Westmoreland County Municipal Authority, Youghiogheny Plant	Fayette	130,000
Westmoreland Municipal Authority, Sweeney Plant	Westmoreland	140,000

SOURCE: Derived from EPA Safe Drinking Water Information System Data, available on-line at *www.epa.gov/safewater/dwinfo/pa.htm.*

TABLE 3-2 Community Water Systems in Southwestern Pennsylvania

County	Number of Active Systems Serving Populations of 10,000 or More	Total Number of Active Community Water Systems
Allegheny	15	41
Armstrong	0	23
Beaver	5	38
Butler	2	64
Fayette	5	29
Greene	1	7
Indiana	1	32
Lawrence	2	29
Somerset	0	45
Washington	2	15
Westmoreland	3	21
Total	36	344

SOURCE: Derived from EPA Safe Drinking Water Information System Data, available on-line at *www.epa.gov/safewater/dwinfo/pa.htm.*

necessary to cope with drinking water regulations that are increasingly complex, and they may have difficulty dealing with problems of source water contamination, should these occur.

An EPA (2001a) survey of drinking water infrastructure needs lists Pennsylvania's statewide need for providing adequate drinking water as $3.148 billion for transmission and distribution, $940 million for treatment, $800 million for storage, $314 million for source needs, and $56 million for other needs. Notably, Pennsylvania's total drinking water infrastructure needs ($5.258 billion) are the highest in EPA Region III and are more than double the dollar needs of Virginia—the second-ranking state in the region ($2.068 billion). Furthermore, these dollar needs are often conservative estimates, because it is difficult to tally comprehensively the small system needs.

Despite the strong reliance in the region's urban core on surface water sources, a substantial population (30 percent, or approximately 800,000 residents) is served by public or private wells. It is important to note, however, that the CWA does not directly address groundwater or water quantity issues (i.e., there is no designated use of groundwater as a source of drinking water). Wellhead protection, required under Section 1428 of the SDWA, was established to protect public groundwater sources from contamination, and Pennsylvania's

Wellhead Protection Program[5] forms the cornerstone of its SWAP. Similar to a SWAP assessment, wellhead protection involves the delineation of the area contributing water and an inventory of potential contaminant sources in that area with the ultimate goal of developing a voluntary, community-based drinking water protection program.

Currently, the only national microbiological standard for groundwater quality is the Total Coliform Rule,[6] which applies only to groundwater used in public water systems and only in the distribution system. However, in 2000, EPA proposed the Ground Water Rule (GWR) in response to the SDWA Amendments of 1996 that mandate the development of regulations for the disinfection of groundwater systems in order to protect human health (EPA, 2000a). The proposed regulation (the final rule has been expected since spring 2003) will establish multiple barriers to protect groundwater drinking water sources from bacteria and virus contamination and will establish a targeted strategy to identify groundwater systems at high risk for fecal contamination. The proposed GWR will apply to public groundwater systems that have at least 15 service connections or regularly serve at least 25 individuals at least 60 days a year. Notably, the GWR does not apply to privately owned wells (nationally approximately 15 percent of Americans rely on private wells; in southwestern Pennsylvania the number is 19 percent), although EPA recommends that private well owners test for coliform bacteria at least once a year. Furthermore, although the state Water Well Drillers License Act (Act 610)[7] requires licensing of water well drillers and filing of well records, the Commonwealth of Pennsylvania does not regulate the construction of or water quality in private wells (PADEP, 2003).

Because construction of and water quality in private wells are unregulated in Pennsylvania, these wells may pose a threat to aquifers due to poor construction and maintenance. Additionally, many older private wells predate the 1956 Act 610, which requires filing of well information with the Pennsylvania Geological Survey; thus, no information is available regarding their construction or location. No regional data were available to assess this potential threat or the public health ramifications posed by unsafe private wells in southwestern Pennsylvania. Anecdotal information about the high rate of on-site sewage treatment and disposal system (OSTDS), or "septic system," failure (described later) suggests that private wells may be at risk of contamination. Similar problems exist in other rural regions of the country, and programs such as the Statewide Rural Wellwater Survey and the Grants to Counties Program in Iowa can serve as a model of cooperative programs designed to protect public health and the environment (see Box 3-1).

Contact Recreation

Because of the importance of outdoor recreation to local economies and social well-being, many waters in Pennsylvania are designated for this purpose and have correspondingly strict water quality criteria. The 2001 Pennsylvania Survey of Fishing, Hunting and Wildlife-

---

[5] Further information on Pennsylvania's Wellhead Protection Program can be found at *http://www.dep.state.pa.us/dep/deputate/watermgt/wc/subjects/srceprot/source/WHPPOVER.htm.*

[6] For further information on the Total Coliform Rule, see NRC (2004) or *http://www.epa.gov/safewater/tcr/tcr.html#coliform.*

[7] The implementing regulations for Act 610 (the Water Well Drillers License Act) are found in 17 Pennsylvania Code § 47 and are available on-line at *http://www.pacode.com/secure/data/017/chapter47/chap47toc.html.*

BOX 3-1
Case Study: Iowa Private Well Programs

Iowa is a rural state with 90 percent of the land under chemically intensive cultivation. The resulting threats to surface and groundwater have become a cause of concern as nitrate levels have risen and pesticides contaminate streams. Recreational waterbodies increasingly fail to meet EPA body contact guidelines for *Escherichia coli*.

Many rural residents live on farms that have multiple wells in various states of repair, including shallow, hand-dug, brick-lined wells more than 100 years old; bored, cement-tile-cased wells; and drilled, steel cased wells. In addition, sand point wells are common along alluvial aquifers. The collective threat to groundwater and risks to individual and public health prompted the State of Iowa to pass the Groundwater Protection Act of 1987. In the 1990s, Iowa's Grants to Counties Program was used to fund the identification and capping of thousands of abandoned wells, upgrades to existing wells requiring maintenance to meet current construction standards, and maintenance and improvements of septic systems adversely impacting groundwater. Counties were encouraged to apply for grant money and to provide oversight for well inspection, sampling, and testing and for sanitary surveys and improvements for septic systems. The program was highly successful because it was administered locally and preceded by an intensive public awareness program to inform stakeholders and potential participants. The program concept served as the basis of a subsequent Nine States Study supported by EPA and the Centers for Disease Control and Prevention to extend the program to the region (see *http://www.cdc.gov/nceh/emergency/wellwater/default.htm* for further information).

When Iowa was settled in the 1800s, there were considerable expanses of wetlands. Farmers sought to recover this land for agriculture by installing drainage tiles that carried surface water to nearby streams or piped surface water to boreholes called agricultural drainage wells. As chemical-intensive farming practices became dominant, these tiled fields became a serious threat to surface and groundwater. Several programs have been implemented to seal these wells; however, substantial numbers of Iowa fields are still tiled to drain into surface streams. This threat to the aquatic environment and groundwater is not unlike combined sewer overflow events in urban southwestern Pennsylvania, and perhaps some of the approaches that have been successful in Iowa could be applied to the Pittsburgh region.

Iowa has an extensive county extension service operated by Iowa State University. The extension service provides a local point of contact for information on health-related issues associated with drinking water and septic systems. Iowa has adopted state-of-the-art requirements for well construction and septic system construction and maintenance, and state law requires these programs be administered through local county health departments, according to regulations and guidelines provided by the Iowa Department of Public Health. The Commonwealth of Pennsylvania seems to lack programs similar to those described above. Existing sanitation regulations are often not enforced or are unenforceable, and there is an apparent need for modernization of sanitation and zoning laws in southwestern Pennsylvania.

Associated Recreation estimates that 1.3 million anglers spent 18.3 million days fishing in the state (DOI and DOC, 2002). Fishing expenditures were estimated at $580 million (DOI and DOC, 2002). Estimates of fishing are not available for counties in the region, but there are extensive resources for fishing, boating, and swimming managed by the Pennsylvania Fish and Boat Commission and Department of Natural Resources and Conservation.[8] Bacteriological indicator data (as described below) are used to assess attainment of contact recreational use criteria in the Commonwealth of Pennsylvania. Sampling is conducted during the swimming

[8] See *http://www.fish.state.pa.us/ and http://www.dcnr.state.pa.us/* for further information about these programs.

season (May 1 through September 30) and is based on indicator organisms that suggest pathogenic organisms may be present and present a health risk to individuals during contact recreation.

Human Health—Fish Consumption

An important activity directly related to recreation is fish consumption, which often drives the specific use designation for surface waters. Water quality impairment can contaminate fish that may be caught from degraded rivers and streams, sometimes to levels that are considered unhealthy for public consumption. The Pennsylvania Fish Tissue Sampling and Fish Advisories Program is responsible for assessment of the attainment of human health use criteria in Pennsylvania waterways. Fish tissue samples are collected during low flow between August and October. Fish tissue concentrations are compared to standards, and decisions regarding fish advisories are made based on a mixture of risk assessment-based methods and U.S. Food and Drug Administration (FDA) Action Levels. Currently, Pennsylvania has a statewide health advisory for recreationally caught sport fish. This advisory recommends no more than one meal of sport fish per week and is based on concerns regarding unidentified contaminants in untested fish. Specific to southwestern Pennsylvania, there are fish advisories in the Ohio River valley related to polychlorinated biphenyls (PCBs), mercury, and chlordane. Advisories cover the main rivers (Allegheny, Monongahela, and Ohio) as well as a number of smaller tributaries, reservoirs, and lakes. Some advisories recommend restricted consumption at one or two meals per month, while others are "do not eat" advisories.

Aquatic Life Use

A final common designated use category, aquatic life use, specifically targets ecosystem health rather than human health and use. Water quality impairment can limit the diversity of aquatic life in an ecosystem, which many states, including Pennsylvania, have determined is of intrinsic importance and also has indirect effects on human health through recreation and fish consumption. Specifically, Pennsylvania uses aquatic life use data (habitat and biological indicator data) to assess the ability of its waterbodies to maintain and/or propagate fish species and additional flora and fauna that are indigenous to aquatic habitats in the state. Habitat is assessed visually using procedures from the Standardized Biological Field Collection and Laboratory Methods manual (as described in PADEP, 2004). Biological indicator data are collected through a biosurvey. Within lakes in the state, aquatic life use attainment decisions are based primarily on the ecological integrity of fish communities.

Water Quality Criteria

Ambient water quality criteria allow states to determine if their surface waters are impaired for designated uses and, if so, to develop total maximum daily loads (TMDLs) for these waters to ensure future attainment of water quality consistent with the designated use (see NRC, 2001, for a full explanation of the TMDL process). Table 3-3 summarizes EPA's published

TABLE 3-3 Selected National Recommended Water Quality Criteria

Priority Pollutant	Freshwater CMC (μg/L)
Arsenic	340
Cadmium	2.0
Chromium (III)	570
Chromium (IV)	16
Copper	13
Lead	65
Mercury	1.4
Nickel	470
Silver	3.2
Zinc	120
Cyanide	22
Pentachlorophenol	19
Aldrin	3.0
gamma-BHC (Lindane)	0.95
Chlordane	2.4
4,4'-DDT	1.1
Dieldrin	0.24
alpha-Endosulfan	0.22
beta-Endosulfan	0.22
Endrin	0.086
Heptachlor	0.52
Heptachlor epoxide	0.52
Toxaphene	0.73

NOTE: A CMC (criteria maximum concentration) is an estimate of the highest concentration of a substance in surface water to which an aquatic community can be exposed briefly without resulting in an unacceptable effect.
SOURCE: EPA, 2002b.

water quality criteria for some chemical constituents. These national criteria were established to provide guidance for states, which are authorized to establish their own water quality standards (no less strict than national standards) to protect human health and aquatic life.

The Commonwealth of Pennsylvania through its Department of Environmental Protection (PADEP) has established numerical ambient water quality criteria for chemical constituents.[9] Pennsylvania's general information on water quality criteria states the following:

> Water may not contain substances attributable to point or nonpoint source discharges in concentration or amounts sufficient to be inimical or harmful to the water uses to be protected or to human, animal, plant or aquatic life. In addition to other substances listed within or addressed by this chapter, specific substances to be controlled include, but are not limited to, floating materials, oil, grease, scum and substances which produce color, tastes, odors, turbidity or settle to form deposits.

As noted previously, water quality criteria are the numeric concentrations, levels, or surface water conditions that must be maintained or attained to protect existing and designated uses. In addition, a few distinct use designations require even more stringent water quality criteria. For example, waters designated by the Commonwealth of Pennsylvania for cold water fish use or for

[9] See 25 PA Code § 93.6 for further information; available on-line at *http://www.pacode.com/secure/data/025/chapter93/chap93toc.html.*

trout stocking as high quality, or as exceptional value waters, must meet the statewide water quality criteria plus lower permissible temperatures and higher standards for dissolved oxygen.

It should be noted that some of the Pennsylvania criteria may be superseded for the Delaware Estuary, Ohio River basin, Lake Erie basin, and Genesee River basin under interstate and international compact agreements with the Delaware River Basin Commission, the Ohio River Valley Water Sanitation Commission (ORSANCO), and the International Joint Commission, respectively. Southwestern Pennsylvania surface water is part of the Ohio River basin and is governed by water quality criteria developed by ORSANCO (see Chapter 6 for further information about ORSANCO). Table 3-4 lists surface water quality criteria as promulgated by ORSANCO. Notably, many of the criteria are stricter than the corresponding national water quality criteria summarized in Table 3-3.

Water quality criteria for bacteria were published by EPA in 1986 and updated in 2002 (EPA, 1986, 2002c). Because of the enormous number and types of pathogens to which humans could potentially be exposed, water quality criteria for human recreational contact specify allowable levels of certain *indicator* organisms, such as fecal coliforms and *Escherichia coli* (described later). The national criteria were selected based on epidemiological work suggesting that body contact at the target level would result in eight gastrointestinal illnesses per 1,000 swimmers in freshwater and 19 illnesses per 1,000 swimmers at marine beaches (EPA, 1986; NRC, 2004). "Excessive amounts of fecal bacteria in surface water used for recreation have been known to indicate an increased risk of pathogen-induced illness to humans. Infection due to pathogen-contaminated recreational waters includes gastrointestinal, respiratory, eye, ear, nose, throat, and skin diseases" (EPA, 2001b).

TABLE 3-4 Water Quality Criteria Promulgated by ORSANCO for Three Common Designated Uses

Conventional Pollutants and Chemical Constituents	Aquatic Life	Public Water Supply	Contact Recreation
Ammonia	Temperature and pH dependent	—	—
Arsenic	—	50 μg/L	—
Bacteria (fecal coliform)	—	GM of 2,000 CFU/100 mL	GM of 200 CFU/100 mL 400 CFU/100 mL in <10% samples
Bacteria (*E. coli*)	—	—	GM of 130 CFU/100 mL 240 CFU/100 mL in any sample
Barium	—	1,000 μg/L	—
Chloride	—	2.5×10^5 μg/L	—
Dissolved oxygen	5,000 μg/L	—	—
Fluoride	—	1,000 μg/L	—
Mercury	—	0.012 μg/L	—
Nitrite + nitrate nitrogen	—	10,000 μg/L	—
Nitrite nitrogen	—	1,000 μg/L	—
pH	6.0-9.0	—	—
Phenolics	—	5 μg/L	—
Silver	—	50 μg/L	—
Sulfate	—	2.5×10^5 μg/L	—
Temperature	Seasonally dependent	—	—

NOTE: GM = monthly geometric mean consisting of at least five samples given in colony forming units (CFUs) per 100 milliliters (CFU/100 mL).
SOURCE: Adapted from ORSANCO, 2002.

Water designated for human contact recreation is considered unimpaired if levels of indicator organisms do not exceed the water quality criteria summarized in Table 3-5. Water containing higher levels of indicator organisms is considered unsafe due to the likely presence of fecal bacteria and other waterborne pathogens, leading to contact recreational risk. Although the EPA recommends the use of *E. coli* and enterococci as indicator organisms, Pennsylvania has retained fecal coliform as the indicator of recreational water pollution. See NRC (2004) for further information on the use of indicators for waterborne pathogens.

Finally, as noted previously, not all water quality criteria are numeric. For many contaminants of concern such as nutrients, the criteria exist as narrative statements, which can make interpretation and thus determinations of attainment difficult (NRC, 2001).

WATER QUALITY MONITORING PROGRAMS IN PENNSYLVANIA

In order to determine the health of its surface waters and the extent to which its water quality standards are being met, each state has developed a comprehensive monitoring program. Section 305(b) of the CWA requires states to compile and summarize water quality information collected by their monitoring programs every two years. In 2002, EPA released the *National Water Quality Inventory: 2000 Report*—the thirteenth installment in a series beginning in 1975 that uses state 305(b) reports to identify widespread water quality problems of national significance and to describe various protection and restoration programs (EPA, 2002d). Furthermore, Section 303(d) of the CWA requires states to list streams and other waterbodies having "impaired" water quality. In 2000, EPA reported that about 21,000 river and stream segments, lakes, and estuaries encompassing more than 300,000 assessed stream-miles and 5 million lake-acres were impaired (EPA, 2000b). In 2004, Pennsylvania's 305(b) and 303(d) reports were published together in a combined document entitled *2004 Pennsylvania Integrated Water Quality Monitoring and Assessment Report* (PADEP, 2004).

TABLE 3-5 Water Quality Criteria for Bacterial Indicators by Recreational Designated Uses (CFU/100 mL)

Bacteria	Steady State, 30-Day Geometric Mean[a]	Single Sample Maximum			
		Designated Beach Area	Moderate, Full Body Contact Recreation	Lightly Used, Full Body Contact Recreation	Infrequently Used, Full Body Contact Recreation
Fecal coliform	200[b]				
Enterococci	35, 33[c]	61	89	108	151
E. coli	126	235	298	406	576

NOTE: CFU = colony forming units.
[a] Five samples in a 30-day period.
[b] Not more than 10% of the total samples may exceed 400 per 100 mL for samples from May through September. For the balance of the year the standard is 2,000 per 100 mL (25 PA Code § 93.7).
[c] The criterion for enterococci is 35 CFU/mL in freshwater and 33 CFU/mL in marine waters.
SOURCE: EPA, 1986.

The PADEP maintains a system of 120 water quality monitoring stations throughout the commonwealth called "routine stations." At these stations, water quality sampling is conducted bimonthly for streamflow, physical analysis (e.g., temperature), and chemical analysis (e.g., dissolved oxygen) and annually for biological evaluation (including macroinvertebrate and fish tissue sampling). Routine stations are located at or near the mouths of streams with drainage areas of about 200 square miles or larger. Another 22 stations, called reference stations, have been established to represent ambient waters with minimal influence from human activity or to represent typical waters having quality similar to that of other waters found in the area. These 22 stations are usually sampled monthly for streamflow and physical and chemical analysis and three times per year for biological parameters. Fish tissue is sampled periodically at about 35 water quality network stations per year. Sampling activity is rotated through the network of stations to give complete coverage over time (PADEP, 2004).

Other than bacterial indicators of waterborne pathogens, the preceding section of this chapter does not list specific water quality criteria for biological parameters because bioassessment is an evolving and burgeoning field, with many states only recently adding new biological parameters to their monitoring programs. In some states, a modified version of EPA's 1989 Rapid Bioassessment Protocol (RBP II)[10] is used to determine if a waterbody is impaired for designated aquatic life use. The assessment is performed in "wadeable" streams and rivers where physical examination of the stream or river and biological sample collection can be conducted. The protocol includes an evaluation of the presence of and identification to the family level of one to three groups of biota: typically periphyton (algae) and/or benthic macroinvertebrates such as crustaceans, insects, snails, and shellfish. A habitat assessment is also performed, which includes characterizing the stream with regard to the nature of the channel, bottom materials, vegetative cover overhead (shade trees), riparian vegetation in general, and aquatic vegetation. Presence of tree trunks and limbs in the channel is also noted, because these constitute habitat.

Assessing the water quality of all the streams and rivers in Pennsylvania is not possible using only the 142 stations described above, so other monitoring programs are also conducted. Intensive surveys of streams and rivers are performed by PADEP for a variety of reasons, including the provision of background water quality data and assessing the effects of pollutant discharges on receiving waters. In addition, PADEP has a program to support volunteer monitoring efforts.[11] The *2004 Pennsylvania Integrated Water Quality Monitoring and Assessment Report* states that more than 180 groups including 11,000 people have taken part in statewide monitoring activities. The PADEP provides workshops and training and quality assurance sessions for volunteer monitors throughout the commonwealth. This kind of volunteer training and education is necessary to help maintain quality control and attain uniformity of reporting when many heterogeneous groups and individuals perform water quality assessments. For the 2004 303(d) process, there were 10 respondents to the PADEP request for data and information from outside sources, and 7 sets of data related to bacteriological monitoring were used to evaluate attainment of recreational uses.

In accordance with the SWAP program, approximately 96 percent of the 14,000 public water systems source waters were assessed by September 2003, with the balance to be completed

---

[10] Details of the Rapid Bioassessment Protocol are available on-line at *http://www.epa.gov/OWOW/monitoring/techmon.html.*

[11] Further information on the volunteer efforts is available on-line *at http://www.dep.state.pa.us/dep/deputate/watermgt/wc/subjects/cvmp/default.htm.*

by September 2004 (PADEP, 2004). In addition, PADEP initiated a Statewide Surface Water Assessment Program (SSWAP) for biological assessment in all of the commonwealth's surface waters. In southwestern Pennsylvania, Watershed Restoration Action Strategy (WRAS) reports[12] have been issued for the Redbank Creek watershed (Allegheny River), the Chartiers Creek watershed (Ohio River), the Raccoon Creek watershed (Ohio River), the Upper Youghiogheny River watershed (Laurel Hill Creek and Indian Creek), the Lower Youghiogheny River watershed, the Stonycreek River and Little Conemaugh River watersheds, the Blacklick Creek and Conemaugh River watersheds, and the Upper Monongahela River watershed (Dunkard, Big Sandy, Georges, and Whiteley Creeks) (see Figure 6-2 for a map of state-delineated watersheds in southwestern Pennsylvania).

In 2001, PADEP initiated a pilot project for monitoring 23 miles of stream segments believed to be at risk for recreational contact use due to bacterial contamination, the results of which were used to inform the 2002 305(b) report. The water contact use support evaluation for the 2002 305(b) report was based on this pilot study of 23 miles, of which 22 miles were found to be impaired. The report notes that this high percentage of impaired streams is due to a selection bias (i.e., the study was targeted at streams where problems were anticipated). The program has since expanded to include 140 miles of streams (PADEP, 2004).

In addition to PADEP, several other governmental and nongovernmental organizations are collecting data related to water quality in southwestern Pennsylvania, including the following:

- U.S. Environmental Protection Agency
- U.S. Geological Survey (USGS)
- U.S. Army Corps of Engineers (USACE)
- Ohio River Valley Water Sanitation Commission
- Allegheny County Health Department (ACHD)
- Allegheny County Sanitary Authority (ALCOSAN)
- Three Rivers Wet Weather Demonstration Program (3RWW)/3 Rivers 2nd Nature (3R2N)
- Pittsburgh Water and Sewer Authority (PWSA) and other drinking water providers
- Water associations, schools, and other nongovernmental organizations

Available water quality data from these sources are discussed below. Some of these sources report their data to EPA's Storage and Retrieval (STORET)[13] computerized environmental data system; others maintain separate hard-copy and computer-based records. Most of these groups focus data collection on physical or chemical parameters of water quality. "Neither US EPA nor the Pennsylvania Department of Environmental Protection nor any other federal or state agency has yet made it a matter of priority to survey the rivers, streams, and creeks of the state for bacterial contamination" (Luneburg, 2004). Private monitoring by volunteer organizations generally does not extend to bacterial testing because of the high cost and professional sophistication of the testing required. Finally, a comprehensive GIS (geographic information system)-linked database of water quality monitoring data for the region does not exist.

---

[12] WRAS Reports are available online *at http://www.dep.state.pa.us/dep/deputate/watermgt/wc/Subjects/Nonpointsourcepollution/Initiatives/Wraslist.htm.*

[13] For further information on STORET, see *http://www.epa.gov/storet/.*

POLLUTANTS

The same broad classes of waterborne contaminants that are of concern in much of the country's water supplies[14] are also a concern in southwestern Pennsylvania. These include pathogenic microorganisms, organic carbon compounds, excessive nutrients, sediment, and toxic compounds. Unfortunately, in southwestern Pennsylvania, as in many parts of the nation, it is difficult if not impossible to determine the individual sources of contaminants found in water supply reservoirs, lakes, rivers, and groundwater because most activities and land uses produce multiple and often similar types of contamination. Furthermore, for financial reasons, monitoring tools are rarely deployed in a way capable of tracing a contamination event back to its source. Thus, this section first briefly summarizes the major classes of contaminants, then catalogs the available water quality data for the region. Chapter 4 discusses the types of activities prevalent in southwestern Pennsylvania that are likely sources of impaired water quality.

Pathogens

Nationally, pathogens (as measured by fecal indicator bacteria; see more below) are the leading cause of impairment in assessed rivers and streams (EPA, 2002d). Most waterborne pathogens of public health concern are not native to the surface waterbodies and groundwater of southwestern Pennsylvania and enter ambient waters from various point and nonpoint sources. Pathogenic microorganisms from human and animal waste have the potential to enter surface and groundwaters via a number of different mechanisms and to affect human health through one or more different exposure routes. *Cryptosporidium parvum* and *Giardia lamblia* are protozoan pathogens receiving increased public health and regulatory attention in the past few decades (NRC, 1999a). Giardiasis is a protozoan parasitic disease infecting primarily children, especially in developing nations and institutional settings (Ali and Hill, 2003). The trophozoite (free-living) form of *Giardia lamblia* does not survive in the ambient environment; however, the cyst stage (sporozite form) is relatively resistant to environmental conditions and persists in an infective state for weeks to months (Rose and Shifko, 1999). A wide variety of both feral and domesticated animals and pets carry *Giardia* spp. (Marshall et al., 1997). Similarly, free sporozites of *Cryptosporidium* may be shed from a variety of mammals, including humans (O'Donoghue, 1995); they are obligate intracellular parasites, requiring a host to reproduce, while the oocyst stage of *Cryptosporidium* spp. is highly resistant to environmental conditions, including disinfection levels typically used in drinking water treatment (Robertson et al., 1992). *Cryptosporidium* spp. infect livestock, humans, and other animals, although host specificity restricts human infections to *Cryptosporidium parvum* Genotypes 1 and 2 (Rose et al., 2002). Although cryptosporidiosis is self-limiting in immunocompetent hosts, it causes devastating disease in immunocompromised individuals, and it is a leading cause of death in AIDS patients (Guerrant, 1997).

Waterborne pathogenic microorganisms of concern in southwestern Pennsylvania also include bacteria. *Escherichia coli* O157:H7 is one of many toxigenic strains of *E. coli* that cause

---

[14] It is important to note that a discussion of drinking water standards (both numeric criteria and treatment/performance requirements) is beyond the scope of this report. For further information on the development and use of drinking water standards and related regulations in the United States, see Pontius (2003).

gastroenteritis in humans. It is frequently detected in livestock, shed in cattle manure, and thus a component of agricultural runoff. *E. coli* O157 has been isolated from deer, and experiments infecting deer with the organism have demonstrated shedding similar to that in cattle; it has also been isolated from dogs and swine (Feder et al., 2003; Hammermueller et al., 1995)

Although disease outbreaks of *E. coli* are usually associated with ground meat products and improper handling or eating of undercooked beef, waterborne outbreaks of this microbe have occurred from sewer contamination of potable water lines, from manure contamination of surface waters used for recreation, and from contamination of well water used for drinking (Geldreich et al., 1992; O'Connor, 2002). Other waterborne bacterial pathogens of public health concern include *Campylobacter*, *Helicobacter*, and *Salmonella*. Shigellosis (from *Shigella* spp.) is one of the most common diseases associated with recreational exposure to untreated surface water. *Shigella* spp. are unique to humans, and their presence in the recreational environment is indicative of human fecal contamination from infected individuals or of sewage or septic tank origin.

A third major class of pathogenic microorganisms important in waterborne disease is viruses. For example, human caliciviruses, which cause diarrheal illness, are thought to be present in sewage and discharged into surface waters in wastewater effluents and during combined sewer overflow (CSO) events, where they could threaten drinking water sources, recreational bathing beaches, and shellfish growing areas (Schaub and Oshiro, 2000). In some cases, viruses have been reported to cause waterborne disease outbreaks, including recreational waterborne disease (Lee et al., 2002; Levy et al., 1998). Lastly, waterborne disease outbreaks resulting in acute gastrointestinal illness of unknown etiology are widely thought to be caused by viruses (NRC, 2004). However, calicivirus concentrations are generally not known in surface water, the efficacy of their removal during sewage treatment has not been determined, and little is known about their persistence and survival in the environment.

Because it is impractical to test waters for all possible pathogenic organisms, the microbiological quality of water is often assessed through the use of indicator microorganisms (usually bacteria) to monitor potable water sources and recreational waters and to determine the treatment efficacy of drinking water and wastewater treatment plants (for a recent comprehensive review of indicators, see NRC, 2004). Two bacterial groups (coliforms[15] and enterococci) are commonly used as indicators of possible fecal contamination. Although these groups are typically not pathogenic, fecal indicator bacteria are used to provide an estimation of the amount of feces and, indirectly, the presence and quantity of fecal pathogens in water. Total coliforms indicate the presence of fecal microorganisms that should not be present in the finished water and must have entered through contamination or failure in the treatment process. However, total coliforms are not typically used to evaluate surface waters because they can come from sources other than fecal contamination. Instead, fecal coliforms are generally used to assess the microbial quality of surface and recreational waters. Enterococci are present in the digestive systems of mammals and have been used as an indicator organism since the 1950s, albeit less frequently than fecal coliforms. EPA (2002c) recommends using *E. coli* and enterococci—the latter also known as fecal streptococci and intestinal enterococci—as better microbial indicator organisms of human health risk when monitoring surface waters, particularly marine recreational waters (WHO, 2000).

---

[15] Coliforms include several genera of bacteria, of which *E. coli* is the most important member. Historically, the definition and subdivision of the group into total and fecal coliforms is based on the methods used for their detection (see NRC, 2004).

Unfortunately, fecal indicator bacteria monitoring is generally considered insufficient to reliably detect viruses and protozoan parasites in drinking water sources or ambient (recreational) waters (NRC, 2004). However, bacteriophages, or viruses that infect bacteria—especially in groundwater—may be useful as indicator organisms and surrogates for viral transport and attenuation. That is, viruses exhibit transport and attenuation behavior distinct from that of bacteria because of size, adsorption characteristics, and other physical factors (Azadpour-Keeley et al., 2003; Higgins et al., 2000; Pesaro et al., 1995). EPA has also supported recent research to determine whether coliphages (viruses that infect *E. coli*) may be a suitable indicator of fecal-contaminated waters.

Nutrients and Organic Carbon

One of the leading water quality issues associated with agriculture, urban stormwater discharges, and domestic wastewater discharges in the United States is nutrient pollution—particularly excessive nitrogen and phosphorus. Nationally, nutrient pollution is the leading cause of impairment in lakes, ponds, and reservoirs (EPA, 2002c). These and other nutrients are routinely applied to cropland in manufactured fertilizers and animal manures to increase yields. In areas with highly intensive livestock production such as animal feeding operations, manure may be applied to cropland primarily to dispose of the waste and secondarily as a fertilizer. Nutrients are also found in some common household products as well as in human waste and can thus end up in wastewater treatment plants, septic systems, and their receiving waterbodies (if not properly treated). Nitrogen in the form of nitrate is easily soluble in water and is transported in runoff, in agricultural tile drainage, and with septic system leachate. Phosphate is only moderately soluble and, relative to nitrate, is not very mobile in soils. However, erosion can transport considerable amounts of sediment-adsorbed phosphate to surface waters.

Excessive nutrients in waterbodies have been shown to cause eutrophication, a process in which increasing nitrogen and phosphorus levels stimulate excessive algae growth, with dramatic (usually adverse) effects on the aquatic ecology. Following an algal bloom, decaying algae are degraded by aerobic microorganisms that deplete the water column of dissolved oxygen, endangering fish and other aquatic life. Algal blooms can also block the sunlight needed by aquatic vegetation, causing the vegetation to die off. This loss in vegetation often leads to subsequent death of fish and other aquatic life higher up the food chain. Eutrophication of freshwater is usually due to phosphates, while nitrates are usually the cause of coastal water eutrophication (NRC, 1992, 1993).

Organic carbon can cause eutrophication via similar mechanisms. When biodegradable organic matter is discharged into receiving waters, bacteria utilize it as a food source and, at the same time, use oxygen in their life processes. As bacteria multiply and more bacteria use the food source, oxygen consumption increases. Eventually, if sufficient biodegradable organic matter is present, all of the oxygen in a stream, river, or lake may be consumed and the water becomes anaerobic. Even if surface water does not become completely anaerobic, oxygen depletion at depth may be sufficient to harm or kill aquatic organisms such as fish. Biodegradable organic matter may occur as dissolved substances, such as sugars or starches from food processing wastes, or in particulate form, such as fecal matter discharged from a malfunctioning septic tank or an overflowing storm sewer or sanitary sewer.

Excessive organic carbon in surface water sources of drinking water can also constitute a public health risk. When certain types of organic carbon react with disinfectants associated with conventional drinking water treatment, potentially carcinogenic disinfectant by-products may result (see EPA, 2003, and NRC, 1987, for further information).

Sediment

A variety of land uses, from agriculture to urban and suburban development, can increase rates of sediment transport from the land to adjacent waterbodies. Disturbing the soil through tillage and cultivation or altering its vegetative cover, especially riparian vegetation adjacent to waterbodies, increases the rate of soil erosion. Dislocated soil particles carried in urban stormwater and agricultural runoff can impair the water quality of streams, rivers, lakes, reservoirs, and wetlands. Excessive sediment causes various types of damage to water resources. For example, accelerated reservoir siltation reduces the useful life of reservoirs. Sediment can clog roadside ditches and irrigation canals, block navigation channels, and increase dredging costs. Sediment can also destroy and degrade aquatic wildlife habitats by covering fish eggs and smothering benthic organisms, reducing diversity and damaging commercial and recreational fisheries. High concentrations of suspended solids can also prevent sunlight from reaching plants in deep water and thus reduce their growth or even result in their death (Livingston et al., 1998). Finally, suspended solids in water cause turbidity, which can increase the cost of water treatment for municipal and industrial water uses. In the United States, sediment is the second leading cause of impairment of rivers and streams and the third leading cause of impairments in lakes, ponds, and reservoirs (EPA, 2002c).

Sediment also provides a delivery mechanism for phosphorus and other pollutants that adhere strongly to sediment particles. Many toxic materials can be tightly bound to clay and silt particles, including some nutrients, pesticides, industrial wastes, metals from mine spoils, and radionuclides (Osterkamp et al., 1998). Depending on the conditions of the receiving water, these compounds may desorb from sediment particles and constitute a threat to both ecological receptors and humans. Furthermore, resuspension of sediment in stream and lake beds can release nutrients and entrained microorganisms (both pathogens and indicator organisms) into the water column (Medema et al., 1998; Schallenberg and Burns, 2004).

Monitored bacterial indicator levels are sensitive to suspended solids, such that indicator concentrations rise sharply with resuspension of sediment (Jensen et al., 2002; VADEQ, 2000). Regrowth of bacteria can occur in sediment and has been documented in several studies conducted in warmer climates (Desmarais et al., 2002; Jensen et al., 2002), and researchers at Gannon University are conducting studies to determine regrowth potential in temperate climate beach sediments on Lake Erie.[16] Concerns about bacterial regrowth potential and impacts of sediment loading are issues raised in the American Society of Microbiology's comments on EPA's proposed policy on National Pollutant Discharge Elimination System (NPDES) permit regulation for wet weather discharges (ASM, 2004).

---

[16] See *http://www.gannon.edu/resource/dept/enviro/research.html* for further information.

Pesticides and Other Chemicals of Concern

Pesticide residues reaching surface water systems may harm freshwater and marine organisms and damage recreational and commercial fisheries (Pait et al., 1992). Aquatic species and their predators can suffer chronic adverse effects from low levels of exposure to pesticides over prolonged periods. Pesticides can also accumulate in the fatty tissue of animals such as shellfish to levels much higher than in the surrounding water (bioaccumulation), and consumption of these animals may lead to chronic effects in predators (biomagnification). These processes are responsible for the damaging effects of dichlorodiphenyltrichloroethane (DDT), which led to its ban in 1972. Herbicides and insecticides in the aquatic environment can also harm birds and other wildlife that feed on the chemicals' target plants and insects.

Many pesticides are probable or possible human carcinogens (Engler, 1993) and could pose risks to human health via ingestion of drinking water. Although the overall state of knowledge about chronic effects of pesticides on human health is quite limited, concerns have been raised about the consequences of low exposures over long periods of time. For example, there is a higher incidence of lung and other types of cancer in farmers and farm workers involved in the handling and application of pesticides (WHO, 1990). In addition to cancer, questions have been raised about other possible effects of pesticide exposure. Thus, regulation of public water supplies requires additional treatment when certain pesticides exceed established health safety levels in drinking water supplies. Box 3-2 describes the effects of some relevant toxic compounds in both fish and humans, which is of considerable concern in the southwestern Pennsylvania region given the significant role of recreation in the regional economy (described later). Box 3-3 summarizes the presence of endocrine disruptors, pharmaceuticals, and personal care products in the aquatic environment.

Understanding the sources, distribution, and control of pesticides in southwestern Pennsylvania waters, as well as potential threats to recreational waters, requires investigation of both current and past pesticide application practices in both agricultural and urban and suburban environments. Southwestern Pennsylvania agricultural production has been primarily dairy, beef, sheep, and other livestock; pasture; hay and forage crops; truck farming and direct marketing of produce; nursery production; orchard production, and woodlot production (USDA, 1997). With the exception of apples and corn, most of these types of agriculture are relatively less intensive in pesticide use. With increasing consumer awareness, increased EPA controls, adoption of integrated pest management (IPM) practices, and the relatively high cost of commercial and agricultural pesticides, pesticide use in nursery and produce farming has become even less intensive in recent years. Residues of older, banned agricultural pesticides such as DDT may pose a public health problem in some areas (e.g., soil residues of lead and arsenic formerly applied in apple orchards).

A potentially greater concern for pesticide contamination, given the Pittsburgh region's significant suburban growth in recent decades, may be urban or suburban use of lawn and landscape pesticides (discussed later; Anderson et al., 2000). Unlike agricultural pesticides—which are controlled by EPA and state regulations regarding purchase, application, storage, disposal, applicator certification, worker protection, and record keeping and are often costly enough for economics to dictate prudent use—oversight of household use of pesticides is considerably less stringent. Although the EPA has recently phased out many of the more toxic products for home use and pesticide labels are the law, the reality is that private consumers can purchase and use pesticides virtually indiscriminately. Unlike agricultural settings, where soil

BOX 3-2
Health Effects of Chemicals in Fish and in Humans Who Eat Them

The health effects caused by ingesting chemically contaminated fish are summarized in fish advisories issued by the EPA, but the text tends to include medical and technical terms that may be unfamiliar to the general public. The Agency for Toxic Substances and Disease Registry (ATSDR) has developed fact sheets (ToxFAQs™) on numerous toxic chemicals, including those found in fish in southwestern Pennsylvania. These fact sheets discuss the toxic chemicals, their sources and effects in the environment, and adverse human health effects. They also summarize the evidence of their carcinogenicity based on information from the Department of Health and Human Services (DHHS), the EPA, and the International Agency for Research on Cancer (IARC). The information on health effects of select chemicals in fish provided below is based on ATSDR fact sheets.

Aldrin and Dieldrin

Aldrin and dieldrin are insecticides with similar chemical structures. Their use in the United States was banned in 1974, with an exception for termite control. In 1987, all uses of these insecticides were banned by EPA. Because aldrin degrades to dieldrin in the environment, the latter is more commonly found. Dieldrin binds to soil particles and breaks down very slowly in soil and in water. It accumulates in the fat of fish. Thus, one means of human exposure is eating fish contaminated with dieldrin.

The ToxFAQs for Aldrin and Dieldrin (ATSDR, 2002a) indicates that persons who ingested large amounts of aldrin or dieldrin have suffered convulsions and some have died. Long-term, moderate exposure by air has led to reports of headaches, dizziness, irritability, vomiting, and uncontrolled muscle movements. The fact sheet indicates that although aldrin and dieldrin have caused liver cancer in mice, the IARC has determined that these chemicals are not classifiable with regard to human carcinogenicity. However, the EPA has determined that they are probable human carcinogens.

Chlordane

Chlordane was used as a pesticide in the United States from 1948 to 1988, after which it was banned. It adheres strongly to soil particles, so stream and river sediments may contain chlordane. Eating fish and shellfish from water contaminated by chlordane is one means of human exposure. In this regard, the U.S. Food and Drug Administration limits the safe amount of chlordane and its breakdown products to less than 100 parts per billion in fish.

The ToxFAQs for chlordane (ATSDR, 1995) states that "chlordane affects the nervous system, the digestive system, and the liver in people and animals." Headaches, weakness, vision problems, vomiting, diarrhea, and jaundice have occurred in people who breathed air containing high concentrations of chlordane or accidentally swallowed small amounts of chlordane. Large amounts of ingested chlordane can cause convulsions and death in people; however, there is no evidence that chlordane exposure causes cancer.

Chlorinated Dibenzo-*p*-dioxins ("dioxins")

Chlorinated dibenzo-*p*-dioxins are not manufactured chemicals but are by-products of processes that use chlorine. Dioxins can form during incineration of some types of solid wastes that contain chlorinated compounds, can attach to soil particles, and can be found in lake and stream sediments. One major form of human exposure is ingesting contaminated fish. In this regard, the FDA recommends against eating fish with concentrations of 2,3,7,8-TCDD (tetrachlorodibenzo-*p*-dioxin) exceeding 50 parts per trillion.

The ToxFAQs for chlorinated dibenzo-*p*-dioxins (ATSDR, 1999a) indicates that chloracne is the most common human health effect, though liver damage may occur in some people. Changes in hormonal levels may be caused by exposure to high concentrations of CCDs. The DHHS has determined that 2,3,7,8-TCDD may reasonably be anticipated to cause cancer, and the World Health Organization has determined that 2,3,7,8-TCDD is a human carcinogen.

DDT, DDE, and DDD

DDT was widely used in the United States for mosquito control and as an agricultural pesticide until its use was banned in 1972 because of harm to wildlife. Commercially produced DDT was contaminated with two similar chemicals, DDE (dichlorodiphenyldichlorethylene) and DDD (dichlorodiphenyldichlorethane). DDT breaks down quickly in sunlight but may remain in soil for years because it adheres strongly to soil particles. It can accumulate in fatty tissue of fish, so eating contaminated fish is a potential route of exposure.

The ToxFAQs for DDT, DDE, and DDD (ATSDR, 2002b) states that "DDT affects the nervous system. People who accidentally swallowed large amounts of DDT became excitable and had tremors and seizures. These effects went away after the exposure stopped. No effects were seen in people who took small daily doses of DDT by capsule for 18 months. A study in humans showed that women who had high amounts of a form of DDE in their breast milk were unable to breast feed their babies for as long as women who had little DDE in the breast milk. Another study in humans showed that women who had high amounts of DDE in breast milk had an increased chance of having premature babies."

The DHHS has determined that DDT may be reasonably anticipated to be a human carcinogen, and the EPA has determined that DDT, DDE, and DDD are probable human carcinogens. The IARC considers that DDT is a possible human carcinogen.

Mercury

Mercury enters the environment as a waste from manufacturing plants, as a result of burning coal and solid waste (if the latter includes products containing mercury), from natural deposits and volcanic emissions, and by waste disposal practices. Bacteria in water and soil can convert inorganic mercury to methylmercury. Eating fish or shellfish contaminated with methylmercury is a public health concern. Older and larger fish may contain more methylmercury because it accumulates in fish tissues over time. The FDA has set a maximum permissible concentration of 1 part per million (ppm) for methylmercury in seafood (ATSDR, 1999b).

Health effects of methylmercury are described as follows (ATSDR, 1999b): "The nervous system is very sensitive to all forms of mercury. Methylmercury and metallic mercury vapors are more harmful than other forms, because more mercury in these forms reaches the brain. Exposure to high levels of metallic, inorganic, or organic mercury can permanently damage the brain, kidneys, and developing fetus. Effects on brain functioning may result in irritability, shyness, tremors, changes in vision or hearing, and memory problems."

"Short-term exposure to high levels of metallic mercury vapors may cause effects including lung damage, nausea, vomiting, diarrhea, increases in blood pressure or heart rate, skin rashes, and eye irritation."

"Mercury's harmful effects that may be passed from the mother to the fetus include brain damage, mental retardation, uncoordination, blindness, seizures, and inability to speak. Children poisoned by mercury may develop problems of their nervous and digestive systems, and kidney damage." The EPA has determined that mercuric chloride and methylmercury are possible human carcinogens, according to the ATSDR.

PCBs

Polychlorinated biphenyls were manufactured in the United States until 1977 and used as coolants and lubricants in electrical equipment especially transformers. They entered the environment as a result of manufacturing, use, and disposal. PCBs do not break down rapidly in the

continues

BOX 3-2 CONTINUED

environment, and they adhere to soil particles. For these reasons, they are found in the sediments of rivers, streams, and lakes. Contaminated sediments can cause contamination of fish, with concentrations in fish being many times greater than concentrations in water. Eating contaminated fish is a means of exposure. The FDA requires that fish contain no more than 0.2 to 3 ppm of PCBs.

The ATSDR ToxFAQs report for PCBs (2001) states that "the most commonly observed health effects in people exposed to large amounts of PCBs are skin conditions such as acne and rashes. Studies in exposed workers have shown changes in blood and urine that may indicate liver damage. PCB exposures in the general population are not likely to result in skin and liver effects."

The fact sheet notes that women who ingested large amounts of fish contaminated with PCBs or who had relatively high levels of exposure at work gave birth to babies weighing slightly less than women who did not have such exposure, and babies born to mothers who ate fish contaminated by PCBs displayed some abnormal responses in infant behavior tests. The DHHS has concluded that PCBs may reasonably be anticipated to be carcinogens, while EPA and IARC have determined that PCBs are probably carcinogenic to humans.

and vegetation may slow the transport of pesticides to streams, the greater extent of impervious surfaces in more urbanized areas may allow pesticides to travel rapidly in runoff overland to streams or into storm sewers. In some areas of the United States (e.g., Fort Worth, Texas), home use of pesticides such as diazinon and malathion has created water pollution problems severe enough to warrant EPA action in the form of large fines and expensive upgrades to treatment plants.[17] As discussed in a later section of this chapter, the USGS National Water Quality Assessment (NAWQA)[18] program indicates that certain pesticides detected in a dominantly urban-suburban watershed in the Pittsburgh area may have originated from lawn care sources.

Metals

The status of metals in the Allegheny and Monongahela River basins was discussed in the USGS report on water quality (Anderson et al., 2000; see also Appendix B). Concentrations of metals in bed sediment were measured because contaminated sediments can adversely affect aquatic life. The results were compared to the probable effect level (PEL) values set in Canada (Canadian Council of Ministers of the Environment, 1995) because standards for metals in bed sediment had not been developed in the United States. Arsenic was detected in bed sediment in all 50 sites, and the concentration exceeded the PEL of 17 μg/g in 12 of 50 sites, with a maximum of 52 μg/g. Anderson et al. (2000) noted that land use did not appear to be a factor in arsenic concentrations in sediment.

The presence of other metals in bed sediments did seem to be affected by land use. Zinc and chromium were found in bed sediments at all 50 sites. Zinc exceeded the PEL of 315 μg/g in 15 sites, while chromium exceeded the PEL of 90.0 μg/g at 5 sites. At four of the sites specifically identified and having zinc concentrations higher than the PEL, land use was either mining or mixed land use. Three of those four sites equaled or exceeded the PEL for chromium

[17] See *http://ci.fort-worth.tx.us/water/perticidewtrqual/pestFAQ.htm* for more information.

[18] Information about the NAWQA Program is available on-line at http*://water.usgs.gov/nawqa/*; see also NRC (2002).

and were among the highest 25 percent of the most degraded sites nationally (Anderson et al., 2000). Other metals found in the sediments and seemingly related to land use were cadmium, lead, and mercury, each of which exceeded the PEL at least once in samples obtained in mixed land use or mined sites. Anderson et al. (2000) noted that although no guidelines exist for cadmium in whole-fish samples, the results from this study were among the highest sampled by NAWQA during 1995-1998.

Anderson and colleagues (2000) noted an effect of surface coal mining on shallow domestic water supply wells, which exceeded secondary maximum contaminant levels for iron and manganese more often than wells in areas not influenced by surface mining. In addition, sulfate concentrations in groundwater usually exceeded the regional background concentrations for sulfate at distances of less than about 1,000 feet from surface coal mines. The importance of coal mining and acid mine drainage to the regions' water quality is discussed more fully in Chapters 2 and 4.

CURRENT WATER QUALITY CONDITIONS

This section reviews the available water quality data for southwestern Pennsylvania. Data on bacterial indicators and protozoan pathogens in both the main stem rivers and their tributaries demonstrate that water quality standards are often unmet and high pesticides levels in fish have been reported. The most recent 305(b) and 303(d) data (PADEP, 2004) confirm that specific uses including fish consumption, drinking water, and recreational use are restricted due to impairments.

Physical and Chemical Parameters

Table 3-6 shows medians and ranges of values for several common water quality parameters for periods from two to six years in the three main stem rivers in the Pittsburgh region. With the exception of pathogen indicators, water quality in the main stem rivers in southwestern Pennsylvania is acceptable (compare with standards in Table 3-4).

A 1993 study found that all groundwater suppliers were in compliance with primary maximum contaminant levels (Chester Engineering, 1996). The primary contaminants cadmium and trichloroethylene were detected at one location, and the secondary contaminants iron, manganese, color, and dissolved solids were detected at five locations. Adequate treatment was in place for these contaminants at the detected locations. Private wells are not routinely sampled in southwestern Pennsylvania. One study in the Upper Mahoning Creek basin of the Allegheny River watershed (USGS, 1996) found that 76 percent of tested wells had at least one constituent concentration that exceeded one primary or secondary maximum contaminant level (SMCL). Concentrations of iron, lead, manganese, pH, bacteria, and radon were commonly detected in excess of standards, though maximum exceedances for all constituents except bacteria, cadmium, and radon were associated with mining activities.

BOX 3-3
Endocrine Disruptors, Pharmaceuticals, and Personal Care Products

In recent years, questions have been raised about the presence of traces of a variety of compounds in natural waters, wastewater, and drinking water. Trace concentrations of endocrine disruptors, pharmaceutical products, and personal care products have been found in the environment. Some are thought to cause problems such as abnormal sexual development in fish. This discussion of trace compounds in water explains what endocrine disruptors are and gives some examples of the kinds of compounds that are known or thought to be endocrine disruptors. In addition, the presence of pharmaceutical products and personal care products in the environment is discussed. Some knowledge of the treatability of these compounds is presented, although information on this is quite limited at present.

Endocrine disruptors are chemicals that interfere with endocrine system function in a variety of ways (Trussell, 2001). The endocrine system functions to regulate growth, behavior, and reproduction in living organisms. The endocrine system includes glands that secrete hormones, the hormones themselves, and other cells in the organism that have hormone receptors. The endocrine system can be disrupted by the presence of other compounds that mimic natural hormones and stimulate some action in a cell. A different form of disruption occurs when a compound prevents or blocks the action of a natural hormone, preventing the action in the cell from taking place. The pesticide DDT behaves as a blocker (Trussell, 2001).

A variety of persistent chemical compounds that are not readily degraded can be found in wastewater and natural waters. Known or potential endocrine disruptors, personal care products, and pharmaceutical compounds such as prescription drugs and over-the-counter medications have been identified in surface waters in the United States (Kolpin et al., 2002). These include the following:

- veterinary and human antibiotics such as erythromycin-H_20, sulfamethizole, and tetracycline;
- prescription drugs such as albuterol, cimetidine, codeine, gemfibrozil, and warfarin; used as antiasthmatic, antiacid, analgesic, antihyperlipidemic, and anticoagulant drugs, respectively;
- over-the-counter medications and nonprescription drugs, such as acetaminophen, ibuprofen, and caffeine used as antipyretic, anti-inflammatory, and stimulant drugs, respectively;
- personal care products such as acetophenone and *N,N*-diethyltoluamide; used as fragrance and insect repellant, respectively.
- steroids and hormones, both natural and synthetic; and
- insecticides, plasticizers, nonionic detergent metabolites, fire retardants.

The presence of such compounds in natural waters, wastewaters, and drinking water has been documented increasingly in recent years as analytical chemistry methods enable scientists to detect ever-decreasing concentrations of such compounds in water. The extent to which the kinds of

Microbiological Parameters

As discussed previously, levels of "pathogens" (as measured by bacterial indicator organisms) are monitored periodically by the state to assess the potential fecal contamination of Pennsylvania waters and to determine if a waterbody meets its designated use for drinking water (as part of human health) or recreational use. The *2004 Pennsylvania Integrated Water Quality Monitoring and Assessment Report* (PADEP, 2004) provides information on impairment of waters related to microbiological contamination; however, it does not provide specific data on concentrations of pathogens or indicator organisms. Statewide, pathogens are implicated in 23 of 1,373 impaired miles for human health use (both drinking water and fish consumption), compounds discussed above act as endocrine disruptors is largely unknown at this time. In 1996 the U.S. Congress directed EPA to develop an endocrine disruptor screening program and to screen endocrine disruptors found in drinking water. The EPA's Endocrine Disruptor Screening Program is relatively new, and EPA estimates that 87,000 chemicals in commerce might have to be evaluated for potential risks. A report by the National Research Council (NRC, 1999b) indicated that much work needs to be done to address the questions related to endocrine disruptors, or hormonally active agents, as these compounds were referred to in that report. The committee's recommendations for screening and monitoring endocrine disruptors were ". . .consistent, in principle, with those of EPA's Endocrine Disruptor Screening and Testing Advisory Committee."

In a high school research project that won the Grand Laureate at the International Stockholm Junior Water Prize competition, Mulroy (2000) detected penicillin, tetracycline, and vancomycin in water samples from a 44 km segment of the Ohio River near Wheeling, West Virginia, and two tributary streams. All *E. coli* cultured from sample sites exhibited acquired antibiotic resistance, with the greatest acquired resistance appearing in samples containing the highest concentrations of antibiotics. Mulroy also detected antibiotics in tap water, at lower concentrations than in the stream and river samples, from three municipalities whose sources are in the study area.

The concept that pharmaceuticals in municipal wastewater might have adverse environmental effects is not new. Snyder et al. (2003) wrote that Stumm-Zollinger and Fair in 1965 and Tabak and Bunch in 1970 expressed concern that natural and synthetic estrogens could become an ecological threat. Snyder et al. (2003) reviewed the treatability of some persistent compounds that are among the categories of pharmaceuticals, endocrine disruptors, and personal care products. They concluded that coagulation would be expected to remove only those compounds that sorb onto particles or colloidal material having a high content of organic carbon. Activated carbon adsorption would be expected to remove hydrophobic compounds very well, but competition for adsorption sites has not been studied thoroughly in the context of the compounds of interest. Oxidation with ozone would occur more rapidly than with chlorine dioxide or chlorine. Although not mentioned by Snyder et al. in their summary of water treatment, oxidation by ozone or other oxidants may result in a partial degradation of complex organic molecules, and the health effects and endocrine-disrupting capabilities of the degradation products are likely to be unknown.

Many pharmaceuticals, endocrine disruptors, and personal care products tend to resist biodegradation and hence have been found in streams, even though they had passed through wastewater treatment plants. This suggests that such compounds may resist degradation in the environment if they are spilled onto the earth's surface, so some of these compounds may also be found in stormwater runoff. Many of the compounds of concern are excreted by humans after they have been ingested for medicinal purposes or otherwise, so they will be found in wastewater, whether in separate sewer systems or in combined sewer systems. Given the present degree of wastewater treatment generally practiced (secondary or biological treatment), one can expect these compounds to be present in most treated wastewater and in natural waters.

representing less than 2 percent of the cause of impairment. Overall, 70 percent of the assessed stream-miles are impaired (PCBs are the leading cause). Pathogens are implicated in all of the 127 stream miles that are impaired for recreational use. Statewide, 90 percent of the assessed miles are impaired, though as noted previously, very few (140) miles have been assessed for recreational use—and those were targeted as being at risk from microbial contamination.

Despite the absence of specific data in the integrated 303(d) and 305(b) report for southwestern Pennsylvania, there are many smaller-scale monitoring efforts and studies in the region for which microbial parameters have been measured and documented as described below.

TABLE 3-6 Surface Water Quality of Major Rivers in Pittsburgh Metropolitan Area

Parameter	Median	Maximum	Minimum
Ohio River at Sewickley: 11/14/2000 through 09/08/2001			
Turbidity, NTU	11.5	50.	2.5
Dissolved oxygen, mg/L	9.8	14.3	7.2
pH, units	7.6	8.0	6.5
Alkalinity, mg/L as $CaCO_3$	40.5	50	28
Dissolved organic carbon, mg/L	2.2	3.1	1.7
Sulfate, mg/L	66.1	107	44.5
E. coli, CFU/100 mL	250	3,100	<5
Allegheny River at New Kensington: 12/6/95 through 09/29/2000			
Dissolved oxygen, mg/L	10.0	14.8	7.2
pH, units	7.5	8.0	5.5
Sulfate, mg/L	53.2	113	22.5
Monongahela River at Braddock: 12/07/95 through 09/27/2001			
Turbidity, NTU	14	75.	4.9
Dissolved oxygen, mg/L	8.2	13.9	6.1
pH, units	7.5	8.1	5.8
Dissolved organic carbon, mg/L	1.7	2.6	1.0
Sulfate, mg/L	94.5	225	27

NOTE: NTU = nephelometric turbidity units; used to measure the clarity of water.
SOURCE: Data from *http://waterdata.usgs.gov/pa/nwis*.

Surface Water

Drinking Water Providers. Water treatment plants routinely monitor source waters for the presence of microbial contaminants and indicator organisms. Of the 23 water producers in Allegheny County 5 were required to participate in the EPA Information Collection Rule (ICR) pathogen and disinfection by-products data collection activity from 1997 to 1998. No other water providers in southwestern Pennsylvania (either inside or outside Allegheny County) were required to collect pathogen data under the ICR because none served at least 100,000 people. In general and as discussed below, the ICR data[19] for Allegheny County show that pathogens are routinely present in source waters for public drinking supplies.

Two water treatment plant sites for the Pennsylvania American Water Company are located on the Monongahela River. Monongahela River source water was found to contain *Cryptosporidium* (20 of 36 samples) at concentrations from 14 to 309 oocysts per 100 L and *Giardia* (31 of 36 samples) at concentrations from 18 to 292 cysts per 100 L. It is important to note, however, that these concentrations—or those report below—are not necessarily for viable/infective (oo)cysts. Coliform bacteria were always detected at concentrations that ranged from 40 to 24,000 per 100 mL while viruses were intermittently detectable (8 of 36 samples) in concentrations from 2.1 to 27.9 (most probable number, MPN) per 100 L.

Two water treatment plants are on the Allegheny River (Pittsburgh Water and Sewer Authority [PWSA] and Wilkinsburg-Penn Joint Water Authority). Allegheny River waterways

[19] The ICR data are available on-line at *http://www.epa.gov/enviro/html/icr/state/PA.html*.

were found to intermittently contain *Cryptosporidium* (10 of 36 samples) at concentrations from 5 to 106 oocysts per 100 L, with all but one of the positive values at the PWSA plant. Allegheny River water intermittently contains *Giardia* (13 of 36 samples) at concentrations ranging from 10 to 263 cysts per 100 L. Coliform bacteria were generally detected (33 of 36 samples) at concentrations ranging from 150 to 23,600 per 100 mL and viruses (20 of 36 samples) at concentrations from 1 to 13.9 MPN per 100 L.

One water treatment plant for the West View Municipal Authority uses Ohio River water as a source. Ohio River water rarely contained *Cryptosporidium* (3 of 18 samples) at concentrations ranging from 31 to 45 oocysts per 100 L. *Giardia* was more commonly detected (8 of 18 samples) at concentrations of 13 to 471 cysts per 100 L. Total coliform counts ranged from 500 to 6,100 per 100 mL in the river, while viruses were usually detectable (12 of 18 samples) in concentrations from 1 to 7.1 MPN per 100 L.

The Municipal Authority of Westermoreland County draws water from a variety of local waterways, including the Youghiogheny River, Indian Creek, and Beaver Run. Data collected under the ICR are available for the Indian Creek Treatment Plant. Although *Cryptosporidium* were not detected in Indian Creek water, *Giardia* were intermittently detected (7 of 18 samples) at concentrations ranging from 10 to 204 cysts per 100 L. Coliform bacteria were always detected, with concentrations between 10 and 800 per 100 mL, and viruses were detected (9 of 18 samples) at concentrations between 1 and 24.1 MPN per 100 L.

3 Rivers 2nd Nature. The 3RWW[20] initiated a project in 2000 entitled 3 Rivers 2nd Nature to assess water quality in surface waters in southwestern Pennsylvania. This study includes dry weather and wet weather monitoring for fecal coliforms and *E. coli* along with several geochemical parameters. Data for 2000 were collected in the Pittsburgh Pool, which includes the areas of the three rivers encompassing the Point and upstream on the Allegheny and Monongahela Rivers to their first locks and downstream on the Ohio to the first lock. In this monitoring context, pool is defined as the water contained within the three rivers up to the first lock and dam (L/D) on each main stem river (L/D 2 on the Allegheny, L/D 2 on the Monongahela, and the Emsworth L/D on the Ohio; see Figure 3-3). In 2001 and 2002 the project focused on the Monongahela and Allegheny Rivers upstream of the Pittsburgh urban area, respectively, and in 2003 it focused on the Ohio River downstream of the City of Pittsburgh. Figure 3-4 shows 3R2N's water quality testing points from 2000 to 2004 for rivers, streams, and watersheds.

In 2000, the 3R2N study concluded that the main river dry weather fecal coliform data were equal to or lower than the 400 CFU/100 mL (EPA, 1986) standard for incidental contact in recreational waters. The *E. coli* level in the main rivers was also less than or equal to the standard of 126 CFU/100 mL. For all three rivers in wet weather conditions, data showed that fecal coliform concentrations increase and remain high for days after a rainfall (Knauer and Collins, 2001, 2002, 2003). The maximum observed fecal coliform concentration in the main stem rivers was 10^3 per 100 mL and in the tributaries was 10^4-10^5 per 100 mL. Maximum *E. coli* levels were 10^2-10^3 per 100 mL in the main stem rivers and 2,500 per 100 mL in the tributaries. It was also observed that the edges of the rivers have higher bacterial concentrations and are slower to recover than the middles of the rivers. This finding is relevant in terms of public

---

[20] For further information on 3RWW, see *http://3riverswetweather.org/index.htm*; for information on 3R2N, see *http://3r2n.cfa.cmu.edu/index.html.*

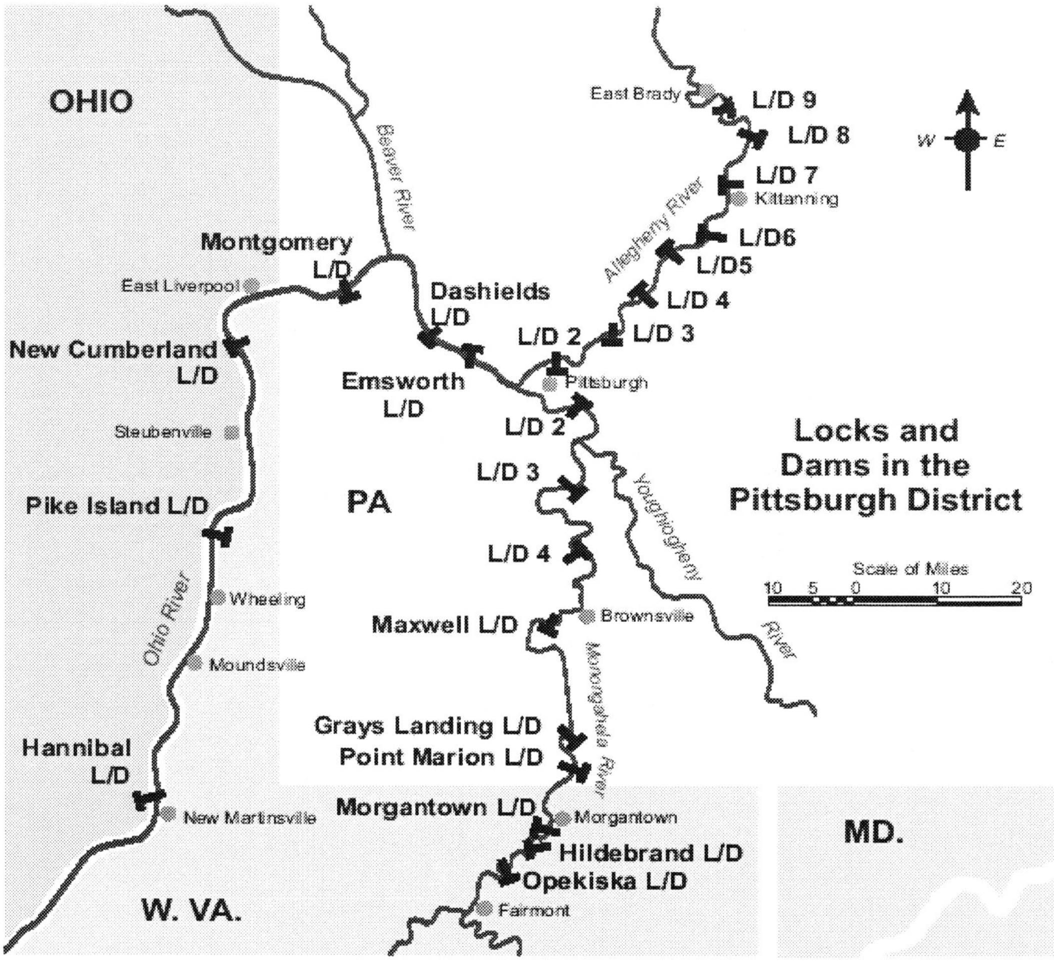

FIGURE 3-3 Lock and dam structure of the Three Rivers.
SOURCE: *http://www.lrp.usace.army.mil/nav/nav.htm.*

access, since most people are in contact with river water while fishing at the edge (Knauer and Collins, 2001).

In 2001, the 3R2N study concluded that during dry weather, sites on the Monongahela River are below regulatory levels for fecal coliforms. During dry weather, tributaries to the Monongahela are higher than regulatory limits, with stream values varying from less than 200 CFU/100 mL to greater than 10^5 CFU/100 mL. Wet weather river water quality generally does not exceed regulatory limits in Pool 3 (the stretch of river between L/D 3, 23.8 miles upstream of the confluence, and L/D 4, 41.5 miles upstream of the confluence [see Figure 3-3]) on the Monongahela; however, Pool 2 (the stretch of the river between L/D 2, 11.2 miles upstream of the confluence, and L/D 3) sites show higher coliform levels in wet weather. The Pittsburgh Pool sites were higher in coliforms than sites in Pools 2 and 3 (Knauer and Collins, 2002).

In 2002, the 3R2N study concluded that during dry weather, sites on the Allegheny River in Pools 2, 3, and 4 are below regulatory levels for fecal coliforms and *E. coli*. However, during dry weather, some tributaries exceed regulatory limits. In Pool 2, Plum Creek, Indian Creek, and Squaw Run exceed the recreational contact standard of 400 CFU/100 mL for fecal coliforms and of 126 CFU/100 mL for *E. coli*. In Pool 3, Baileys Run and Clarks/Crawford Run exceed fecal

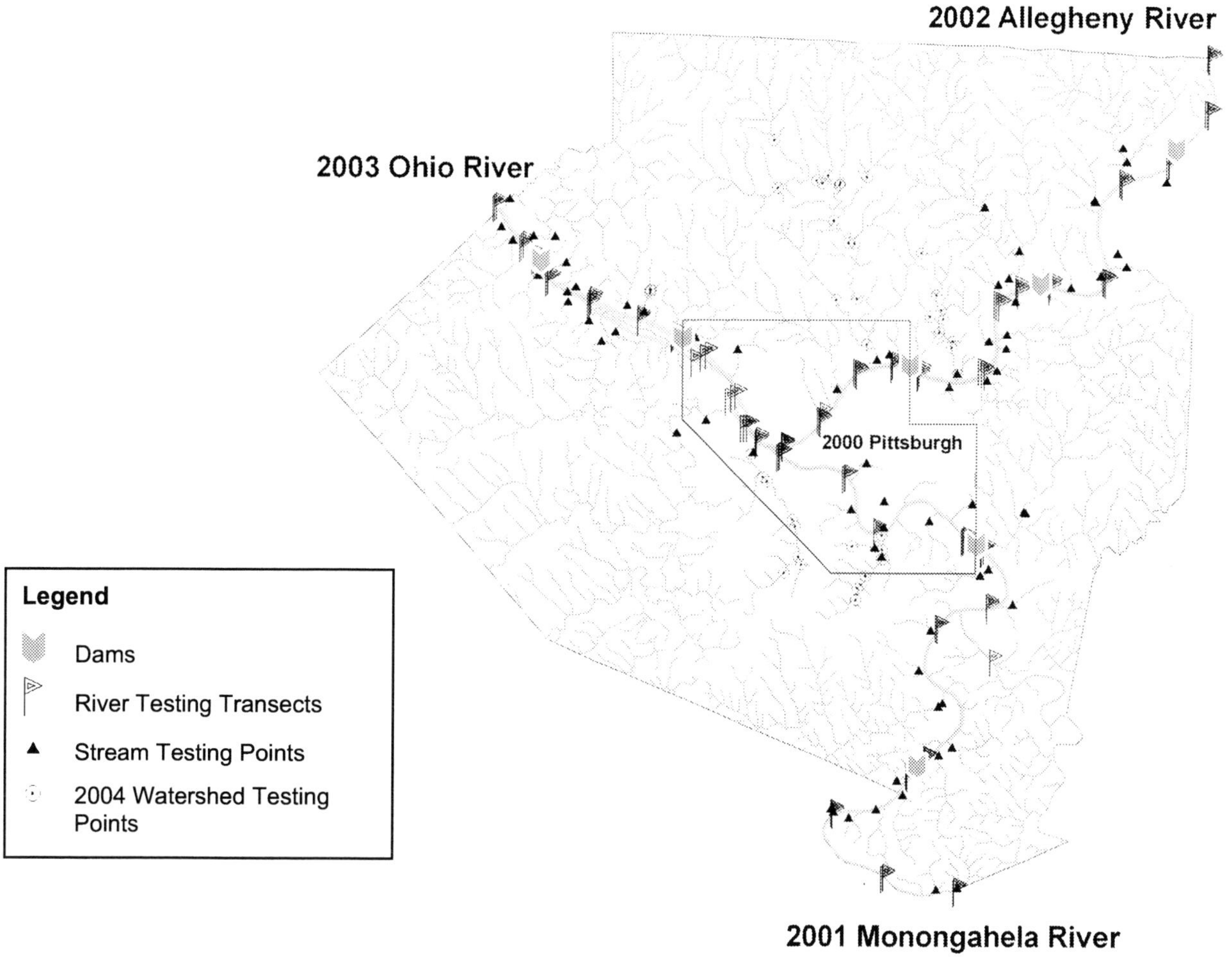

FIGURE 3-4 3 Rivers 2nd Nature integrated water quality testing points, 2000-2004
SOURCE: Adapted and reprinted, with permission, from 3 Rivers 2nd Nature. © 3 Rivers 2nd Nature.

coliform and *E. coli* standards in dry weather. Although all Pool 3 tributaries exceed the *E. coli* standard in dry weather, the single tributary monitored in Pool 4 (Buffalo Creek) was below regulatory limits during dry weather. Wet weather event sampling indicated that locations in the Pittsburgh Pool have high concentrations of fecal coliforms after rain events. In the upper pools, locations near CSO structures have higher indicator concentrations following precipitation (Knauer and Collins, 2003).

In 2003, the 3R2N evaluated sites on the Ohio River and its tributaries, but the results have not yet been released.

Allegheny County Sanitary Authority. ALCOSAN has collected data on bacterial indicator organisms in surface water in southwestern Pennsylvania. From 1993 to 1996, the Allegheny River, Monongahela River, Ohio River, Chartiers Creek, Saw Mill Run, Turtle Creek, and Thompson Run were monitored both upstream and downstream of the ALCOSAN service area. These data are not directly available; however, they were reviewed by the Third Party Review Committee (TPRC, 2002; see Chapter 5 for further information). According to the TPRC report

(2002), downstream fecal coliform levels are greater than upstream by a factor of 10 for these rivers. Upstream levels are greater than the maximum standard of 400 CFU/100 mL 30 percent of the time in the Allegheny River and 50 percent of the time in the Monongahela River. Data exceeded the geometric mean of 200 CFU/100 mL at both upstream locations. The creek data are similar in that downstream data are higher than upstream data, but not all values meet standards. Maximum upstream fecal coliform levels are approximately 10^3-10^4 CFU/100 mL in the main rivers and 10^4-10^5 CFU/100 mL at the downstream stations. In the tributaries, maximum levels were reported to be 10^4-10^5 CFU/100 mL upstream and 10^5-10^6CFU/100 mL at downstream stations. The TPR report concluded that high dry-weather bacteria levels in the tributaries may indicate dry weather overflow, illegal direct connection of sanitary sewers to stormwater systems, or groundwater and/or upstream sources (lying outside of the ALCOSAN service area) (TPRC, 2002). High downstream and wet weather levels indicate SSOs and CSOs related to wet weather collection system overloads. However, in most cases there is insufficient information to determine the sources of the bacterial indicator organisms.

Allegheny County Health Department and U.S. Geological Survey. In 2001, the ACHD in collaboration with the USGS (Water Resources Division, Pittsburgh, Pennsylvania) began collecting data on fecal indicators in the three main stem rivers near the City of Pittsburgh. The goal of this data collection effort is to develop a sampling protocol, to sample indicator bacteria in accordance with 25 PA Code § 93.7[21] (which requires a specific number of samples within a defined time frame for validity) and to develop a decision tool for issuing and lifting river advisories. Three sets of indicator organisms were evaluated—fecal coliforms, *E. coli,* and enterococci. The samples were collected weekly and for several successive days after three wet weather events in the summer of 2001 at five sampling locations (two on the Monongahela; two on the Allegheny; one on the Ohio). Raw data for this sampling are available on-line.[22] While the ACHD analysis of the data is still ongoing regarding conclusions that can be made about the bacterial quality of the three rivers in Allegheny County, preliminary analysis of the raw data shows that all indicator organism levels exceed water quality standards some of the time in these rivers, with total counts increasing in response to wet weather.

U.S. Army Corps of Engineers. Pittsburgh District Office. The USACE has conducted two studies related to bacterial water quality (Koryak and Reilly, 2000; USACE, 1997). For Montour Run (USACE, 1997), fecal coliform were collected in September 1996 at four locations on Montour Run and at the mouths of 14 of its largest tributaries. The Montour Run watershed is in western Allegheny County, with much of the basin within 10 miles of Pittsburgh. Notably, the basin contains the Pittsburgh International Airport and was under significant development during the USACE study. Conversion of agricultural and woodland uses to suburban housing and retail complexes was significant. Fecal coliform bacterial concentrations were highest in the western portion of the watershed (in the headwaters) and ranged from 650 to 4,600 CFU/100 mL. In the eastern part of the watershed, fecal coliform concentrations were all below 20 CFU/100 mL. The report speculates that high levels in the western portion of the watershed are related to malfunctioning on-site septic systems discharging partially treated effluent.

---

[21] Section 93.7 is available on-line at *http://www.pacode.com/secure/data/025/chapter93/s93.7.html.*

[22] For further information see *http://pa.water.usgs.gov/ar/wy02/pdfs/cso-proj.pdf.*

Nine Mile Run is a tributary of the Monongahela River in eastern Allegheny County. It enters the Monongahela at river mile 7.6 and encompasses a 7.5 square mile urban watershed with much of the flow through culverts. Total coliform, fecal coliform, and *E. coli* bacterial samples were collected from six locations along Nine Mile Run in 1999. Results indicate that Nine Mile Run is unsafe for human contact during dry and wet weather; it is seriously degraded by sewage because it is a corridor for SSOs and CSOs for eastern suburbs as well as the east end of Pittsburgh (Koryak and Reilly, 2000). Total coliform counts ranged from 101,000 to 1,311,000 CFU/100 mL, while fecal coliform counts ranged from 125 to 1,051,200 CFU/100 mL. *E. coli* ranged from 125 to 1,009,800 CFU/100 mL. Wet weather values increase by orders of magnitude. The authors conclude that dry weather exceedances are related to chronic SSOs, while wet weather exceedances are related to CSO events.

Watershed Associations, Schools, and Other Nongovernmental Agencies. Many other groups collect water quality data in southwestern Pennsylvania for a variety of reasons. For example, many public schools participate in Creek Connections.[23] This program involves local schools in hands-on, inquiry-based learning through water quality monitoring in local creeks and streams, and data are available on-line. Monitoring parameters include temperature, pH, total dissolved solids, dissolved oxygen, nitrate, phosphorus, alkalinity, turbidity, and biological index. Some studies also include flow measurements and iron, and one high school (North Hills) has collected a few samples for *E. coli.* As an example, Figure 3-5 shows sampling locations for the Pittsburgh region. The North Hills *E. coli* sampling program found levels in Girty's Run that exceeded acceptable levels for contact recreation in seven samples taken from October 2003 to January 2004. The highest levels were seen during wet weather.

The Pennsylvania Organization for Watersheds and Rivers (POWR) is a statewide nonprofit organization that includes individuals, watershed associations, conservation agencies and organizations, and corporations that support watershed science and education. Its web site[24] includes a directory of watershed groups active in each county and links to other programs such as the Keystone Watershed Monitoring Network. The directory does not list recorded data directly, but links to individual organizations may provide lists of projects and data for specific programs. In many cases, local organizations in POWR may be the only groups conducting water quality monitoring in rural headwater streams. The extent of monitoring and the available data for bacteria in headwaters streams are uncertain and difficult to ascertain. In the farthest southwestern counties of the study area (Washington and Greene), technical advisers to local watershed alliances were not aware of any local watershed groups conducting bacterial monitoring (Ben Stout, Wheeling Jesuit University, personal communication, 2003; Mary Joy Haywood, Carlow College [retired], personal communication, 2003). Box 5-3 discusses an example of a headwaters study in a West Virginia watershed that provides important evidence for watershed conditions both locally and on the larger, main stem stream.

Water quality data collected by independent watershed associations and nongovernmental agencies are poorly catalogued within the region. Data exist, but they may not meet the rigid standards of quality assurance and quality control required for utilization by state regulators. For example, in preparing the *2004 Pennsylvania Integrated Water Quality Monitoring and Assessment Report*, the PADEP requested information on available water quality data from

[23] See *http://merlin.alleg.edu/group/creekconnections/* for data and further information about the program.
[24] The POWR website can be found on-line at *http://www.pawatersheds.org/index.asp/*.

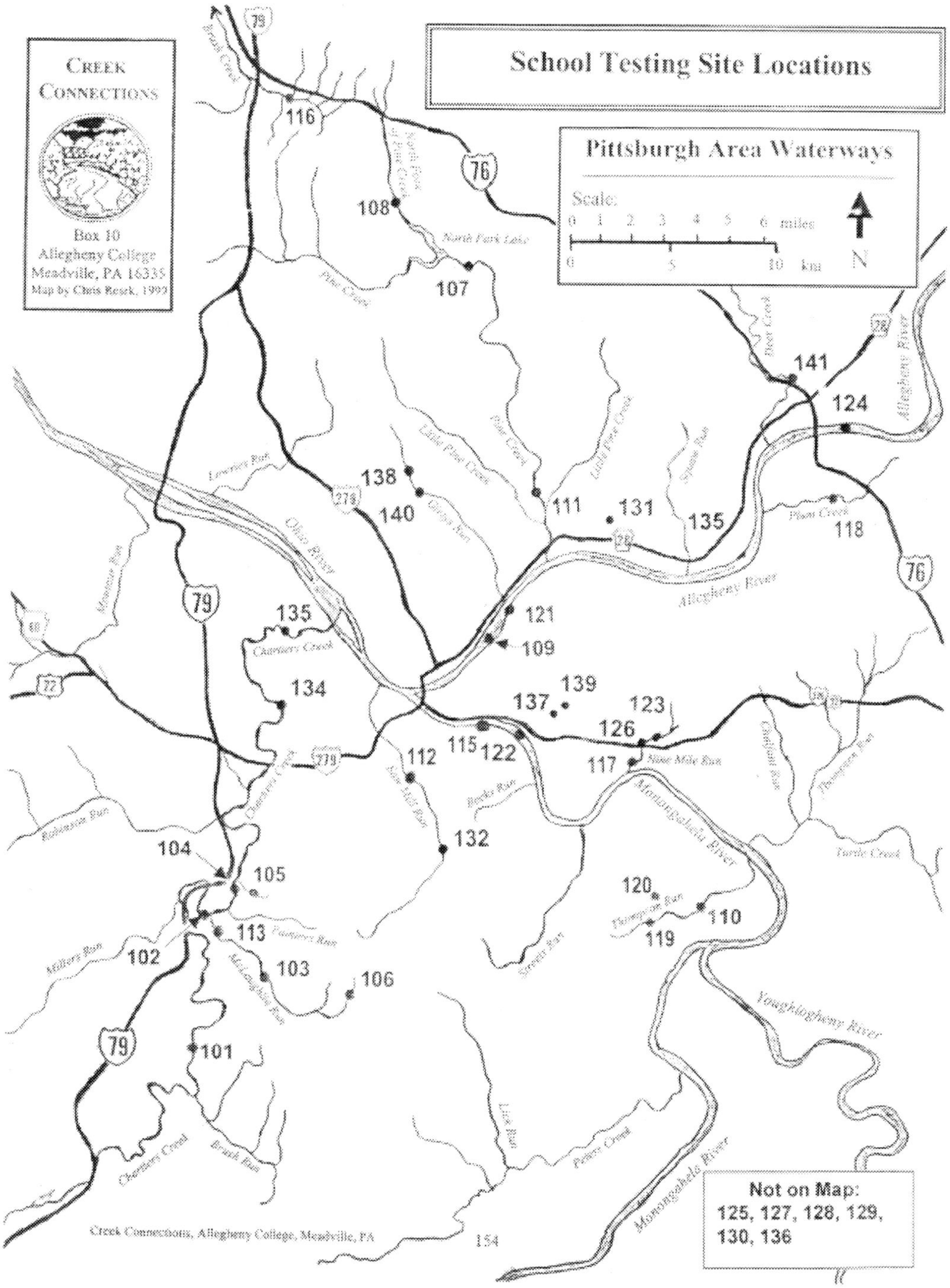

FIGURE 3-5 School testing locations in and around the City of Pittsburgh for the Creek Connections program.
SOURCE: Reprinted, with permission, from *http://creekconnections.allegheny.edu/swpamap.html*. © 1999 Allegheny College.

hundreds of independent organizations (PADEP, 2004). Ten organizations responded and data from seven of these were used to evaluate the attainment of recreational uses. Specifically, four local community-based water associations provided bacteriological data: Broadhead Creek Watershed Association, Brush Creek Watershed Association, the Schuylkill Chapter of the Pennsylvania Senior Environment Corps, and the Tri-County Conewago Creek Association (PADEP, 2004).

The PADEP encourages citizens' groups to become trained to conduct water quality monitoring programs. The Citizens' Volunteer Monitoring Program (CVMP) provides technical support and assistance to organizations that want to collect data for use in the 305(b) listing process. Independent sources of data that do not meet PADEP standards for clear delineation of stream segment and for adequate site location and collection protocols are not used in the 305(b) listing process.

Specific Independent Studies. In addition to large agency-based or sponsored efforts, individual researchers have collected data regarding the water quality in southwestern Pennsylvania. Many of these studies were focused on a specific water quality impairment (e.g., CSO events) and thus are discussed in Chapter 4 with the relevant impairment. The results are summarized here. States et al. (1997) reported on protozoa levels in the Allegheny River, which is the source water for the PWSA treatment plant. In the Allegheny River, they found that 63 percent of the samples were positive for *Giardia* cysts, with concentrations ranging from 0 to 421 cysts per 100 L of water, and 63 percent of the samples were positive for *Cryptosporidium* oocysts, with concentrations ranging from 0 to 2,233 oocysts per 100 L of water. In the Youghiogheny River, similar percentages and concentrations were observed (see complete data summarized in Table 4-8). Gibson et al. (1998) found that *Giardia* and *Cryptosporidium* were routinely present in Saw Mill Run (an urban stream which is a tributary to the Ohio River) during dry weather, with concentrations of 5-105 oocysts per 100 L and 13-6,579 cysts per 100 L (see complete data summarized in Table 4-9). Both studies found higher levels of protozoa (oo)cysts in CSOs and in surface waters downstream of CSOs during wet weather. Collins et al. (1998) conducted a series of dry weather bacteriological surveys of Nine Mile Run in 1997 using multiple sample locations on a single day. Data were provided at six sampling locations. Fecal coliforms ranged from more than a million per 100 mL to 125/100 mL. *Escherichia coli* likewise ranged from 125 to more than a million per 100 ml. Total coliforms ranged from 100,000 to 1.3 million per 100 mL.

Groundwater

As part of a joint study by the USGS and PADEP, indicator organisms (total coliforms and *E. coli*) and specific pathogens (e.g., culturable viruses, the bacterium *Helicobacter pylori*) were monitored in noncommunity water supply wells in Pennsylvania (Lindsey et al., 2002). Overall, the study found that 62 percent of the wells were positive for total coliforms, while 10 percent were positive for *E. coli*. Seventeen percent of the samples that were positive for total coliform were also positive for *E. coli*. Of 60 wells, 4 were positive for *H. pylori* (although one of these samples had no other indicators or pathogens detected). Culturable viruses were detected in 5 of 60 wells (two of these locations had no other indicators or pathogens detected).

However, these wells were distributed throughout the state (but predominantly in southeastern Pennsylvania) and may not be indicative of well water conditions in southwestern Pennsylvania (see Figure 3-6).

Sharpe et al. (1985) sampled private wells in Pennsylvania and found that 42 percent were positive for coliform bacteria. The USGS (1996) completed a study of 50 wells or springs in the Upper Mahoning Creek basin in Pennsylvania. Mahoning Creek runs through Jefferson, Indiana, and Clearfield Counties in western Pennsylvania and flows into the Allegheny River about 60 miles north of Pittsburgh. Seventy-four percent of the samples were positive for coliform bacteria. The report does not identify specific sources for the contaminated wells; however, it implicates improperly functioning septic systems, improperly sealed wells, and runoff from barnyards as potential sources in general.

Specific Organic Compounds in Water: USGS Studies

Pesticides and Herbicides

Analysis of water samples for pesticides and herbicides is somewhat expensive, so testing for these contaminants is typically conducted less frequently in ambient surface waters than in finished drinking water—especially for the pesticides and herbicides for which enforceable maximum contaminant levels (MCLs) exist in federal drinking water regulations. From 1996 to 1998, the USGS conducted a comprehensive study of water quality in the Allegheny and

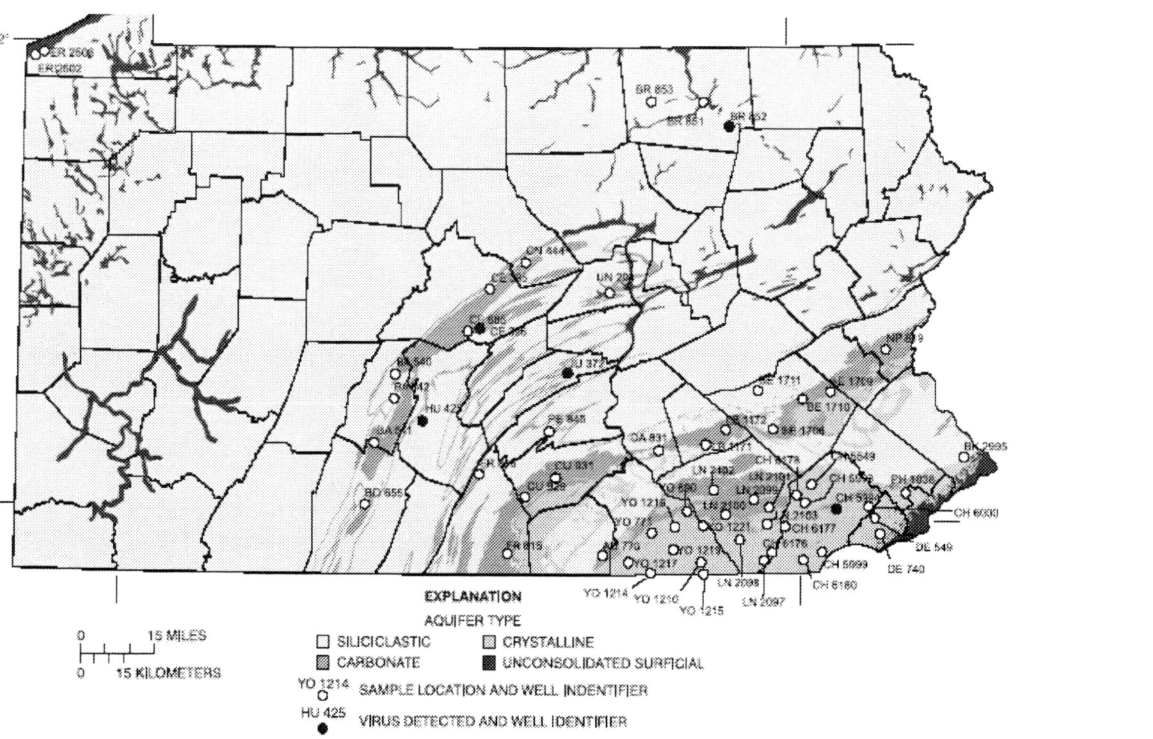

FIGURE 3-6 Sampling locations and detection of viruses in noncommunity supply wells in Pennsylvania. SOURCE: Lindsey et al., 2002.

Monongahela River basins (called the ALMN Study Unit) as a part of its NAWQA program. A summary of the study unit's water quality is available in printed form (Anderson et al., 2000; see also Appendix B), and detailed data are available on-line at *http://pa.water.usgs.gov/ almn/*. In this study unit, one surface water site on Deer Creek (see Figure 3-5) near Dorseyville in Allegheny County was selected as an "Intensive Urban/Residential Indicator Fixed-Site." Samples from this site were collected for comprehensive analysis of water quality, including 83 dissolved pesticides and 87 volatile organic compounds (VOCs), as summarized and discussed below.

South Branch Plum Creek at Five Points, Pennsylvania, an agricultural area about 40 miles northeast of Pittsburgh, was the only other site in addition to Deer Creek that was selected for intensive sampling of pesticides and organic chemicals in the study unit. Although those data are not presented in this discussion, the use of only one other intensive sampling site in the ALMN Study Unit suggests that comprehensive water quality sampling for large numbers of synthetic organic chemicals is not conducted frequently. Anderson et al. (2000) reported that of the 83 pesticides tested for, 25 were detected at least once in Deer Creek, but no pesticide was detected in a concentration greater than its drinking water MCL set by EPA. The authors noted that atrazine, simazine, and metolachlor appeared seasonally, during herbicide applications periods and when increased rainfall occurred in the spring. These chemicals tended to peak in Deer Creek in May and June.

A review of detailed Deer Creek pesticide data from the web site for the Allegheny and Monongahela River basin study was conducted by the committee (see Table 3-7). The herbicide detected in the highest concentrations in water flowing from Deer Creek, a residential-urban watershed, was 2,4-D. Some of the trade names under which this chemical is sold suggest that it finds uses not only in agriculture but also in lawn care, which may explain why it was found at higher concentrations than other herbicides in a watershed where land use was not primarily agricultural (Anderson et al., 2000).

TABLE 3-7 Pesticides Found in Water of Deer Creek During USGS Study[a,b]

Pesticide (Trade Name)	Concentration Range (μg/L)	Months Detected/ Months Sampled
Herbicide		
Atrazine (AAtrex, Atrex, Atred, Gesaprim)	0.004-0.250	12/14
Cyanazine (Bladex, Fortrol)	0.007-0.14	1/14
2,4-D (Aqua-Kleen, Lawn-Keep, Weed-B-Gone)	0.17-1.16	5/11
Metolachlor (Dual, Pennant)	0.005-0.099	11/14
Prometon (Pramitol, Princep)	0.025-0.355	4/14
Simazine (Princep, Caliber 90)	0.008-0.092	5/14
Insecticide		
Diazinon (Basudin, Diazotol, Neocidol, Knox Out)	0.006-0.096	6/14

[a] Sampling conducted from June 1996 through September 1998.
[b] Table contains only data presented on web site with no qualifiers (i.e., only data ≥ method detection limit included, and no data included in which quantitative determination was uncertain).
SOURCE: Adapted from data on USGS web site (*http://pa.water.usgs.gov/almn/*).

Volatile Organic Compounds (VOCs)

Anderson et al. (2000) reported that VOCs related to gasoline and cleaning solvents were found in 24 of 25 samples from Deer Creek. Of the 87 VOCs for which analysis was conducted, 22 were detected at least once, and slightly more than half of those VOCs were compounds related to gasoline. However, no VOC was found at a concentration exceeding its drinking water MCL. Anderson et al. (2000) also noted that VOCs can accumulate on impervious surfaces and be washed off during precipitation events. Analysis of a set of five samples collected from Deer Creek during a storm showed that VOC concentrations peaked as streamflow was increasing, while the lowest VOC concentrations were measured as flow decreased. Anderson and colleagues concluded that the relationship between flow and VOC concentration showed an effect of flushing VOCs from the land surface and into streamflow.

EVIDENCE OF WATERBORNE DISEASE

In addition to analyzing water quality data for chemical and biological parameters and noting water quality impairments, another way to gauge water quality conditions in the southwestern Pennsylvania region is to document cases of waterborne disease. Most states and territories participate in the National Waterborne Disease Outbreak Surveillance System maintained by the Centers for Disease Control and Prevention (CDC), EPA, and the Council of State and Territorial Epidemiologists, by reporting cases of salmonellosis, shigellosis, *Escherichia coli* O157:H7, hepatitis A, giardiasis, and cryptosporidiosis (NRC, 2004). Some states collect surveillance data on other enteric diseases, such as yersiniosis and campylobacteriosis. Repeated efforts by the committee to procure surveillance data for southwestern Pennsylvania from the state epidemiologist and the Allegheny County Health Department were unsuccessful. In lieu of those data, overall Pennsylvania data on reported drinking water and recreational waterborne diseases were obtained from CDC and are discussed below. It is important to note, however, that these CDC reports do not include all known waterborne disease outbreaks in Pennsylvania (e.g., a drinking water-related outbreak of giardiasis in McKeesport, Pennsylvania, in winter 1983-1984 was not included in these reports; see Box 3-4 for further information).

Between 1971 and 1985, Pennsylvania reported a total of 90 waterborne disease outbreaks resulting in 29,380 cases of illness. This represented 19 percent of reported outbreaks and 26 percent of cases of illness between 1971 and 1985 in the United States, leading the nation in the number of waterborne disease outbreaks and number of cases of illness reported. In a review of waterborne disease surveillance activity between 1946 and 1980 (Lippy and Waltrip, 1984), Pennsylvania reported 0.6 to 1.9 waterborne disease outbreaks per 100 community water systems, with 1 to 20 cases of illness per 10,000 population served by water utilities. The number of illnesses exceeded 10,000 cases during the review period. Only Pennsylvania, New York, Florida, Texas, and California experienced more than 10,000 cases of such illness. Finally, Frost et al. (1995) summarized waterborne disease surveillance practices based on a survey of state and territorial epidemiology programs. Pennsylvania ranked with New York, New Hampshire, Colorado, Missouri, and Minnesota for the highest number of waterborne disease outbreaks reported between 1986 and 1992. Table 3-8 summarizes waterborne illness

TABLE 3-8 Pennsylvania Waterborne Disease Surveillance Data from CDC for Outbreaks Associated with Drinking Water Use: 1983 to 2000

Year	Agent	Cases	Deficiency	Class	Type of Location	Source
1983	*Giardia*	366	3	C	16 communities	Sewage-contaminated watershed
1983	*Giardia*	135	3	C	Community	Stream
1983	AGI	11	2	IND	Camp	Well
1983	AGI	11,400	3	NC	Religious festival	Well
1983	AGI	25	2	NC	Recreation area	Well
1983	AGI	200	2	NC	Resort	Well
1983	AGI	146	2	NC	Recreation area	Well, spring
1983	AGI	298	3	C	Community	River
1984	*Giardia*	8	2	IND	Picnic	Well
1984	AGI	34	2	IND	Bicycle race	Private well
1984	AGI	18	2	IND	Industry	Well
1984	AGI	98	2	NC	Resort	Well
1985	AGI	70	3	NC	Restaurant	Well
1985	AGI	275	2	NC	School	Well
1985	AGI	11	3	NC	Restaurant	Well
1985	*Shigella*	27	1	NC	Camp	Well
1986	AGI	213	3	NC	Restaurant	Well
1987	AGI	53	5	NC	Resort	Well
1987	AGI	22	2	NC	Camp	Well
1987	AGI	?	2	IND	Home	Well
1988	*Giardia*	172	3	C	Community	Lake
1988	AGI	26	2	NC	Camp	Well
1989	AGI	50	2	NC	Camp	Well
1990	Hepatitis A	22	2	IND	Homes	Well
1990	Hepatitis A	3	3	C	Community	Well
1990	AGI	63	5	C	Inn	Lake
1991	AGI	8	3	NC	Restaurant	Well
1991	AGI	170	3	NC	Picnic area	Well
1991	*Giardia*	13	3	NC	Park	Well
1991	*Cryptosporidium*	551	3	NC	Picnic area	Well
1991	AGI	300	3	NC	Camp	Well
1992	AGI	5	3	NC	Restaurant	Well
1992	AGI	28	5	C	Park	River
1992	AGI	38	2	IND	Home	Well
1992	AGI	42	3	NC	Camp	Well
1992	AGI	50	3	NC	Camp	Well
1992	AGI	57	3	NC	Camp	Well
1992	AGI	80	3	NC	Camp	Well
1993	*Giardia*	20	3	NC	Trailer park	Well
1993	AGI	65	3	NC	Ski resort	Well
1994	AGI	200	3	NC	Resort	Well
1995	AGI	19	2	NC	Inn	Well
1996-2000	None Reported	—	—	—	—	—

NOTE: AGI = acute gastrointestinal illness of unknown etiology; NC = noncommunity, C = community, IND = individual; 1 = untreated surface water, 2 = untreated groundwater, 3 = treatment deficiency, 4 = distribution system deficiency, 5 = unknown or miscellaneous deficiency.
SOURCES: Adapted from various CDC reports available on-line at *http://www.cdc.gov/ncidod/dpd/healthywater/publications.htm*.

outbreaks reported to CDC in Pennsylvania due to drinking water sources from 1983 to 2000. Notably, no waterborne disease outbreaks were reported during 1996 to 2000.

The preceding studies and reviews focused on drinking water as the vector of contamination. Identifying and summarizing epidemiological studies of waterborne illness contracted due to recreational exposure is more challenging. Most exposed individuals develop symptoms days after recreational contact, and most infected individuals attribute their illness to food poisoning or flu. Underreporting of gastrointestinal illness associated with recreational water contact (and, indeed, drinking water exposure) is expected to be high (NRC, 2004). Table 3-9 shows waterborne illness outbreaks reported to CDC in Pennsylvania due to recreational water contact during the 1980s and 1990s.

Box 3-4 describes the one well-documented case of waterborne disease (in this case, giardiasis) known to have occurred in southwestern Pennsylvania, an incident related to operational failures at a treatment plant. Because no other specific outbreak data for southwestern Pennsylvania were obtained, it is difficult to extend the preceding findings on waterborne disease to the entire region. For example, no data were produced to suggest that southwestern Pennsylvania experienced more gastrointestinal illness that other parts of the state. Although the fecal indicator and pathogen concentrations released into the environment during wet weather events have exceeded federal guidelines in the past, no related illnesses have been identified. The only documented evidence of a drinking water-related public health problem for the entire state is the number of disease outbreaks in noncommunity drinking water supply wells and through recreational exposure that were caused by acute gastrointestinal illness of unknown etiology (AGI [likely of viral source]) and *Shigella*. The high viral host specificity indicates that these are likely caused by wells contaminated with human waste.

TABLE 3-9 Pennsylvania Waterborne Disease Surveillance Data from CDC for Outbreaks Associated with Recreational Water Use

Year	Etiologic Agent	Cases	Illness	Source	Setting
1982	*Pseudomonas*	127	Dermatitis	Whirlpool	Motel
1982	*Pseudomonas*	36	Dermatitis	Whirlpool	Motel
1982	*Pseudomonas*	68	Dermatitis	3 pools	Hotel
1982	*Pseudomonas*	14	Dermatitis	Pool	Motel or hotel
1987	*Pseudomonas*	22	Dermatitis	Hot tub	Motel
1988	*Shigella sonnei*	138	Gastroenteritis	Lake	Recreation area
1990	AGI	60	Gastroenteritis	Lake	Camp
1991	*S. sonnei*	203	Gastroenteritis	Lake	Park
1995	AGI	17	Gastroenteritis	Lake	Park
1995	*S. sonnei*	70	Gastroenteritis	Lake	Beach
1998	*Cryptosporidium parvum*	8	Gastroenteritis	Lake	State park

SOURCES: Adapted from various CDC reports available on-line at *http://www.cdc.gov/ncidod/dpd/healthywater/publications.htm.*

BOX 3-4
The McKeesport Outbreak

The Outbreak

During the winter of 1983-1984, more than 340 cases of waterborne giardiasis occurred in McKeesport, Pennsylvania. In late December 1983, water demand in McKeesport was very high, depleting distribution system water storage and preventing effective backwash of the filters (Logsdon et al., 1985). Demand increased on December 24, and reservoir levels dropped to about half capacity by the 26th. On December 27 and 28, breaks in 3-inch and 6-inch fire lines were discovered at a U.S. steel mill (Stoecker, 1985). An elevated backwash tank at the water treatment plant was out of service at the time, so all backwash water had to come from the distribution system. Filters were run for several days without backwashing until December 31, when backwashing was performed for the first time in a week. (A filter typically is backwashed after operating times of one to three days when filtered water quality goals are met and when head loss through the filter has not reached the maximum allowed). On January 3, the plant had been pumping finished water at the rate of 13 to 13.5 million gallons per day (mgd) since Christmas, whereas normal winter pumping was about 10 mgd (Stoecker, 1985).

This high rate of water production and failure to backwash filters for a week led to a large-scale turbidity breakthrough and significant deterioration of the finished water turbidity. Data entered in the plant report from December 25, 1983, through January 14, 1984 (MMWA, 1983-1984), indicate that composite turbidity from the plant rose to 5 nephelometric turbidity units (ntu) on December 29 and was 2.0 ntu or higher for 10 days during this 21-day period. For comparison, the weekly average turbidity of combined filter effluent was 0.24, 0.28, and 0.37 ntu, respectively, during the three weeks before the treatment problems started. After January 11, 1984, filtered water turbidity was generally 1.0 ntu or lower. From December 25 through January 14, free chlorine residual at the plant ranged from 0.7 to1.9 mg/L and was below 1.0 mg/L only twice in 21 days, which is typical of water treatment practices at that time. Total chlorine residual was generally about 0.3 mg/L higher than the free residual at the plant. During the three-week period from November 27 to December 17, 1983, one treatment plant effluent sample had a confirmed total coliform MPN count of 2.2 per 100 mL. From December 27, 1983, through January 30, 1984, five samples had an MPN count of 2.2 per 100 mL, and one sample had an MPN count of 5.1 per 100 mL (Logsdon et al., 1985).

Two studies (Jarroll, et al., 1981; Rice et al., 1982) had shown that at 5°C (the approximate temperature of source water at McKeesport at this time) free chlorine is not very effective for inactivating *Giardia* cysts, but this information had not been publicized widely by the end of 1983 nor did drinking water regulations at that time reflect the findings that chlorination practices that were adequate for inactivation of bacteria were likely to be inadequate for inactivation of *Giardia* cysts.

The high-turbidity episode was followed by an unusually high incidence of giardiasis that became apparent in the second week of February 1984. Microscopic analysis of sediment from large samples of finished drinking water, sampled by cartridge samplers during the week of February 27, 1984, by the EPA's Health Effects Research Laboratory, later confirmed the presence of *Giardia* cysts in the raw water, the finished water at the plant, and the water in the distribution system (Logsdon et al., 1985). These results showed that the treatment plant was not performing as expected.

Logsdon et al. (1985) reported that a "boil water" notice was issued on February 22, 1984, by the Allegheny County Health Department to all consumers in the affected communities, based on evaluations of 15 cases of giardiasis. The conclusion that the drinking water was the source of the giardiasis in McKeesport was strengthened because the time between the turbidity problem in late December and early January and the subsequent disease outbreak was similar to the incubation period for giardiasis.

continues

BOX 3-4 CONTINUED

Water Quality

After the McKeesport giardiasis outbreak, a monitoring program for *Giardia* cysts in surface water was undertaken (Sykora et al., 1986). From November 1984 through September 1986, 37 samples were collected from the Youghiogheny River at the McKeesport treatment plant. All samples were positive for *Giardia* cysts. Seventeen of the samples contained between 11 and 100 cysts per 100 gallons (378 L). Ten samples had 101 to 438 cysts per 100 gallons. Ten samples had cysts in the range of 1 to 10 per 100 gallons. At wastewater treatment plants in the Youghiogheny River watershed, samples of raw wastewater and secondary effluent were obtained and analyzed. Raw sewage samples were reported to have *Giardia* cysts in the range of about 10^4 to 10^5 per 100 gallons at five plants, and in activated sludge effluent cysts were in the range of 10^2 to10^3 per 100 gallons. The data of Sykora et al. demonstrate that discharge of treated wastewater to surface waters is very likely to contribute *Giardia* cysts to those waters.

Engineering Aspects

The McKeesport Municipal Water Authority operated a 9 mgd conventional treatment plant that was constructed in 1907-1908 (Trax, 1916) and served about 51,000 residents in the communities of McKeesport, Versailles, Port Vue, and White Oak. Raw water was obtained from the Youghiogheny River about 0.5 mile (0.8 km) upstream from its confluence with the Monongahela River. Raw water typically had a turbidity of 2.5 to 200 ntu. Clarification processes included hydraulic mixing, baffled flocculation with no direct power input, sedimentation, and filtration.

The sand filters at the plant were converted to monomedia anthracite filters in 1960. In the spring of 1984, when filters contained about 30 inches (0.75 m) of media, a core sample was obtained. The EPA's Drinking Water Research Division performed three separate sieve analyses on the media that showed a mean effective size of 0.92 mm (range 0.89 to 0.93 mm) (Logsdon et al., 1985). This effective size was considerably larger than the size generally used for sand filters (0.5 mm) and was also larger than the fine media used in dual-media or mixed-media filters. Thus, although the bed depth was typical, the grain size was not.

In response to the potential risk to recreational users of direct body contact with contaminated surface water, the ACHD initiated in 2003 a cross-sectional survey of recreational and competitive rowing organizations in the Three Rivers region. The total affected population is estimated at 10,000 rowers. The goal of the project is to assess whether there is any increased health risk associated with direct contact with river water during periods of wet weather when the microbiological water quality standards in the rivers are not being met. Results of the study are expected in early 2005.

The committee concludes that a thoughtful analysis of the relationship between environmental conditions and disease incidence and morbidity in southwestern Pennsylvania is hampered by the unavailability of both public health and environmental data. Existing public health surveillance systems and environmental water quality monitoring programs lack the sophistication to adequately characterize surface water and groundwater microbial quality in the region.

Engineering investigations after the outbreak revealed a number of problems with the treatment plant (ACHD, 1984). Some were related to operation and maintenance, whereas others were related to facilities. Problems included the following:

- Excessive accumulation of sludge occurred in mixing and flocculating chambers and sedimentation basins.
- Some valves needed repairs so filter backwashing and sedimentation basin blow-off could be done properly.
- Filter media was dirty and needed to be removed and cleaned.
- Filter rate controls were not operable.
- Filters had no flow rate indicators.
- Filters had no loss-of-head gauges.
- Backwash water storage tank had to be repaired and returned to service to increase amount of water available for filter washing.
- Rate of flow during filter backwash had to be increased.

Summary

A combination of factors resulted in the outbreak at McKeesport. Treatment process equipment, including mixing, flocculation, sedimentation, and filtration process facilities had not been maintained adequately. The filter media being used was not as effective as the more commonly used media designs. Certain very important filter control and filter performance monitoring equipment had not been installed or was not working. The plant was unable to produce safe water at the excessive production rates needed to prevent depressurization of the distribution system after the water line breaks had occurred in late December. The chlorination practice was not sufficient for thorough inactivation of *Giardia* cysts, although this was not known at the time because only limited research data on chlorination of *Giardia* cysts had become available by the end of 1983. The confluence of numerous adverse factors led to the serious outbreak at McKeesport early in 1984.

SOURCES: Based on Stoecker (1985) and Logsdon et al. (1985).

SUMMARY

Surface waters in southwestern Pennsylvania are impaired for a variety of uses including recreational use due to microbiological indicators and pathogens in surface water, fish consumption due to organic (PCBs) and inorganic (Hg) contamination, and aquatic life use due to metal concentrations and low pH. Inadequacies in the type and extent of water quality dataavailable in the region prevented the committee from assessing the full extent of adverse effects due to pollution. Almost all of the water quality data available to the committee were derived from single studies in specific areas for limited durations. Recently, a variety of agencies have expanded water quality data collection in the region; however, these activities do not appear to be coordinated. As a result, it is difficult to say how extensive and significant the water quality contamination is.

Groundwaters in southwestern Pennsylvania, especially those used directly for drinking through private wells, have not been completely assessed. Limited data suggest that pathogen contamination is not unusual in groundwater. Whereas groundwater used for public drinking supplies generally meets water quality guidelines, private wells show significant variability, and the effects of mining are apparent.

Geochemical parameters in waters in southwestern Pennsylvania do not identify significant problems in the rivers. Water in the three main rivers in southwestern Pennsylvania generally shows adequate dissolved oxygen, is at near-neutral pH, and does not exceed water quality standards for inorganic constituents. Pesticides and volatile organic compounds were detected at levels below maximum contaminant levels in waters in southwestern Pennsylvania.

Microbiological parameters indicate a wet weather contamination problem for the main rivers and a continual microbial problem in tributaries. Wet-weather microbiological water quality in the main stem rivers is demonstrably worse than dry weather microbiological water quality. Microbiological water quality in many tributaries does not meet standards in either wet or dry weather, suggesting the potential for multiple sources of pollution. Pathogenic protozoa and indicator organisms are routinely detected in surface waters used as drinking water sources in the region.

Despite high levels of pathogens and indicators in regional waters, there is no evidence that southwestern Pennsylvania has recently experienced any waterborne disease that would link impaired source water quality with human health effects. However, as with water quality data, significant gaps exist in public health monitoring, thus preventing an adequate assessment of possible endemic waterborne disease occurrences.

REFERENCES

ACHD (Allegheny County Health Department). 1984. Engineering Evaluation of the McKeesport Water Treatment Plant. Draft. Pittsburgh, PA: ACHD.

Ali, S., and D. Hill. 2003. *Giardia intestinalis*. Current Opinion in Infectious Diseases 16(5):453-460.

Anderson, R., K. Beer, T. Buckwalter, M. Clark, S. McAuley, J. Sams, and D. Williams. 2000. Water Quality in the Allegheny and Monongahela River Basins: Pennsylvania, West Virginia, New York, and Maryland (1996-98). Denver, CO: United States Geologic Survey.

ASM (American Society of Microbiology). 2004. Comments on EPA's Proposed Policy for National Pollutant Discharge Elimination Systems Permit Requirements for Wastewater Treatment Discharges During Wet Weather. Available on-line at *http://www.asm.org/Policy/index.asp?bid=24834*. Accessed June 15, 2004.

ATSDR (Agency for Toxic Substances and Disease Registry). 1995. ToxFAQs™ for Chlordane. Available on-line at *http://www.atsdr.cdc.gov/tfacts31.html*. Accessed March 23, 2004.

ATSDR. 1999a. ToxFAQs™ for Chlorinated Dibenzo-*p*-dioxins (CDDs). Available on-line at *http://www.atsdr.cdc.gov/tfacts104.html*. Accessed March 23, 2004.

ATSDR. 1999b. ToxFAQs™ for Mercury. Available on-line at *http://www.atsdr.cdc.gov/tfacts46.html*. Accessed March 23, 2004.

ATSDR. 2001. ToxFAQs™ for PCBs. Available on-line at *http://www.atsdr.cdc.gov/tfacts17.html*. Accessed March 23, 2004.

ATSDR. 2002a. ToxFAQs™ for Aldrin/Dieldrin. Available on-line at *http://www.atsdr.cdc.gov/tfacts1.html*. Accessed March 23, 2004.

ATSDR. 2002b. ToxFAQs™ for DDT, DDE, and DDD. Available on-line at *http://www.atsdr.cdc.gov/tfacts35.html*. Accessed March 23, 2004.

Azadpour-Keeley, A., B. Faulkner, and J. Chen. 2003. Movement and longevity of viruses in the subsurface. EPA 540/S-03/500. Cincinnati, OH: EPA, National Risk Management Research Laboratory.

Canadian Council of Ministers of the Environment. 1995. Protocol for the Derivation of Canadian Sediment Quality Guidelines for the Protection of Aquatic Life. CCME EPC-98E. Winnipeg, Manitoba: Task Group on Water Quality Guidelines.

Chester Engineering. 1996. Comprehensive Water Supply Plan: Allegheny County Pennsylvania. Pittsburgh, PA: Chester Engineering.

Collins, T., D. Dzombak, J. Rawlins, K. Tamminga, and S. Thompson. 1998. Nine Mile Run Watershed Rivers Conservation Plan: Appendix IV: Water Quality Issues in the Nine Mile Run Watershed. Pittsburgh, PA: Studio for Creative Inquiry, Carnegie Mellon University.

Desmarais, T., H. Solo-Gabriele, and C. Palmer. 2002. Influence of soil on fecal indicator organisms in a tidally influenced subtropical environment. Applied Environmental Microbiology 68(3):1165-1172.

DOI (U.S. Deaprtment of the Interior) and DOC (U.S. Department of Commerce). 2002. 2001 National Survey of Fishing, Hunting, and Wildlife-Associated Recreation. FHW/01-NAT. Washington, DC: DOI, U.S. Fish and Wildlife Service, and DOC U.S. Census Bureau.

Engler, R. 1993. Memorandum: Lists of Chemicals Evaluated for Carcinogenic Potential. Washington, DC: EPA Office of Prevention, Pesticides, and Toxic Substances.

EPA (Environmental Protection Agency). 1986. Ambient Water Quality Criteria for Bacteria—1986. EPA 440/5-84-002. Washington, DC: Office of Water.

EPA. 2000a. National Primary Drinking Water Regulations: Ground Water Rule; Proposed Rules. Federal Register 65(91). Available on-line *at http://www.epa.gov/safewater/ gwr/gwrprop.pdf.* Accessed June 16, 2004.

EPA. 2000b. Atlas of America's Polluted Waters. EPA 840-B-00-002. Washington, DC: Office of Water.

EPA. 2001a. Infrastructure Needs Survey: Second Report to Congress. Washington, DC: Office of Water.

EPA. 2001b. Protocol for Developing Pathogen TMDLs. EPA 841-R-00-002. Washington, DC: Office of Water.

EPA. 2002a. Consider the Source: A Pocket Guide to Protecting Your Drinking Water. EPA 816-K-02-002. Washington, DC: Office of Ground Water and Drinking Water.

EPA. 2002b. National Recommended Water Quality Criteria: 2002. EPA 822-R-02-047. Washington, DC: Office of Water, Office of Science and Technology.

EPA. 2002c. Public Review Draft: Implementation Guidance for Ambient Water Quality Criteria for Bacteria. EPA 823B-02-003. Washington, DC: Office of Water.

EPA. 2002d. National Water Quality Inventory: 2000 Report. EPA 841-R-02-001. Washington, DC: Office of Water.

EPA. 2003. Proposed Stage 2 Disinfectant and Disinfection Byproducts Rule. EPA-815-F-03-006. Washington, DC: Office of Ground Water and Drinking Water.

Feder, I., F. Wallace, J. Gray, P. Fratamico, P. Fedorka-Cray, R. Pearce, J. Call, R. Perrine, and J. Luchansky. 2003. Isolation of *Escherichia coli* O157:H7 from intact colon fecal samples of swine. Emerging Infectious Diseases 9:380-383.

Frost, F., R. Calderon, and G. Craun. 1995. Waterborne disease surveillance: Findings of a survey of state and territorial epidemiology programs. Journal of Environmental Health 58:6-11.

Geldreich, E, K. Fox, J. Goodrich, E. Rice, R. Clark, and D. Swerdlow. 1992. Searching for a water supply connection in the Cabool, Missouri disease outbreak of *Escherichia coli* O157:H7. Water Research 26:1127-1137.

Gibson, C., K. Stadterman, S. States, and J. Sykora. 1998. Combined sewer overflows: A source of *Cryptosporidium* and *Giardia*. Water Science and Technology 38(12):67-72.

Guerrant, R. 1997. Cryptosporidiosis: An emerging, highly infectious threat. Emerging Infectious Diseases 3(1):51-57.

Hammermueller, J., S. Kruth, J. Prescott, and C. Gyles. 1995. Detection of toxin genes in *Escherichia coli* isolated from normal dogs and dogs with diarrhea. Canadian Journal of Veterinary Research 59:265-270.

Higgins, J., G. Heufelder, and S. Foss. 2000. Removal efficiency of standard septic tank and leach trench septic systems for MS2 coliphage. Small Flows Quarterly 1(2):26-27, 57.

Jarroll, E., A. Bingham, and E. Meyer. 1981. Effect of chlorine on *Giardia lamblia* cyst viability. Applied and Environmental Microbiology 41:483-487.

Jensen, P., Y. Su, K. Lee, A. Boer, H. Rifai, S. Payne, R. Stein, and T. Running. 2002. The 2002 Texas Water Monitoring Congress Proceedings: Indicator bacteria monitoring on Houston Bayous. Available on-line at *http://www.txwmc.org/2002_TWMC_Proceedings.pdf*. Accessed June 16, 2004.

Knauer, K., and T. Collins. 2001. 3 Rivers 2nd Nature Aquatic Report: Pittsburgh Pool, Phase 1—2000, Water Quality. Pittsburgh, PA: STUDIO for Creative Inquiry, Carnegie Mellon University.

Knauer, K., and T. Collins. 2002. 3 Rivers 2nd Nature Aquatic Report: Monongahela River Valley, Phase 2—2001, Water Quality. Pittsburgh, PA: STUDIO for Creative Inquiry, Carnegie Mellon University.

Knauer, K., and T. Collins. 2003. 3 Rivers 2nd Nature Aquatic Report: Allegheny River Valley, Phase 3—2002, Water Quality. Pittsburgh, PA: STUDIO for Creative Inquiry, Carnegie Mellon University.

Kolpin, D., E. Furlong, M. Meyer, E. Thurman, S. Zaugg, L. Barber, and T. Buxton. 2002. Pharmaceuticals, hormones, and other organic wastewater contaminants in U.S. streams, 1999-2000: A national reconnaissance. Environmental Science and Technology 36:1202-1211.

Koryak, M., and J. Reilly. 2000. Nine Mile Run, Allegheny County, Pennsylvania: Aquatic Ecosystem Restoration Water Quality and Aquatic Life Report. Pittsburgh, PA: U.S. Army Corps of Engineers.

Lee, S., D. Levy, G. Craun, M. Beach, and R. Calderon. 2002. Surveillance for waterborne-disease outbreaks—United States, 1999-2000. Morbidity and Mortality Weekly Report 51(SS-8):1-47.

Levy, D., M. Bens, G. Craun, R. Calderon, and B. Herwaldt. 1998. Surveillance for waterborne-disease outbreaks—United States, 1995-1996. Morbidity and Mortality Weekly Report 47(5):1-34.

Lindsey, B., J. Rasberry, T. Zimmerman. 2002. Microbiological Quality of Water from Noncommunity Supply Wells in Carbonate and Crystalline Aquifers of Pennsylvania. Water Resources Investigations Report 01-4268. Washington, DC: DOI, USGS.

Lippy, E., and S. Waltrip. 1984. Waterborne disease outbreaks—1946-1980: A thirty-five year perspective. Journal of the American Water Works Association 6(2):60-67.

Livingston, R., S. McGlynn, and X. Niu. 1998. Factors controlling seagrass growth in a gulf coastal system: Water and sediment quality and light. Aquatic Botany 60(2):135-159.

Logsdon, G., V. Thurman, E. Frindt, and J. Stoecker. 1985. Evaluating sedimentation and various filter media for removal of *Giardia* cysts. Journal of the American Water Works Association 77(2):61-66.

Luneburg, W. 2004. Where the Three Rivers Converge: Unassessed Waters and the Future of EPA's TMDL Program: A Case Study (Water Quality Policy and Regulation in Allegheny County, PA). A report for 3 Rivers 2nd Nature, STUDIO for Creative Inquiry, College of Fine Arts, Carnegie Mellon University, Pittsburgh, PA.

Marshall, M., D. Naumovitz, Y. Ortega, and C. Sterling. 1997. Waterborne protozoan pathogens. Clinical Microbiology Reviews 10:67-85.

Medema, G., F. Schets, P. Teunis, and A. Havelaar. 1998. Sedimentation of free and attached *Cryptosporidium* oocysts and *Giardia* cysts in water. Applied and Environmental Microbiology 64:4460-4466.

MMWA (McKeesport Municipal Water Authority). 1983-1984. Unpublished Purification Plant Report Data Sheets from McKeesport for Weeks Ending December 31, 1983; January 7, January 14, January 21, and January 28, 1984. McKeesport, PA: MMWA.

Mulroy, A. 2000. Correlating residual antibiotic contamination in public water to drug-resistant *Escherichia coli*: Is remediation an option? Alexandria, VA: Water Environment Federation 2000 U.S. Stockholm Junior Water Prize.

NRC (National Research Council). 1987. Drinking Water and Health, Volume 7 Disinfectants and Disinfectant By-Products. Washington, DC: National Academy Press.

NRC. 1992. Restoring Aquatic Ecosystems. Washington, DC: National Academy Press.

NRC. 1993. Managing Wastewater. Washington, DC: National Academy Press.

NRC. 1997. Safe Water from Every Tap. Washington, DC: National Academy Press.

NRC. 1999a. Setting Priorities for Drinking Water Contaminants. Washington, DC: National Academy Press.

NRC. 1999b. Hormonally Active Agents in the Environment. Washington, DC: National Academy Press.

NRC. 2001. Assessing the TMDL Approach to Water Quality Management. Washington, DC: National Academy Press.

NRC. 2002. Opportunities to Improve the U.S. Geological Survey National Water Quality Assessment Program. Washington, DC: National Academy Press.

NRC. 2004. Indicators for Waterborne Pathogens. Washington, DC: National Academies Press.

O'Connor, D. 2002. Report of the Walkerton Inquiry: The Events of May 2000 and Related Issues. Part One: A Summary. Toronto: Ontario Ministry of the Attorney General, Queen's Printer for Ontario.

O'Donoghue, P. 1995. *Cryptosporidium* and cryptosporidiosis in man and animals. International Journal for Parasitology 25:139-195.

ORSANCO (Ohio River Valley Water Sanitation Commission). 2002. Biennial Assessment of Ohio River Water Quality Conditions for Water Years 2000 and 2001. Cincinnati, OH: ORSANCO.

Osterkamp, W., P. Heilman, and L. Lane 1998. Economic considerations of a continental sediment-monitoring program. International Journal of Sediment Research 13 (4):12-24.

PADEP. 2003. Frequently Asked Questions About Private Water Wells in Pennsylvania. Fact Sheet 3800-FS-DEP2657. Available on-line at *http://www.dep.state.pa.us/dep/deputate/watermgt/wc/Subjects/SrceProt/well/questions/default.htm*. Accessed March 18, 2004.

PADEP. 2004. 2004 Pennsylvania Integrated Water Quality Monitoring and Assessment Report. Available on-line at *http://www.dep.state.pa.us/dep/deputate/watermgt/wqp/wqstandards/303d-report.htm*. Accessed June 16, 2004.

Pait, A., A. DeSouza, and D. Farrow. 1992. Agricultural Pesticide Use in Coastal Areas: A National Summary. Washington, DC: U.S. Department of Commerce, National Oceanic and Atmospheric Administration.

Pesaro, F., I. Sorg, and A. Metzler. 1995. In situ inactivation of animal viruses and a coliphage in nonaerated liquid and semiliquid animal wastes. Applied and Environmental Microbiology 61(1):92-97.

Pontius, F.W. (ed.) 2003. Drinking Water Regulation and Health. New York: John Wiley & Sons.

Rice, E., J. Hoff, and F. Schaefer, III. 1982. Inactivation of *Giardia* cysts by chlorine. Applied and Environmental Microbiology 43(1):250-251.

Robertson, L., A. Campbell, and H. Smith. 1992. Survival of *Cryptosporidium parvum* oocysts under various environmental pressures. Applied and Environmental Microbiology 58(11):3494-3500.

Rose, J., and T. Shifko. 1999. *Giardia, Cryptosporidium*, and *Cyclospora* and their impact on foods: A review. Journal of Food Protection 62(9):1059-1070.

Rose, J., D. Huffman, and A. Gennaccaro. 2002. Risk and control of waterborne cryptosporidiosis. Federation of European Microbiological Societies Microbiology Reviews 26:113-123.

Schallenberg, M. and C. Burns. 2004. Effect of sediment resuspension on phytoplankton production: Teasing apart the influences of light, nutrients, and algal entrainment. Freshwater Biology 49(2):143-159.

Schaub, S., and R. Oshiro. 2000. Pubic health concerns about caliciviruses as waterborne contaminants. Journal of Infectious Diseases 181:S374-S380.

Sharpe, W., D. Mooney, and R. Adamis. 1985. An analysis of ground water quality data obtained from private individual water systems in Pennsylvania. North Eastern Journal of Environmental Science 4:155-159.

Snyder, S., P. Westerhoff, Y. Yoon, and D. Sedlak. 2003. Pharmaceuticals, personal care products, and endocrine disruptors in water: Implications for the water industry. Environmental Engineering Science 20(5):449-469.

States, S., K. Stadterman, L. Ammon, P. Vogel, J. Baldizar, D. Wright, L. Conley, and J. Sykora. 1997. Protozoa in river water: Sources, occurrence, and treatment. Journal of the American Water Works Association 89(9):74-83.

Stoecker, J. 1985. Memorandum: McKeesport Water Quality. Philadelphia, PA: EPA.

Stumm-Zollinger, E., and G. Fair. 1965. Biodegradation of steroid hormones. Journal of Water Pollution Control Federation 37(11):1506-1510.

Sykora, J., S. States, W. Bancroft, S. Boutros, M. Shapiro, and L. Conley. 1986. Monitoring of water and wastewater for *Giardia*. In Proceedings of the American Water Works

Association Water Quality Technology Conference, Portland, Oregon. Denver, CO: AWWA.

Tabak, H., and H. Bunch. 1970. Steroid hormones as water pollutants. Developments in Industrial Microbiology 11:367-376.

TPRC (Third Party Review Committee). 2002. Third Party Review of the ALCOSAN Regional Long Term Wet Weather Control Concept Plan. Pittsburgh, PA: ALCOSAN.

Trax, E. 1916. A new raw water supply for the City of McKeesport, Pennsylvania. Journal of the American Water Works Association 3:947-958.

Trussell, R. 2001. Endocrine disruptors and the water industry. Journal of the American Water Works Association 93(2):58-65.

USACE (U.S. Army Corps of Engineers). 1997. Montour Run Watershed, Allegheny County, Pennsylvania: Water Quality and Aquatic Life Resources. Pittsburgh, PA: U.S. Army Corps of Engineers.

USDA (U.S. Department of Agriculture). 1997. 1997 Census of Agriculture: Pennsylvania State and County Data. Available on-line at *http://www.census.gov/prod/ac97/ac97a-38.pdf.* Accessed June 16, 2004.

USGS (U.S. Geological Survey). 1995. The Allegheny-Monongahela River Basin NAWQA Project. Available on-line *at http://pa.water.usgs.gov/almn/.* Accessed June 16, 2004.

USGS. 1996. Quality of ground water at selected sites in the upper Mahoning Creek basin, Pennsylvania. Fact Sheet 176-96. Available on-line at *http://pa.water.usgs.gov/reports/fs_176-96/report.html.* Accessed June 16, 2004.

VADEQ (Virginia Department of Environmental Quality). 2000. Fecal Coliform TMDL for Muddy Creek, Virginia: Revised Final Report. Available on-line at *http://www.deq.virginia.gov/tmdl/apptmdls/shenrvr/muddyfe.pdf.* Accessed November 30, 2004.

WHO (World Health Organization). 1990. The WHO Recommended Classification of Pesticides by Hazard and Guidelines to Classification. Geneva, Switzerland: WHO.

WHO. 2000. Monitoring Bathing Waters: A Practical Guide to the Design and Implementation of Assessments and Monitoring Programmes, J. Bartram and G. Rees (eds.). London: E. and F.N. Spon.

4

Causes of Water Quality Impairment

Chapter 3 provides an overview of water quality standards and the types of water quality problems in southwestern Pennsylvania. In this chapter the causes of impairments to waters that prevent their designated uses are discussed with an emphasis on those caused by improper human waste disposal methods.

Waters can be impaired for any of their designated uses and by a variety of contaminants. Any waterbody that does not meet ambient water quality standards pertaining to its designated use after elimination of point source pollution through applicable technology is considered impaired. Table 4-1 summarizes designated uses and the parameters used for evaluating impairment in Pennsylvania. It should be noted that some causes of aquatic life use impairment are not susceptible to physical or chemical analysis of water samples, including siltation, other habitat alteration, and flow alteration or variability. When these are the primary causes of water quality degradation, bioassessment protocols (discussed in Chapter 3; see also Box 5-2) must be used in lieu of chemical analyses. For other causes of aquatic life impairment that might be measured by physical or chemical analysis of water samples, no water quality criteria exist (e.g., suspended solids, turbidity, oil and grease). Conversely, some water quality criteria exist for which no related cause of impairment is obvious, including color and alkalinity. Thus, there is rarely a one-to-one correlation between impairments and parameters that can be measured by chemical analyses.

Whereas the list of potential pollutants and conditions that cause impairment for aquatic life use is long, single causes of impairment exist for recreational use and for human health use in Pennsylvania. The recreational use of water is impaired by the presence of microbial pathogens, for which fecal and total coliforms are used as indicators.[1] Because surface waters are presumed to be treatable for the production of potable water, the health aspect for human use is not drinking water but rather fish consumption. In other words, waters are not considered "impaired" for use as drinking water sources because of the extensive treatment that is routinely performed on these sources, particularly surface water. It should be noted that private well owners are not required currently by state or federal regulations to monitor for contaminants or to treat their drinking water.

---

[1] Waters containing high levels of indicator organisms are sometimes termed "impaired for pathogens," although this usage can be misleading. The intent is that these waters exceed the surface water standards for a specific stream segment classification based upon fecal indicator monitoring, such as fecal coliforms, *Escherichia coli*, and/or enterococci. The relationship of indicator counts to pathogen presence depends on the source of pollution resulting in the indicator standard exceedence, such as sewage, agricultural runoff, or nonpoint source contamination from feral animals (see NRC, 2004, for further information).

TABLE 4-1 Parameters Found in Water Quality Standards to Measure Impairment of Aquatic Life, Human Health, and Recreational Water Use in Pennsylvania

Designated Use	Impairment Causes in 303(d) Report	Water Quality Parameter to Determine Impairment
Aquatic life	Metals	Iron, manganese
	Acidity	pH
	Nutrients	Nitrite plus nitrate
	Salinity, total dissolved solids, chlorides	Total dissolved solids, chlorides, osmotic pressure
	Other inorganics	Sulfate
	Organic enrichment	Dissolved oxygen
	Thermal modification	Temperature
	Toxicity	Ammonia nitrogen, fluoride, phenolics
	Excess algae growth	Not listed
	Siltation	Not listed
	Other habitat alteration	Not listed
	Suspended solids	Not listed
	Turbidity	Not listed
	Flow alteration; water or flow variability	Not listed
	Oil and grease	Not listed
	Chlorine	Not listed
	Pesticides	Not listed
Human health	Chemicals in fish	Federal food and drug standards used
Recreation	Pathogens (surrogate)	Fecal coliforms, total coliforms

SOURCE: PADEP, 2002.

As explained above, impairment of waters for designated uses can be determined by comparison of water quality to standards for specified uses or by bioassessment protocols. Causes of impairment may be determined by physical, chemical, or biological analysis of water samples; bioassessment protocols; or observation of environmental conditions in a reach of a stream or river. Causes of impairment are identified by descriptors such as siltation, metals, pH, low dissolved oxygen, and nutrients. However, the *2004 Pennsylvania Integrated Water Quality Monitoring and Assessment Report* (PADEP, 2004a) does not present data on concentrations of impairment-causing contaminants or data on the degree to which water quality is impaired. Without detailed data for each reach of stream or river, it is not possible to determine how severely surface waters are impaired, even when impairment is caused by one or more pollutants.

According to the most recent assessment of surface water by the Commonwealth of Pennsylvania (PADEP, 2004a), 83,161 miles of streams (82 percent of total) in Pennsylvania have been assessed, and the remaining 15,182 stream-miles (18 percent) are scheduled for assessment. Of those assessed, 57,801 river-miles (82 percent) support their aquatic life and fish use designations, and 10,762 (18 percent of the assessed and 13 percent of the total stream-miles) are classified as impaired. The two largest sources of impairments specified are abandoned mine drainage (4,040 miles impaired) and agriculture (3,903 miles impaired). Three sources related to runoff (urban runoff, road runoff, and small residential runoff) account for an additional 3,007 miles impaired. Four sources related to wastewater handling and treatment (municipal point sources, on-site wastewater, combined sewer overflows, and package plants) account for 744 miles impaired.

Pennsylvania has 215 significant lakes totaling 98,942 acres.[2] A total of 75,543 acres have been assessed, with 30,346 (40 percent) supporting their designated aquatic life use and 45,197 (60 percent) showing impairment. Of 28,665 lake-acres assessed for fish consumption use, 99 percent are impaired with unacceptable concentrations of mercury and/or polychlorinated biphenyls (PCBs). A total of 64,588 lake-acres have been assessed for contact recreation, with 98 percent supporting this use. Only 1,237 lake-acres are impaired due to pathogens (1,150) and nutrients (174). Major sources of impairments for lakes were agriculture, "other," and "unknown." Human waste-related sources including on-site sewage treatment and disposal systems (OSTDSs or "septic systems"), municipal point sources, and runoff were responsible for smaller areas of lake impairment.

In addition to assessing streams and lakes for specific impairments, Pennsylvania also monitors the concentrations of several toxicants in fish tissue. Nationally, 28 percent of assessed lake-acres and almost 14 percent of river-miles in the United States are under restricted consumption advisories (EPA, 2003a). In Pennsylvania, a statewide advisory for all surface waters was recently issued for mercury in fish.[3] Specific fish consumption advisories in southwestern Pennsylvania include an advisory on high PCB levels in fish caught in the region's three main stem rivers as well as in the Beaver River, Chartiers Creek, Little Chartiers Creek, and the Mahoning and Shenango Rivers. Advisories on the pesticide chlordane were issued for the Monongahela River, Chartiers Creek, Little Chartiers Creek, and the Cheat River.

SOURCES OF POLLUTION

As described above, the major causes of surface water quality impairment within the state of Pennsylvania are (1) acid mine drainage, (2) agriculture, (3) urban and stormwater runoff, and (4) wastewater. Likewise, the U.S. Geological Survey (USGS) National Water Quality Assessment (NAWQA) program (as reported in Anderson et al., 2000; see also Appendix B) identified the following causes of surface water quality impairment in the Allegheny and Monongahela River basins: (1) coal mine discharge, (2) urban runoff, and (3) agricultural runoff. However, the NAWQA study did not include pathogen or indicator organism data collection, which could have implicated wastewater as a major source of surface water contamination. Evaluation of water quality data in the region by this committee (discussed in Chapter 3) indicates that wastewater-related impairment of surface waters is a significant source of microbial loading to surface waters.

HUMAN AND OTHER WASTE DISPOSAL

As discussed in Chapter 2, human waste disposal methods in southwestern Pennsylvania have developed over time to include individual home on-site sewage treatment and disposal systems (OSTDSs), decentralized small systems, and centralized collection sewers and treatment

---

[2] An additional 146 public waterways are used as lakes but do not have the 14-day retention time required for designation as a "lake." These are monitored but not included in the totals above.

[3] For further information, see *http://www.dep.state.pa.us/dep/deputate/watermgt/Wqp/WQStandards/FishAdvis/fishadvisory04.htm.*

systems. All of these systems, if not operating or maintained properly, have the potential to adversely affect the region's water quality. The most critical effect of improperly treated human waste is the release of pathogenic microorganisms to waterways; however, human waste also contains suspended solids, biochemical oxygen demand (BOD), and nutrients that can adversely affect water quality if released improperly. Centralized wastewater treatment plants (WWTPs) are utilized in the region's urban areas, while OSTDSs are commonly used in more sparsely populated (rural) areas where the cost of conveyance to a centralized treatment plant is often prohibitive. Figure 4-1 shows the distribution of WWTPs and OSTDSs in southwestern Pennsylvania. Of the 1,172,274 households in southwestern Pennsylvania reported in the 1990 U.S. census,[4] 76 percent were on public sewers connected to WWTPs, while 23 percent were utilizing some sort of on-lot septic systems. Notably, 11,289, or 1 percent, reported neither sewers nor on-lot disposal; these homes may use cesspools or straight pipes that discharge directly to surface water or groundwater, they may be served by older substandard or unknown treatment systems that pre-date permitting programs, or they may simply represent the residents' lack of knowledge about the treatment system.

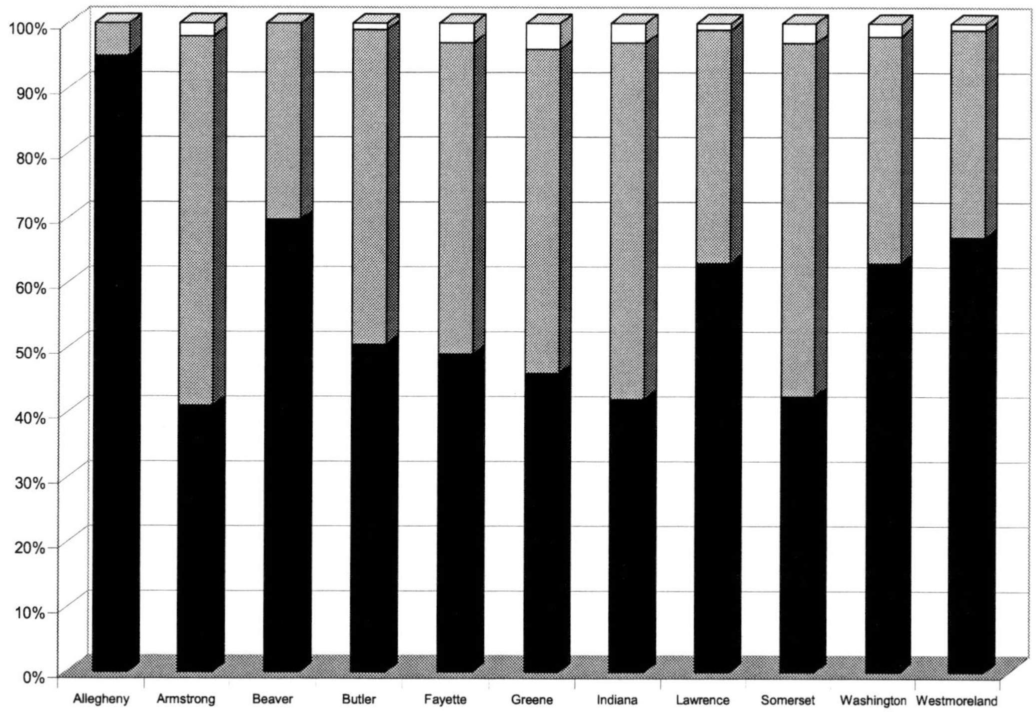

FIGURE 4-1 Human waste disposal methods by county in southwestern Pennsylvania. NOTE: Black (centralized WWTP); gray (on-site systems); white ("other;" e.g., cesspools, straight pipes). SOURCE: Data from WSIP, 2002.

[4] Notably, the 2000 census did not include a question regarding the mode of wastewater treatment for the household as in previous years.

As shown in Table 4-2, there are more than 1,100 industrial, municipal, and nonmunicipal sanitary sewage dischargers in the area that have been granted National Pollutant Discharge Elimination System (NPDES) permits by the Pennsylvania Department of Environmental Protection (PADEP). Of these, 96 are classified as "major sources," including 32 industrial sources, 63 municipal sources, and 1 nonmunicipal sanitary sewage source. Industrial sources are classified as major based on a combination of flow and quality criteria as established by the U.S. Environmental Protection Agency (EPA). A nonindustrial source is classified as major if the permitted discharge rate exceeds 1 million gallons per day. About 50 percent of the NPDES permits are for nonmunicipal sanitary sewage wastewater disposal, although only one of these is classified as a major source. How well these systems are operated would require further investigation, but the general problem of operating small wastewater disposal systems in compliance with permit requirements is well known (see, for example, EPA, 2003b).

Wastewater Treatment Plants

Centralized wastewater collection and treatment systems convey wastewater from individual homes and businesses to a wastewater treatment plant where multiple-stage treatment removes total suspended solids (TSS), BOD, nutrients, and pathogenic microorganisms (and their microbial indicators), such that the effluent from wastewater treatment plants is significantly less contaminated than the influent and meets NPDES discharge requirements. Nutrient levels can be further reduced with tertiary biological treatment when necessary. Table 4-3 summarizes influent and effluent parameters for typical domestic sewage through primary (settling) and secondary (biological) treatment.

In addition to indicator organisms (such as fecal coliforms, included in Table 4-3), specific pathogens are common in raw sewage and in wastewater effluents. For example, Chauret et al. (1995) reported an arithmetic mean of 1,730 *Cryptosporidium* oocysts per 100 L in untreated domestic sewage at Ottawa, Canada. Madore et al. (1987) examined sewage in

TABLE 4-2 Number of Permitted Wastewater Dischargers in Southwestern Pennsylvania by County[a] and Classification

	Industrial		Municipal		Nonmunicipal		Other	
County	All	Major	All	Major	All	Major	All	Major
Allegheny	97	10	48	21	80	0	225	31
Armstrong	21	2	14	3	35	0	70	5
Beaver	51	8	21	14	51	0	123	22
Fayette	19	0	28	2	62	0	109	2
Greene	8	1	11	1	22	0	41	2
Indiana	23	3	28	1	46	0	97	4
Somerset	17	0	25	1	52	0	94	1
Washington	32	5	30	9	75	0	137	14
Westmoreland	53	3	37	11	134	1	224	15
Total	321	32	242	63	557	1	1,120	96

[a] Does not include Butler and Lawrence Counties of southwestern Pennsylvania (see Box 1-2).
SOURCE: Compiled from data provided by Renee Larry, PADEP, July 2004.

TABLE 4-3 Influent and effluent Parameters for Typical Domestic Sewage

Source	5-Day BOD_5(mg/L)	TSS (mg/L)	Total N (mg/L)	Total P (mg/L)	Fecal Coliforms (CFU/100 mL)
Influent	110-400	100-350	20-85	4-15	10^7-10^9
Effluent from secondary treatment	<5-30	<5-30	15-25	<1-5	<200

NOTE: CFU = colony forming units.
SOURCES: Adapted from Metcalf and Eddy, 1991 and WEF, 1999.

Arizona and found 85,000 to 1,370,000 oocysts per 100 L in raw sewage. Studies have also demonstrated that although conventional sewage treatment can achieve greater than 90 percent removal of *Cryptosporidium* oocysts, treated wastewater effluent can still contain measurable concentrations of oocysts (Chauret et al., 1995; Desvousges et al., 1987). Similarly, *Giardia* cyst removal can be greater than 90 percent in conventional sewage treatment (Caccio et al., 2003; Sykora et al., 1986).

Sewage typically contains a variety of chemical contaminants disposed of by consumers and industries. Toxic chemicals released to the sewage system are believed to be partially removed through wastewater treatment; however, the EPA estimates that 25 percent of these toxic substances pass through sewage treatment to receiving waters (EPA, 1997). More than 1.4 billion pounds of toxic chemicals were sent to sewage treatment plants in the United States between 1992 and 1996—50 percent more than the amount directly released to waterways during that same period (Puchalsky and LaPlante, 1998).

The preceding discussion suggests that although treated wastewater effluent continues to contain indicator microorganisms, pathogens, and chemicals that may escape treatment or removal, the major water quality concern is when downstream drinking water treatment fails or such systems become overloaded. When this occurs, as it did in Mckeesport, Pennsylvania, in 1983-1984 (see Box 3-4), untreated or inadequately treated drinking water can be supplied to consumers, which constitutes a distinct public health threat.

Small Wastewater Treatment Facilities and Failures

Many small wastewater treatment entities in isolated, low-income rural counties in the Central Appalachians struggle to operate and maintain facilities, make necessary repairs, and maintain financial health. In many of these rural areas, water quality problems are in part attributable to these facilities. For example, Kentucky's PRIDE assessment project[5]—a coalition of government, academic, and volunteer groups addressing regional water pollution problems in 38 counties in southeastern Kentucky—has identified small municipal wastewater treatment plants and small package treatment plants, in addition to straight pipes and failing septic systems, as major pathogen sources in the area's watersheds. Many of these facilities have malfunctioned because of improper operation and maintenance. South of southwestern Pennsylvania, the Morgantown (West Virginia) Utility Board is beginning a sewer extension project to serve White's Run, a tributary to Cheat Lake, a popular recreational waterbody in the area.[6] Water quality in White's Run has been adversely affected not only by failing OSTDSs but also by nine

[5] For further information, see *http://pride.uky.edu/pollutionsources.cfm.*
[6] More information on this project is available on-line at *http://www.mub.org/eng.htm.*

privately owned package wastewater treatment plants, many of which were failing or overloaded beyond capacity.

Throughout Pennsylvania, many small communities experience problems with antiquated sewage treatment facilities and, particularly in southwestern Pennsylvania, older systems that are overloaded during heavy rainfalls (Strawley, 2002). Within the Redbank Creek watershed in the Allegheny River basin, a package plant was implicated in water quality impairment of Fivemile Run (WRAS, 2003a). The Pennsylvania infrastructure Investment Authority (PENNVEST) lists a number of loans to small communities during 2002 and 2003 to correct these types of problems.[7] Even with new or upgraded facilities, communities must be able to support adequate operation and maintenance of their systems, a need that can be difficult for small utilities to meet. For this reason, several programs have been targeted for assisting small wastewater facilities in Pennsylvania.

Section 104(g)(1) of the Clean Water Act (CWA) authorizes funding for the Wastewater Treatment Plant Operator On-Site Assistance Training Program. This program provides on-site operator training, financial management, troubleshooting, and other operation and maintenance assistance to small underserved communities through a network of operator training personnel, peer trainers, EPA regional office coordinators, state and regional training centers, and state programs. The PADEP's Division of Technical Training and Outreach has administered not only this program but also a Drinking Water and Wastewater Operator Information Center, resources to assist local governments, and a variety of financial assistance programs. Due to budgetary and staffing concerns, PADEP plans to discontinue regularly scheduled basic operator training courses during 2004; instead, these will be offered by approved academic, association, and private sector providers. The division will continue to provide outreach and technical assistance for small community systems.

Several nonprofit organizations help small rural utilities with many of the aforementioned challenges. The Pennsylvania Rural Water Association (PRWA), a member-supported nonprofit organization and state associate of the National Rural Water Association, provides training, technical assistance, and "circuit-rider" assistance in operation, maintenance, and management to its small water and wastewater utility members. The PRWA also assists communities in developing and adopting groundwater protection plans and is a partner in the Pennsylvania Water Well Owner Network. The Rural Community Assistance Program (RCAP) provides guidance, training, and technical assistance for a variety of rural concerns, from community and leadership development to rural housing and health care. Through a partnership agreement with EPA, RCAP supports assistance programs for small water and wastewater facilities through the Small Community Wastewater Project.[8]

In a study of small, rural Appalachian communities in Tennessee, teams from the University of Tennessee and Tennessee Technological University found that inadequate wastewater treatment was associated with a variety of socioeconomic factors distinctive of isolated communities (MTAS, 1997). Box 4-1 reviews these factors, many of which are commonly found in southwestern Pennsylvania's rural areas, and thus are important to consider in crafting solutions to the region's wastewater problems (Ann Bargerstock, Greene County Planning Commission, personal communication, 2002; Barbara McMillen, U.S. Department of Agriculture, Rural Utilities Service, personal communication, 2002, 2003).

---

[7] See PENNVEST press releases 11/20/02, 05/07/03, 07/23/03, 11/19/03, available on-line at *http://www.pennvest.state.pa.us/ pennvest/cwp/browse.asp?A=11&BMDRN=2000&BCOB=0&C=43125.*

[8] Further information about this program can be found on-line at *http://www.epa.gov/owm/mab/smcomm/1rcap.htm.*

BOX 4-1
Small Rural Communities Study

In an EPA-funded study examining environmental justice issues in wastewater services, research teams from the University of Tennessee's Municipal Technical Advisory Service and Tennessee Technological University's Center for Management, Utilization, and Protection of Water Resources conducted an intensive in-field study of socioeconomic factors in 12 small, rural Tennessee communities (including 2 control communities), selected because of one or more negative wastewater impacts of health risks, water quality violations, or aesthetic problems. Results of the study dispel many common myths about isolated rural communities, while documenting in detail the distinctive challenges faced in achieving proper wastewater management in similar communities.

Populations in the study communities ranged from 200 to 6,000. Local employment and economic base varied widely: manufacturing, timber, food processing, and commuting to jobs located from 12 to 70 miles away. One community had a high percentage of elderly residents; another had a very high local unemployment rate. Local water supplies included private wells, community wells and springs, a community-owned and operated water treatment plant (surface water), and purchase of water from another utility. Sewage treatment included on-site systems, central activated sludge plants, a central trickling filter plants, central oxidation ditch plant, and central aerated lagoon plants.

Lessons learned from this study provide a snapshot of characteristics common to many small rural communities in the Central Appalachians. Residents were mostly white, and only one of the communities had a high school graduation rate that met the state average of 67 percent (most ranged from slightly above 40 percent to slightly more than 50 percent). Seven of the communities had poverty rates above the state average of 15.7 percent (ranges were from 20 to more than 40 percent). All but one of the communities met or exceeded the statewide average (17.1 percent) for population over 60.

Researchers found that six of the eight communities with central wastewater treatment systems charged rates below the statewide average. In addition, the state's Wastewater Financing Board had cited these six communities for failure to adequately fund sewer operations. Communities in the survey used grants to build and upgrade systems, not because of advance planning and budgeting, but because a grant funding source was brought to their attention. Furthermore, centralized wastewater operations were understaffed; in five of the eight plants surveyed, operators worked alone. Three of these five operators were responsible for the water treatment and distribution system, and several also handled lab work and grounds maintenance. Such understaffing may reflect attempts to save money or difficulty in finding and retaining qualified staff to work in isolated rural locations. Operators had few opportunities for adequate training or for sharing ideas with other operators, because they had to travel a long distance for training and often had trouble getting away from the plant to attend training. Because of these factors, some had difficulty completing the training needed to retain certification. However, local operators were often very resourceful in coping with limited budgets, aging facilities, and design oversights; they kept plants functioning by rebuilding parts and devising repairs and improvements from locally available materials.

Engineers' design mistakes and lack of attention to community resources and concerns accounted for many of the problems that communities and operators experienced with their wastewater systems. For example, Inflow/infiltration, overflows, and bypasses were found to be widespread problems.

Community residents did not like the government and outsiders telling them what to do, and often felt that outsiders did not bother to understand their needs. Growth and change came slowly to these communities and residents sometimes lived with problems (e.g., odor, sewage on the ground) because they felt helpless to change the situation. Many residents stated that their communities were good places to live and did want to change the community itself—even if wastewater problems presented a documented health threat. These residents did without many common services and amenities; as a result, wastewater-related problems ranked low on their list of concerns, and those affected tended to blame, deny, minimize, or not recognize the problems. In several communities

continues

BOX 4-1 CONTINUED

with failing and improperly operating septic systems, residents stated that "people take care of their septic systems," that the house was on public sewer, are that they did not know about any problems. The mayor of one of these communities confided that there were many problems with septic systems, but that there were concerns that small lots allowed no room for repairs, that the community lacked the money for centralized treatment, and that even if funds could be obtained for a central system there would be nobody to oversee it.

Citizens and local government leaders were uneducated about how to solve their wastewater problems. They did not know how to access help, obtain funding, select an engineer who understands small community issues, select treatment technologies compatible with community resources, and estimate long-term operation and maintenance costs. Many existing wastewater treatment problems have resulted from one or more of these deficiencies in education and preparation for making decisions. Although communities may have learned how to obtain grants, they often do not have the knowledge to target the funding toward correcting the problem. Finally, community leaders, consultants, and technical assistance providers had difficulty obtaining information, at both state and local levels, to identify problems and craft solutions. Records were typically inconsistent, poorly organized and maintained, and not in electronic form.

Recommendations related to this study, for policy makers, regulators, educators, and technical assistance providers who work with similar communities, are discussed in Chapter 5 (see also Boxes 5-3 and 5-5).

SOURCE: MTAS, 1997.

Combined Sewer Overflows

While small treatment systems may suffer from maintenance and operational failures, all systems have the potential to suffer from problems associated with excessive flows. When flows exceed the capacity of the collection or treatment system, overflow events are triggered. Seven hundred and seventy-two communities in 32 states throughout in the United States have combined sewer systems or dedicated sanitary sewer systems that are linked hydraulically to combined systems. These systems were designed in the nineteenth and early twentieth centuries to convey wastewater to treatment plants and to convey mixed stormwater and wastewater partially to the treatment plant and partially to nearby waterways (rivers and streams) during significant precipitation events (see also Chapter 2 and 5). Inherent in the design were diversion structures that release mixed rainwater and sewage to surface waters when flows exceed the capacity of the sewer collection or wastewater treatment system (combined sewer overflow structures, or CSOs). When heavy rains lead to high flows in the collection system, a gate or relief system diverts flow away from the sewage treatment plant and to surface waters. This diversion of high flows prevents sewage from backing up in the collection system (e.g., flooding basements or streets) and prevents damage to the wastewater treatment plant from flows that exceed design capacity. Figure 4-2 shows a schematic of combined and separate sewers and how they contribute to wet weather-related surface water pollution.

EPA has estimated that there are CSO discharges of 1,260 billion gallons per year in the United States through 9,471 outfall locations regulated through 859 NPDES permits (EPA, 2001a). Notably, Pennsylvania has the highest number of CSO structures of any state (1,671). Figure 4-3 shows the distribution of CSOs in Pennsylvania and their concentration in

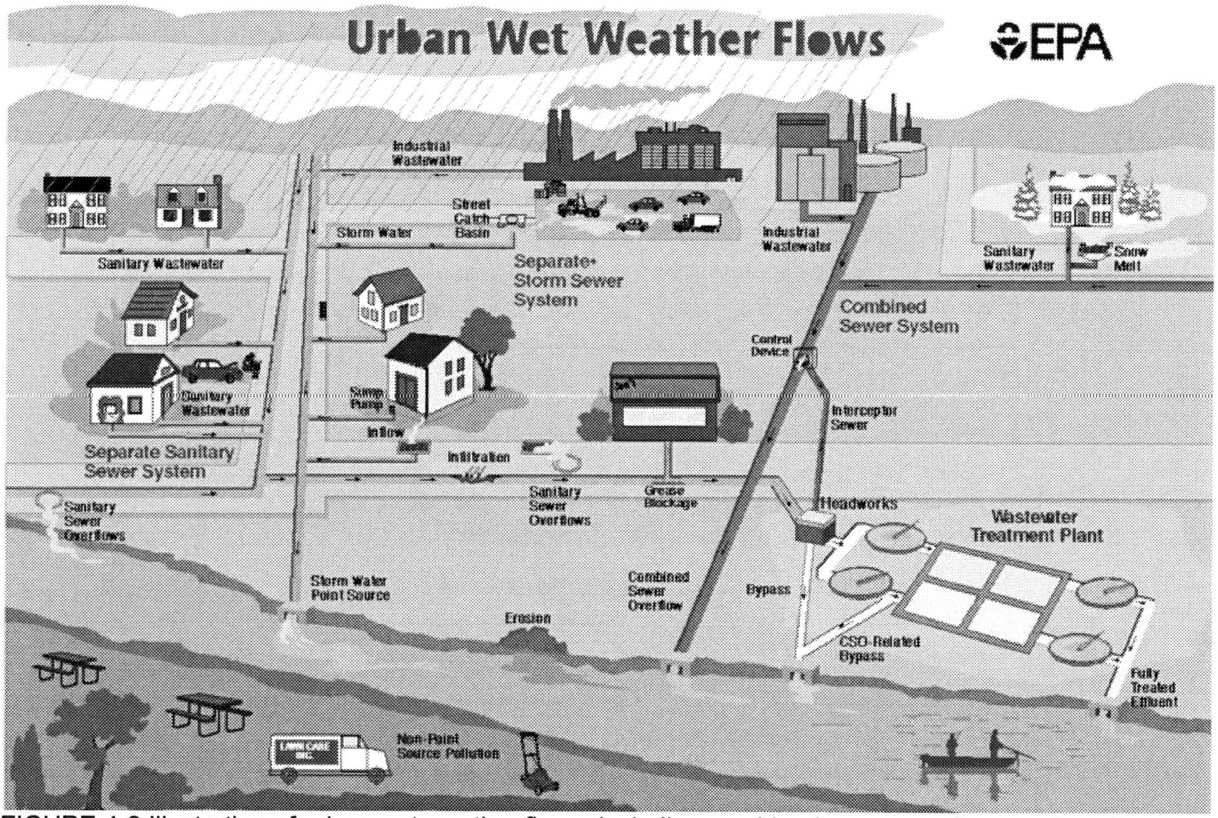

FIGURE 4-2 Illustration of urban wet weather flows, including combined sewer overflows and sanitary sewer overflows.
SOURCE: *http://www.epa.gov/reg3wapd/cso/images/uww.jpg.*

southwestern Pennsylvania. Tables 4-4 and 4-5 summarize the distribution of CSOs in southwestern Pennsylvania by county and by major authority or agency in the region.

Figure 4-4 shows specific CSO outfall locations and water intake locations for the major drinking water providers in Allegheny County. CSOs are considered point sources of pollution and are subject to NPDES permitting, compliance, and enforcement. In May 1997, PADEP revised its CSO strategy and began implementation following EPA approval in July 1997. As noted previously, water released during CSOs is a mixture of dilute raw sewage and primarily urban surface runoff. The chemical and microbial constituents of CSOs have been studied less than those of stormwater. Initially during a rainfall event (the "first flush"), CSO effluents resemble raw sanitary sewage. When high flows scour the sewer pipes the first flush can contain higher levels of suspended solids and BOD than typical raw sewage (Larsen et al., 1998). After the first flush, stormwater dilutes the raw sewage, and pollutant concentrations in CSO effluents decline.

The major contaminants in CSO discharges include suspended solids, BOD, chlorides (typically in winter months from the application of road salt), nutrients (nitrogen and phosphorus), fecal bacteria and other microorganisms, and various chemicals. A recent study of Canadian waters found stormwater and CSOs to be similar in solids concentrations, but CSOs had higher concentrations of BOD and nutrients and lower concentrations of heavy metals and anthropogenic organic compounds (e.g., pesticides) (Chambers et al., 1997). Table 4-6 compares typical concentrations of several chemical and biological parameters for CSOs and urban runoff. With respect to microorganisms, Burm and Vaughan (1966) found that a combined sewer system

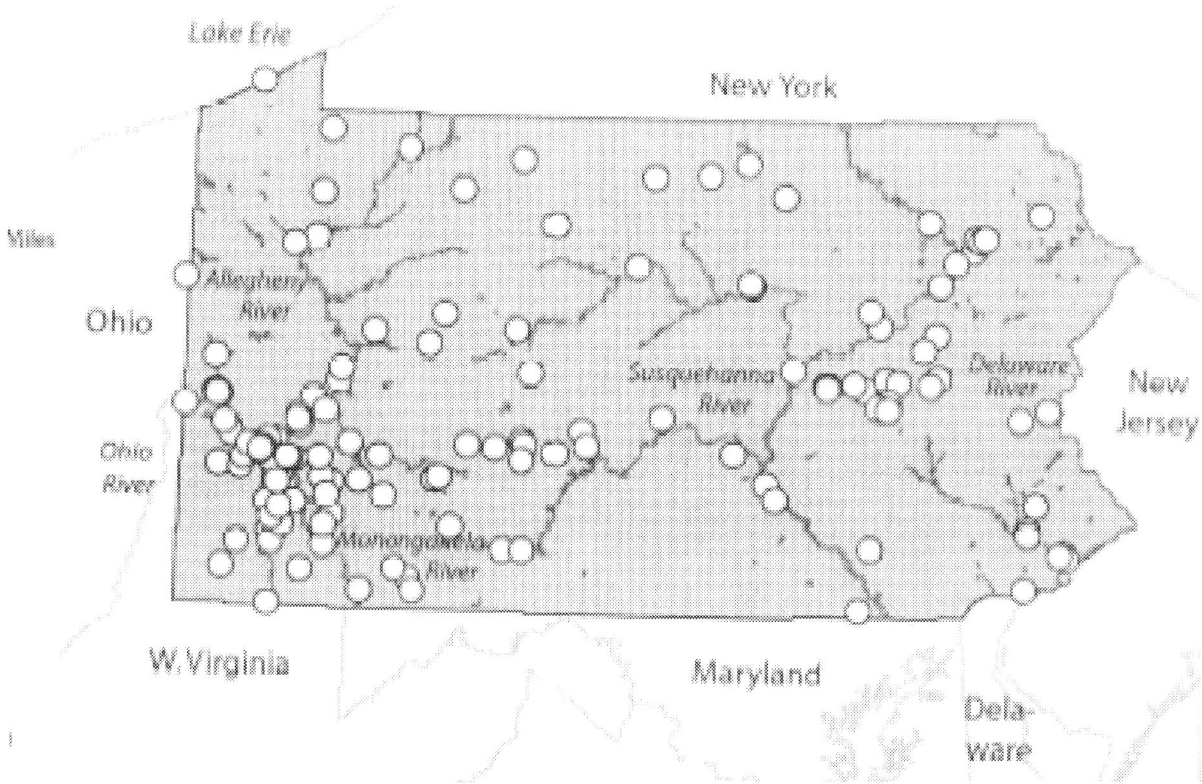

FIGURE 4-3 Distribution of CSOs in Pennsylvania.
NOTE: This figure shows the locations of the agencies that hold the 155 CSO permits.
SOURCE: EPA, 2002a, Appendix B.

TABLE 4-4 Distribution of CSOs in Southwestern Pennsylvania by County

County	CSO Structures/Outfalls
Allegheny	414
Armstrong	18
Beaver	14
Butler	0
Fayette	72
Greene	2
Indiana	22
Lawrence	1
Somerset	7
Washington	79
Westmoreland	126
Total	755

SOURCE: Adapted from PADEP Combined Sewer Overflow Listing, *http://www.dep.state.pa.us/eps/docs/extras/TG/Finals/wswm/CSO_LIST.xls.*

TABLE 4-5 Distribution of Select CSOs in Southwestern Pennsylvania by Major Authority or Agency

Community	CSO Structures/Outfalls
City of Pittsburgh	216
McKeesport City Municipal Authority	28
ALCOSAN	21
Upper Allegheny Joint Sanitary Authority	19
Turtle Creek	10
Borough of Wilmerding	9
Girty's Run Joint Sewer Authority	9
Braddock	8
Etna	8
Total	328

SOURCE: *http://cfpub2.epa.gov/npdes/cso/demo.cfm.*

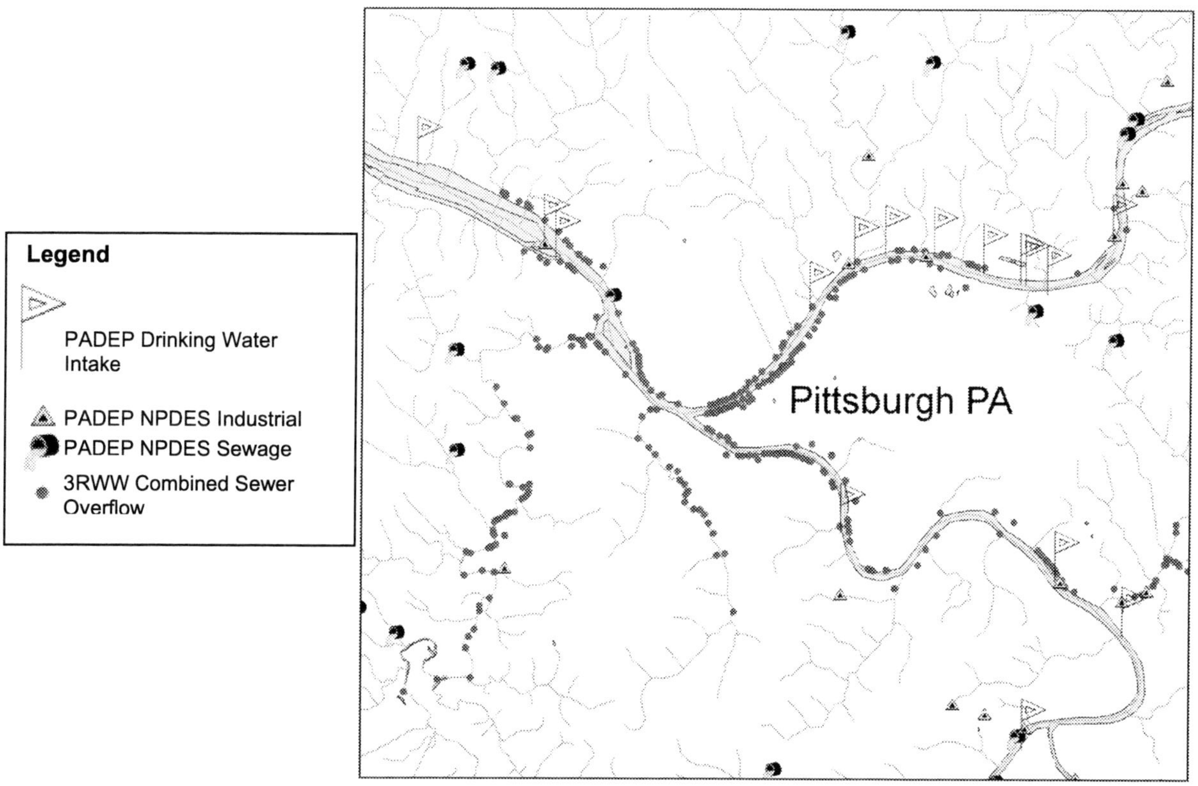

FIGURE 4-4 Illustration of the proximity of public surface water intakes and CSO outfalls in Allegheny County in the vicinity of the City of Pittsburgh. SOURCE: Adapted and reprinted, with permission, from 3 Rivers 2nd Nature. © 3 Rivers 2nd Nature.

contained an order of magnitude more indicator microorganisms than similarly sized stormwater systems. Typical fecal indicator concentrations in CSO discharges are provided in Table 4-7. Because CSOs are characterized by high flows over short time periods during wet weather, the effects of CSO are more pronounced in small streams and lakes than in large rivers with extensive dilution.

The public health impact of CSOs has been difficult to measure directly. Box 4-2 summarizes recent studies that have attempted to correlate waterborne disease outbreaks with precipitation events, which may indicate a role for CSOs. In August 2002, EPA convened an experts workshop (EPA, 2002a) in order to inform a report that it is preparing for Congress on the health impacts of sewer overflows (overdue as of December 15, 2003, as required in P.L. 106-554). Notably, preparation for that workshop yielded very little information linking waterborne diseases to CSOs or SSOs (sanitary sewer overflows) through a literature survey. However, a preliminary list of pathogens expected to be present in sewer overflows was developed and is summarized in Table 4-8. Although EPA's report to Congress was still pending at the time this report was nearing completion, the results of the workshop indicate the following (EPA, 2002a):

- CSOs are likely to contain pathogens and chemicals of public health concern.
- In the majority of cases in waterborne disease outbreaks, the etiologic agent is not identified. Furthermore, many cases are self-medicated and not reported. For these and other reasons, health monitoring data are expected to be significant underestimates.
- Few waterborne disease outbreaks are known to have resulted from sewer overflows. The largest outbreak (cryptosporidiosis) in the United States took place in Milwaukee, Wisconsin, in 1993 and was associated with sewage in the Lake Michigan source water and a failure at the drinking water treatment plant (see Edwards, 1993 and MacKenzie et al., 1994, for further information).
- Although elimination of CSOs and SSOs would likely have real public health benefit, it is unlikely under the current measuring and tracking surveillance system for waterborne illness that the corresponding reduction in illness levels would be detectable.

TABLE 4-6 Concentration Ranges of Select Constituents in CSOs and Urban Runoff

	5-Day BOD (mg/L)	TSS (mg/L)	Total N (mg/L)	Total P (mg/L)	Fecal Coliforms (CFU/100 mL)
CSO	25-100	150-400	3-24	1-10	10^5-10^7
Urban runoff	10-250	67-101	0.4-1.0	0.7-1.7	10^3-10^7

SOURCE: WEF, 1999.

TABLE 4-7 Concentration Ranges of Bacterial Indicators During CSOs

Indicator Organism(s)	Concentration Range (CFU/100 mL)
Fecal coliforms	1.0×10^5 to 3.5×10^7
Escherichia coli	3×10^4 to 2.2×10^7
Enterococci	2.9×10^3 to 7.2×10^6

SOURCE: EPA, 2001a.

BOX 4-2
Wet Weather Events and Human Health Risk

Rose et al. (2000) reviewed U.S. waterborne disease outbreaks from 1971 through 1994 for an association with high total monthly precipitation. The months that were examined for outbreak occurrence were those having total precipitation ranked in the highest 10 percent, the highest 5 percent, and the highest 2.5 percent of monthly totals, or those that followed the months with high precipitation. For systems using surface water sources, it was shown that high amounts of precipitation can affect source water quality and lead to outbreaks. Similarly, Curriero et al. (2001) found a statistically significant association between outbreaks and precipitation events. They reviewed 548 reported waterborne outbreaks between 1948 and 1994 and reported that 51 percent of outbreaks were associated with precipitation events above the 90th percentile ($p = .002$) and 68 percent were associated with outbreaks above the 80th percentile ($p = .001$). Surface water outbreaks occurred during the month of the precipitation event, while groundwater outbreaks demonstrated a two-month lag time.

TABLE 4-8 Pathogenic Microorganisms That Can Be Found in Untreated Domestic Wastewater

Pathogen	Pathogen
Bacteria	Protozoa
Shigella	*Entamoeba histolytica*
Salmonella	*Giardia lamblia*
Vibrio	*Balantidium coli*
Escherichia coli	*Cryptosporidium parvum*
Yersinia enterocolitica	
Leptospira	Viruses
Campylobacter jejuni	Enteroviruses
	Hepatitis A
Helminths (worms)	Adenovirus
Ascaris lumbricoides	Rotavirus
Ancylostoma duodenale	Parvovirus
Necator americanus	Norovirus
Strongyloides stercoralis	Reovirus
Trichuris trichiura	Astrovirus
Taenia spp.	Calicivirus
Enterobium vermicularis	Coronavirus
Echinococcus granulosus	

SOURCE: EPA, 2002b.

CSOs are specifically implicated as the cause of impairment in 109 of 10,762 miles of streams impaired for aquatic life support in Pennsylvania (PADEP, 2004a). The watershed restoration action strategy (WRAS, 2003b) state water plan for the Upper Youghiogheny River watershed, Laurel Hill Creek, and Indian Creek in Fayette, Somerset, and Westmoreland Counties indicates that CSOs are known contributors to water quality impairment in Deadman Run and three unnamed tributaries to Laurel Hill Creek. The WRAS state water plan for Chartiers Creek watershed (Ohio River; see Figure 6-2) in Washington and Allegheny Counties indicates that combined sewer overflows carry considerable urban runoff into the streams of this watershed. PENNVEST awarded $1.9 million to the City of Washington for a stormwater control project in this watershed. The Chartiers Creek watershed is part of the Allegheny County

Sanitary Authority (ALCOSAN) collection systems and is working in consultation with EPA, PADEP, and ALCOSAN on CSOs in the region (WRAS, 2003c).

In addition to the specific areas described above, there is also some direct evidence for the role of CSOs in degrading water quality in other parts of southwestern Pennsylvania. Several studies that have considered geographical or weather-related differences in pathogen loading suggest that these impairments are related to sewage handling within the region, and their correlation with precipitation events specifically targets CSOs. States et al. (1997) reported on protozoa levels in the Allegheny River, while Gibson et al. (1998) evaluated a tributary. Details of their results are presented in Boxes 4-3 and 4-4 and Tables 4-9 and 4-10, respectively. Both studies found higher levels of protozoa in CSOs and in surface waters downstream of CSOs during wet weather.

BOX 4-3
Protozoa in Surface Waters in Southwestern Pennsylvania

States et al. (1997) surveyed source waters for drinking water in the Pittsburgh area for protozoa in a two-year program of monthly sampling and analysis. They collected monthly samples from the Allegheny River, from a stream flowing through a dairy farm that had about 20 to 25 head of cattle, and from secondary wastewater treatment plant effluent that flowed into the Allegheny River. In addition, they occasionally collected samples from overflowing combined sewer outfalls. Samples were collected from the Youghiogheny River for comparison to water quality in the Allegheny River. Monitoring results indicated similar quality for the two rivers, as shown by the geometric means (see Table 4-9 below). Protozoa were found to be much more abundant in CSO samples than in river water.

TABLE 4-9 Protozoa Monitoring Results on Main Stem Rivers in the Pittsburgh Area

Source Sampled	Number of Samples	*Giardia* cysts per 100 L			*Cryptosporidium* oocysts per 100 L		
		Range	Geom. Mean	Percent Positive	Range	Geom. Mean	Percent Positive
Allegheny River	24	ND-421	34	63	ND-2,233	31	63
Youghiogheny River	24	ND-526	118	54	ND-1,473	58	63
Combined sewer overflow	5	3,750-114,000	28,681	100	ND-3,000	2,013	80
Small stream flowing through dairy farm	24	ND-1,527	82	55	ND-2,290	42	82
Wastewater treatment plant effluent	24	ND-4,614	664	83	ND-4,927	924	33

NOTE: ND = not detected.
SOURCE: States et al., 1997.

BOX 4-4
Studies on the Effects of CSOs in Tributaries in Southwestern Pennsylvania

Gibson et al. (1998) followed up on the continuing water sampling and analysis program of States et al. (1997) with sampling and testing in the Pittsburgh area at Saw Mill Run, a tributary of the Ohio River in late 1996 and 1997. Saw Mill Run is about 12 miles long, with 26 CSO sites along the stream. The Upper Saw Mill Run site was upstream of the reach of the stream influenced by CSOs. The sampling site for Lower Saw Mill Run was downstream of the CSOs, in a location closest to the Ohio River where the stream water quality was not affected by backwater from the Ohio River. Dry weather sampling was carried out at times when no precipitation had occurred for at least 72 hours. Wet weather sampling conditions occurred when 0.1 inch or more of rainfall had occurred after at least 72 hours without rainfall. During five wet weather events, a total of 11 CSO samples were collected. Pathogens were detected in CSO end pipes at concentrations of 250-40,000/100 L for *Cryptosporidium* and 9,000-283,000/100 L for *Giardia*, which suggests a public health risk to recreational water users and for drinking water. Fecal coliform concentrations ranged from 3,000 to 85,000 colony forming units (CFUs) per 100 mL from the end pipe. Summary data from this study are shown in Table 4-10. Gibson et al. (1998) concluded that combined sewer overflows can contribute to the load of protozoa in ambient waters.

In another study, total coliform, fecal coliforms, and *Eschericia coli* were sampled at six locations along Nine Mile Run in 1999. Nine Mile Run is a tributary of the Monongahela in eastern Allegheny County. There are six documented CSO outfalls that discharge directly into Nine Mile Run. The study results indicated that Nine Mile Run is unsafe for human recreational contact during dry and wet weather; it is seriously degraded by sewage. Total coliform counts ranged from 101,036 to 1,311,000 CFU/100 mL, while fecal coliform counts ranged from 125 to 1,051,200 CFU/100 mL. *Escherichia coli* ranged from 125 to1,009,800 CFU/100 mL. Wet weather values are orders of magnitude higher than dry weather values. The authors concluded that "it appears that dry weather bacteriological data documented mostly the influences of chronic sanitary sewage, while the wet weather data show the overwhelming impacts of CSOs on Nine Mile Run" (USACE, 2000).

TABLE 4-10 Microbiological Monitoring Results in Vicinity of Saw Mill Run

	Upper Saw Mill Run			Lower Saw Mill Run		
Condition and Parameter	*Crypto-sporidium*	*Giardia*	Fecal Coliforms	*Crypto-sporidium*	*Giardia*	Fecal Coliforms
Dry weather						
Range	5-39	<13-66	170-6,500	<33-105	21-6,579	280-13,300
Geometric mean	18	36	642	78	343	1,137
Geometric standard deviation	2.6	1.8	4.3	1.4	8.0	3.7
Mean	24.6	42.0	1,637	81.2	1,539	2,686
Standard deviation	20.4	19.1	2,503	22.7	2,579	5,122
Median	13.0	42.0	370	91.0	356	845
Wet weather						
Range	<39-72	67-288	89,000-127,000	429-1,667	429-5,800	6,100-87,000
Geometric mean	70	225	107,203	754	2,653	18,328
Geometric standard deviation	1.1	2.2	1.2	2.0	4.9	4.0
Mean	70	133	108,333	899	4,576	34,900
Standard deviation	3.5	125	19,009	671	3,691	45,204
Median	69.5	78	109,000	600	5,800	11,600

NOTE: *Cryptosporidium* = oocysts per 100L, *Giardia* = cysts per 100 L, fecal coliforms = colony forming units (CFU) per 100 mL.
SOURCE: Gibson et al. (1998).

There are indications that the CSO problem has worsened in the last decade in southwestern Pennsylvania. Table 4-11 provides the total number of advisories and days affected by advisories for the summer recreational season in Allegheny County. River advisories are issued when rainfall in the region is high enough to potentially cause sewer overflows and lead to unsafe bacterial concentrations in the river. When an alert is in effect, marinas and docks fly an orange and black CSO sign to alert recreational users of potentially unsafe conditions. The Allegheny County Health Department (ACHD) recommends restricted recreational exposure during advisories (e.g., anglers are advised to wash their hands after fishing and not to cut fish line with their teeth). As shown in Table 4-11, there has been a steady rise in the number of days that the water is considered impaired and restriction of body contact recreation is recommended by the ACHD. As noted previously, the ACHD is conducting a study of the river quality and human health by evaluating the health status of rowers who use the river for practice and competition (ACHD, 2004). Initial data are expected in October 2004.

One of the reasons for increasing CSOs in Allegheny County is the aging collection system and the problem of infiltration. Although many wet weather stormwater flows are directed into sewer pipes in the combined systems, many additional flows find their way into the system during wet and dry weather. Figure 4-5 shows possible sources of infiltration and inflow (I/I) into a collection system. The source of infiltration and inflow in sewer systems is site specific. In some locations, foundation drains are connected to the sewer lines. In other areas, rain leaders (roof gutter drains and areaway drains) are connected to the sewer lines. In many areas, house laterals (the component of the system owned by individual homeowners) show significant deterioration, allowing groundwater to enter sewer pipes.

Sanitary Sewer Overflows

Unlike combined systems, dedicated sanitary sewer systems were designed to carry only sanitary waste. However, pipe cracks and illegal connection of "French drains" or roof collection systems can add stormwater to sanitary systems. When significant infiltration occurs, sanitary sewer overflows can take place, especially during rain events. SSOs are illegal in the United States under the federal CWA. Since the sanitary system was not designed to overflow into local waterways, SSOs result in groundwater contamination, backups of sewage into basements, and overflows through manhole covers (see Figure 4-6).

TABLE 4-11 Water Quality Advisories in Allegheny County: 1994-2003

Year	No. of Advisories	No. of Days	Portion of season with Advisory (percent)
1994	11	33	24
1995	12	46	33
1996	12	62	45
1997	12	46	33
1998	10	50	36
1999	11	33	24
2000	13	71	51
2001	15	68	49
2002	13	83	60
2003	8	109	79

SOURCE: Charles Vukotich, ACHD, personal communication, 2004.

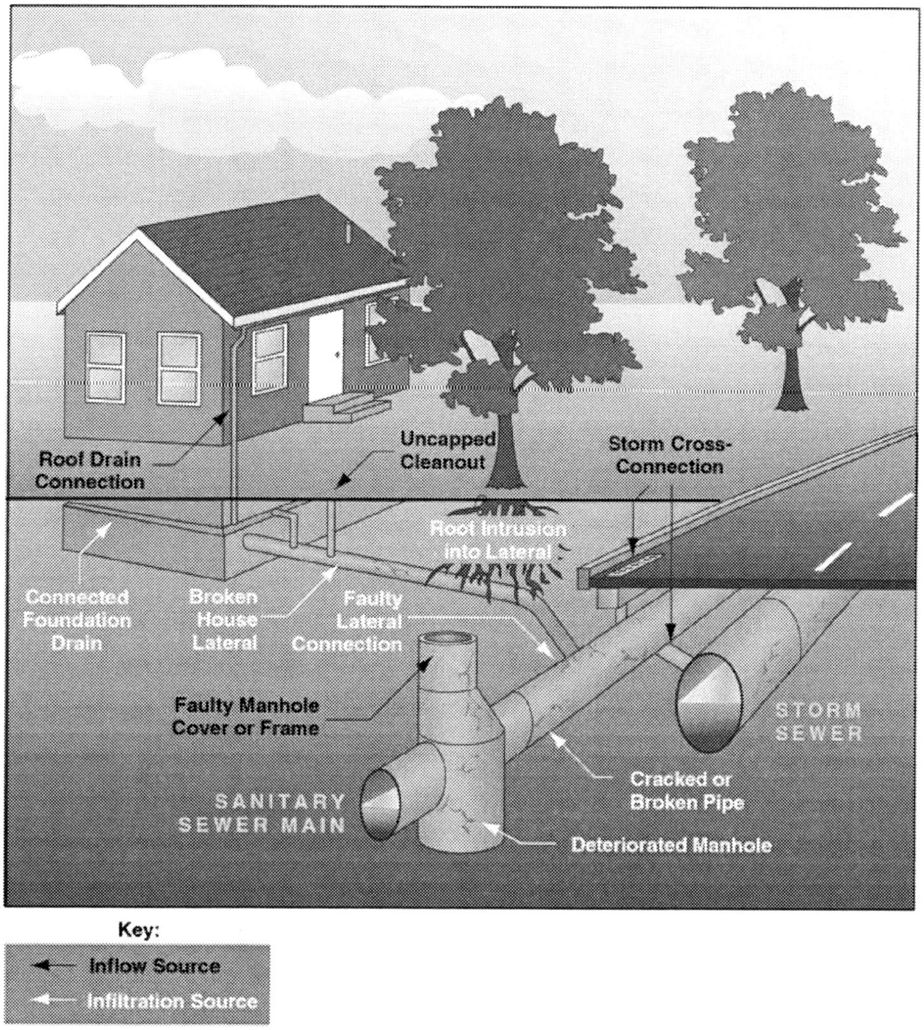

FIGURE 4-5 Infiltration and inflow sources.
SOURCE: Adapted from *http://dnr.metrokc.gov/WTD/i-i/whatis.htm.*

Nationally, the EPA estimates that there are at least 40,000 overflows of sanitary sewers each year (EPA, 2001b). The WSIP (2002) report estimates more than 600 SSOs a year in southwestern Pennsylvania. Of the 76 percent of homes in southwestern Pennsylvania that are on public sewer systems (see Figures 1-2 and 4-1), 11 percent are served by combined sewer systems and 48 percent are served by separate systems whose pipes connect to combined sewer systems or whose waste is treated at a plant that also serves combined sewer systems. These hydraulic interconnections allow overflows of raw sanitary sewage in regions with separate systems when total flow to the plant receiving both types of sewage exceeds the plant capacity. Some separate sewer systems include overflow structures because they were built before these structures were illegal. Many systems do not have these structures, and when excessive flows cannot be handled, these systems back up in basements, at manholes, or underground. Only one reference to sewage backups in state Watershed Restoration Action Strategy (WRAS) plans

FIGURE 4-6 Sanitary sewer overflow from an elevated manhole in the Pittsburgh. region. SOURCE: Photograph courtesy of 3RWW.

was found (WRAS, 2003d). The Stonycreek River and Little Conemaugh River watersheds report a PENNVEST project in 1998 to Windber Borough to eliminate backup of sewage into basements during wet weather and to replace collection lines. Additional sanitary sewer projects in this watershed were also undertaken.

On-Site Sewage Treatment and Disposal Systems

Individual on-lot septic systems (more accurately referred to as on-site sewage treatment and disposal systems) are frequent alternatives to wastewater treatment plants in sparsely populated areas of the country where the costs of constructing centralized treatment systems are prohibitive. If properly sited and functioning, OSTDSs can receive, treat, and dispose of wastes in a manner that is comparable to wastewater treated in a central facility (EPA, 1980, 1997, 2002c). Table 4-12 provides some ranges for typical contaminants from septic tank effluent and from downgradient in the leach field. Both the tank and the leachfield must operate properly for treatment to be complete. OSTDSs can be designed to provide waste treatment from a single house, business, or groups of structures. The 1997 Response to Congress on Use of Decentralized Wastewater Treatment Systems[9] states that "adequately managed decentralized wastewater systems are a cost-effective and long-term option for meeting public health and water quality goals" (EPA, 1997).

---

[9] This report is a response to the congressional House Appropriations Committee's request that EPA report on the benefits of decentralized wastewater systems alternatives; the potential savings and/or costs associated with the alternatives; and the ability and any plans of EPA to implement the alternatives during the 1997 fiscal year. The full report is available on-line at *http://www.epa.gov/owm/mtb/decent/response/*.

TABLE 4-12 Case Study: Septic Tank Effluent and Soil Water Quality[a]

Parameter (units)	Statistics	Septic Tank Effluent Quality	Soil Water Quality[b] at 0.6 Meter	Soil Water Quality[b] at 1.2 Meters
BOD	Mean	93.5	<1	<1
(mg/L)	Range	46-156	<1	<1
	No. of samples	11	6	6
TOC	Mean	47.4	7.8	8.0
(mg/L)	Range	31-68	3.7-17.0	3.1-25.0
	No. of samples	11	34	33
TKN	Mean	44.2	0.77	0.77
(mg/L)	Range	19-53	0.40-1.40	0.25-2.10
	No. of samples	11	35	33
NO_3-N	Mean	0.04	21.6	13.0
(mg/L)	Range	0.01-0.16	1.7-39.0	2.0-29.0
	No. of samples	11	35	32
TP	Mean	8.6	0.40	0.18
(mg/L)	Range	7.2-17.0	0.01-3.8	0.02-1.80
	No. of samples	11	35	33
TDS	Mean	497	448	355
(mg/L)	Range	354-610	184-620	200-592
	No. of samples	11	34	32
Cl	Mean	70	41	29
(mg/L)	Range	37-110	9-65	9-49
	No. of samples	11	34	31
Fecal coliforms	Mean	4.57	ND[c]	ND
(log of No. per	Range	3.6-5.4	<1	<1
100 mL)	# samples	11	24	21
Fecal	Mean	3.60	ND	ND
streptococci (log	Range	1.9-5.3	<1	<1
of No. per 100 mL)	# samples	11	23	20

NOTE: [a] The soil matrix consisted of a fine sand; the wastewater loading rate was 3.1 cm per day over 9 months. TDS = total dissolved solids; TKN = total Kjeidahl nitrogen; TOC = total organic carbon.
[b] Soil water quality measured in pan lysimeters at unsaturated soil depths of 2 feet (0.6 m) and 4 feet (1.2 m).
[c] ND = Not detected.
SOURCES: Adapted from Anderson et al., 1994; EPA, 2002c.

The most common OSTDS is the septic tank and drainfield, the former of which is essentially a settling basin in which the suspended solids are separated from the liquid fraction of waste. The solids settle to the bottom of the tank where they are degraded by anaerobic bacteria, while the lighter material, including fats, oils, and grease accumulates at the liquid surface. The liquid portion of the waste flows from the tank through an outlet near the top and is distributed through perforated pipes into a subsurface drainfield or infiltration system. Within the drainfield, the soil filters out pathogenic microorganisms from the OSTDS effluent before it reaches groundwater. Because aerobic conditions enhance destruction of pathogens, drainfields are placed within the unsaturated portion of the soil profile. Properly designed, sited, and maintained septic tanks and drainfields can remove 90 percent of BOD, 85-95 percent of total phosphorus, 99-99.99 percent of fecal coliforms, but as little as 10-40 percent of total nitrogen (EPA, 2002b).

Other types of OSTDSs vary from older, substandard cesspits (no longer legal under modern codes) to more innovative and efficient sand (and other media) filters and alternative drainfields such as low-pressure pipe systems. Media filters are constructed beds of sand or other suitable granular material usually two to three feet deep. Partially treated wastewater (e.g., from a septic tank) is applied to the filter surface and receives treatment as it slowly trickles through the media; the wastewater then collects in an underdrain and flows to further treatment and/or soil dispersal. Low-pressure pipe (LPP) systems use shallow-placed, pressure-dosed, perforated pipes in narrow trenches for controlled, periodic release of effluent into the soil. LPPs allow more even distribution of effluent and may overcome problems of localized overloading of the soil and anaerobic conditions due to continuous saturation. Because they are small scale, are widely dispersed, and discharge to relatively large subsurface areas, OSTDSs constitute a nonpoint source of pollution within a watershed, unlike centralized WWTPs. A major consequence of this fact is that measuring the impact of OSTDSs on nearby water quality can be extremely difficult. To the committee's knowledge, environmental monitoring of OSTDS effluent in the southwestern Pennsylvania area has not occurred.

OSTDS Failures

Contamination of groundwater by failing or substandard septic systems is a considerable risk in much of Pennsylvania because of the state's geology, soils, land development patterns, and large numbers of aging or unknown treatment systems. As of 1990, 1.2 million Pennsylvania homes used on-lot septic systems (Fleeger, 1999). For southwestern Pennsylvania, Figure 4-1 shows the percentage of homes in each county using OSTDSs. A total of 264,408 households were serviced by OSTDSs, or 23 percent of the total households in the 11-county region. The percentage of homes on "other" means declined from 2.5 percent in 1970 to 1.2 percent in 1990. (As noted previously, the 2000 census did not include a question regarding wastewater treatment method.) The percentage of homes utilizing centralized treatment (sewers) increased slightly from 1970 to 1990. The percentage of homes on septic systems changed little during this period (1970-1990).[10]

National failure rates for OSTDSs are reported at 10 percent annually based on self-reported failures in the three months prior to the housing survey (DOC and HUD, 1997; Knowles, 1998). The EPA (2000) reviewed a series of studies and concluded that failure rates were 10-20 percent; however, failure definitions varied and were not systematically linked to water quality impacts. A survey of on-site treatment in 28 states found failure rates from 0.5 to

[10] Data from SepticStats available on-line at *http://www.nesc.wvu.edu/nodp/nodp_index.htm.*

70 percent. Pennsylvania was not surveyed, but nearby states had high failure rates (Ohio, 25-30 percent; West Virginia, 60 percent) (Nelson et al., 1999).

Assuming a 10 percent annual failure rate would suggest 26,000 failing septic tanks in the southwestern Pennsylvania region *annually*. However, several regional reports (see Appendix B) have estimated much higher failure rates by considering the suitability of soils in the region for on-site treatment systems. In general, soils in southwestern Pennsylvania are not well suited to conventional septic systems. The PADEP reports that less than 5 percent of permits for new systems are for conventional in-ground systems (William Davis, PADEP, personal communication, 2004). There are known areas in southwestern Pennsylvania where most OSTDs are failing (designated "mass malfunction areas") (see below). The *2004 Pennsylvania Integrated Water Quality Monitoring and Assessment Report* (PADEP, 2004) indicates that septic systems are among the eight highest-priority sources of groundwater contamination in the commonwealth. Failed and failing septic systems may be associated with pollution of groundwater by nitrates, microbial indicators and pathogens, and excessive dissolved solids.

In Pennsylvania, an OSTDS is considered failed if a confirmed malfunction is documented in one or more of the following ways: (1) dye testing, (2) lab results, (3) observation by a sewage enforcement officer (SEO) or other experienced professional, and/or (4) seasonally wet adsorption areas, where drainfield soils are saturated by system overload or seasonal high water tables or both. Utilization of best technical guidance (BTG) repair that does not meet the technical standards of the existing code for a new system also leads to classification as a failed system. A failed system requires response by the Pennsylvania Association of Sewage Enforcement Officers (PASEO) and repair by the homeowner. There is no pre-sale inspection protocol in place for on-site systems during property transfers, and regular inspections are not required by the state.[11] In reality, many mortgage lenders require inspections as part of the home loan process to protect their own financial interests (Gil Longwell, Pennsylvania Septage Management Association, personal communication, 2003). However, these inspections, many of which are conducted by SEOs or Pennsylvania Septage Management Association-certified inspectors, are not tracked or recorded in the OSTDS and water quality programs.

Nationally, septic systems are reported as major sources of groundwater contamination by 31 of 52 states (EPA, 2002a), including Pennsylvania (PADEP, 2004a). On-site system failures are implicated in 149 impaired stream miles and 6,110 impaired lake-acres in the Commonwealth of Pennsylvania (PADEP, 2004a). Within the study region, many of the WRAS state plans implicate individual sewage systems in water quality impairment. For the Lower Youghiogheny River watershed in Westmoreland and Fayette Counties, 95-99 percent of the soils are not suitable for conventional septic systems and less than one-quarter of the watershed had municipal sewage treatment as of 1994 (WRAS, 2001). A PENNVEST loan within the watershed of $1.4 million to East Huntington Township was used to construct 10 miles of collection sewers and a treatment facility to eliminate raw sewage discharges from on-lot septics. The Chartiers Creek watershed in Washington and Allegheny Counties reported malfunctioning on-lot septic systems and wildcat sewer discharges into yards and streams; a PENNVEST loan of $350,000 was awarded to Midway Sewage Authority to design a new treatment plant, pump station, and collection lines to mitigate this problem. The Center Township Sewer Authority received a PENNVEST loan in 1997 to eliminate malfunctioning OSTDSs and raw sewage

---

[11] From Title 25 Environmental Resources Chapter 71, 72, and 73 Standards for Sewage Disposal Facilities (October 2, 1999).

discharges. The Blacklick Creek and Conemaugh River watersheds are impaired by numerous discharges of untreated sanitary wastes from municipalities and private OSTDSs. Significant PENNVEST funding was received by these watersheds, including $22 million to the Jackson and East Taylor Sewer Authority, to construct sewers and pumping stations to eliminate discharges from malfunctioning OSTDSs into tributaries of Hinkston Run. Lastly, the Stonycreek River and Little Conemaugh River watersheds report on the Forest Hills Sewer Project that will connect unsewered communities and eliminate discharges of raw sewage (WRAS, 2003d).

Even though specific monitoring data are not available on the performance of OSTDSs, and not all watersheds have completed WRAS reports summarizing specific problems, documentation of existing OSTDS sewage problems, straight pipes, and wildcat sewers, through Pennsylvania Act 537 needs identification, is required when competing for need-prioritized funding of sewage treatment—for example, PENNVEST or federal Rural Utilities Service (RUS) assistance (PADEP, 2003). Information from PENNVEST[12] (Lawrence Gasparato, PENNVEST, personal communication, 2003) indicates that OSTDS failures have been documented extensively in the region and, during 2003 alone, resulted in construction or planned construction of community wastewater treatment systems in 10 of the 11 counties in the study area. For example, two pending PENNVEST projects in Fayette County were planned for communities with OSTDS failure rates of 62 percent and 65 percent. In Armstrong County, a project was pending at the end of 2003 for a community in which OSTDSs had a 77 percent failure rate.

PENNVEST and RUS have also funded community water system projects in areas where pervasive bacterial contamination affects numerous private water well supplies. In some of these cases, failing septic systems have been documented. For example, in 2003, the Bentleyville Municipal Authority in Washington County received a PENNVEST loan to install water lines to 28 homes that were served by shallow wells contaminated by nearby malfunctioning septic systems.[13]

The complete absence of wastewater treatment facilities of any sort in portions of southwestern Pennsylvania is also indicated by activities in the PENNVEST program (Lawrence Gasparato, PENNVEST, personal communication, 2003). In late 2003, ten projects were completed, under way, or pending to eliminate wildcat sewer systems that served or are serving populations ranging from 200 to 2,000 in six counties in southwestern Pennsylvania. "Wildcat" systems are those that discharge raw sewage directly to nearby streams without treatment (see Appendix C). The extent of wildcat systems within the region is not known precisely, but anecdotal reports to the committee indicate that thousands of homes having no treatment system for human waste still exist throughout the region.

These projects indicate the pervasive and serious nature of the problems associated with wastewater treatment in portions of southwestern Pennsylvania that are not served by effective community wastewater treatment systems or appropriate septic systems.

---

[12] See PENNVEST press releases 11/20/02, 05/07/03, 07/23/03, 11/19/03, available on-line at *http://www.pennvest.state.pa.us/pennvest/cwp/browse.asp?A=11&BMDRN=2000&BCOB=0&C=43125.*

[13] See *http://www.pennvest.state.pa.us/pennvest/cwp/view.asp?A=11&Q=175087* for 07/23/03 PENNVEST press release.

Contamination of Surface and Groundwater by Failing OSTDSs

No studies have been conducted that demonstrate the contribution of failing OSTDSs to pathogen loading in surface waters of southwestern Pennsylvania. As discussed in Chapter 3, fecal coliform bacteria samples were collected and analyzed by the ACHD from 4 stations on Montour Run and the mouths of 14 of its largest tributary streams in September 1996 (USACE, 1997). Fecal coliforms were highest in the western, headwater portion of Montour Run and in the tributaries in the western portion of the watershed. The authors suggest that higher levels are likely caused by improperly operating septic systems discharging partially treated wastewater effluents. Land use in this region was in rapid transition at the time of the study—from agricultural lands and woodlands to suburban housing and retail development. The contribution of common headwaters sources, such as wildlife and/or livestock, to measured bacterial loads cannot be excluded.

No known scientific studies correlating malfunctioning septic systems and fecal contamination of water wells were available for the study area. However, several national studies suggest potential risks of groundwater contamination. In a surveillance summary of incidents in 1989 to 1990, the U.S. Centers for Disease Control and Prevention (CDC) reported that 13 of 26 drinking waterborne disease outbreaks in the United States were due to contaminated groundwater, with viruses as the main agents. In most of these outbreaks, contamination originated from malfunctioning on-lot sewage systems (Herwaldt, 1991). In a surveillance summary of incidents in 1999-2000, CDC stated that 29 of the 39 reported drinking water outbreaks were associated with groundwater, with most of unknown etiology and associated with private or noncommunity wells. In a shift from previous reports, most of the outbreaks were not associated with distribution systems or treatment failure. Rather, most outbreaks were associated with drinking groundwater, suggesting that this is an increasing risk in the United States (NRC, 2004). Although no specific causes were discussed, the need to increase public awareness of the risks of direct consumption of untreated water from any source was discussed in Lee et al. (2002).

Scandura and Sobsey (1997) seasonally seeded four on-site wastewater treatment systems (three conventional drainfields and one LPP system) in sandy soils of the North Carolina coast with known amounts of a model enterovirus (BE-1). They studied the survival and transport of BE-1, fecal coliforms, and other wastewater constituents in groundwater by sampling from drainfield monitoring wells. At one site—a conventional drainfield system in soils with clay content of 15 percent, a vadose (unsaturated) zone of 1 meter or more, and no seasonal submergence of drainfield lines—they observed extensive reduction of viruses, fecal coliforms, and nitrates. However, at the remaining sites, they detected contamination of groundwater by viruses, bacteria, and nutrients under conditions of coarse, sandy soils; shallow water tables; and drainfield lines submerged by seasonal high water tables. One of these latter sites was the LPP system, suggesting that soil absorption system design is a less important factor in contaminant reduction than soil properties and relative location of the water table. Rapid and extensive movement of contaminants in groundwater at one site appeared to be related to the steep hydraulic gradient and land slope at that site.

In the first study in the United States to systematically sample private water wells for human enteric viruses, Borchardt et al. (2003) investigated their incidence in 50 single home wells in Wisconsin, with half in locations near septage land application sites and the others in rural subdivisions served by septic systems. Wells were selected to represent seven of the state's hydrogeologic districts and were sampled four times during a year, once each season. All but one well was isolated from surface water influence. In addition to viruses, wells were sampled for several other water quality parameters (total coliforms, *Escherichia coli*, enterococci, coliphages, nitrate, and chloride). Four wells (8 percent), and 5 of 194 samples (3 percent) were positive for viruses, including hepatitis A, rotovirus, poliovirus, or Norwalk-like virus. None of the wells were virus positive for two successive samplings, suggesting that contamination was transient. All four wells were relatively new and constructed according to Wisconsin state code, with a minimum casing depth of 40 feet. All virus-positive wells were in subdivisions served by septic systems, suggesting that septic systems were more likely to be a contamination source (although the authors cautioned against precluding land application sites as a potential source). Three of the four wells were located in coarse-textured soils. Two of the contaminated wells (drilled and constructed to code and cased to 52-meter depth) were in the Door County Peninsula, in an area of extensively fractured bedrock overlain by shallow soils. (This latter observation suggests the potential for similar risks in the shallow soils and highly fractured bedrock of southwestern Pennsylvania.) Chloride was the only indicator with a comparatively high true-positive rate (i.e., when virus was present, chloride concentration was elevated), suggesting that the virus-positive wells were in a fecal plume. Chloride, however, had low positive predictive value at 15 percent.

Summary

The data examined in the preceding sections suggest that surface water and groundwater in southwestern Pennsylvania experience significant fecal contamination from practices related to treatment of human waste. Fecal pollution of this magnitude has the potential to cause outbreaks of waterborne gastroenteritis or other diseases associated with enteric pathogens when contaminated water is ingested either for drinking or during contact recreation. However, properly maintained drinking water treatment of surface source water appears to be sufficient to prevent waterborne diseases in users of public water supplies (although, as noted previously, many drinking water-related outbreaks go unreported). Limited surveillance data (see Chapter 3) do not suggest that water contaminated by insufficiently treated human waste is causing widespread illness in southwestern Pennsylvania.

URBAN STORMWATER

The main constituents of concern in urban stormwater are suspended solids, nutrients (particularly phosphorus), heavy metals, toxic organic chemicals, and fecal bacteria. A recent review of 140 studies from the United States, Europe, and Canada identified 28 water quality parameters[14] in stormwater with the potential to affect aquatic life or human health through drinking water (Chambers et al., 1997; Environment Canada, 1999; Makepeace et al., 1995).

Several other studies have characterized the pollutants present in urban stormwater (Ellis and Wang, 1995; Van Metre and Mahler, 2003; Walker et al., 1999). Urban stormwater

---

[14] Total solids; total suspended solids; metals Al, Be, Cd, Cl, Cr, Cu, Fe, Pb, Mn, Hg, N, Ag, Zn; low dissolved oxygen; PCBs; bis(2-ethylehexyl) phthalate; γ-BHC; chlordane; heptachlor; heptachlor epoxide; total polyaromatic hydrocarbons; benzo[*a*]pyrene; tetrachloroethylene; fecal coliforms; fecal streptococci; enterococci.

discharges are associated with numerous effects on receiving waters. Flooding, erosion, and sedimentation are amenable to engineering solutions (Marsalek and Chocat, 2002). The physical and chemical composition of urban stormwater may adversely affect aquatic life. Nationwide urban stormwater quality data have been collected for the past 20 years, although few parameters are related to public health (Smullen et al., 1999). While stormwater management has received much attention in recent decades, the impact of stormwater on public health has largely been ignored.

Indicator Organisms and Sources

Concentrations of fecal coliform bacteria in separate urban stormwater systems are frequently in excess of levels considered safe for recreational water contact (Novotny et al., 1985; see also Chapter 3). Anecdotal information about discharges from separate storm sewers in the Pittsburgh region was presented in testimony to the committee late in the study, and this information was consistent with prior experiences of committee members. On this basis, the committee believes there is reason to consider that discharges from separate storm systems represent a potentially significant source of microorganisms that should be anticipated in the design of water quality monitoring programs and the formulation of management options. As described previously, sources of fecal pollution may include failing sanitary sewer lines, OSTDSs, or illegal discharges; however the role of nonpoint source pollution depends largely on the amount of animal (pets, livestock, wildlife) fecal material accumulating between rainfall events (Schiff and Kinney, 2001). In addition, recent development of recreational trails in urban-suburban riverside greenbelts and parks introduces a new relative concentration of a contaminant source directly adjacent to streams (i.e., dog waste) (Hamilton, 2001; Rodricks, 2003).

Nonhuman sources can be major contributors of fecal contamination in urban and suburban watersheds as well (see also Boxes 5-6 and 5-7). In a New York City water supply watershed, Alderiso et al. (1996) found that 95 percent of fecal coliform in urban stormwater was of nonhuman origin. A similar study in the Four Mile Run watershed, in densely developed northern Virginia suburbs of Washington, D.C., found that humans and canines contributed approximately 25 percent, while waterfowl were the source of 37 percent, of *E. coli* in the watershed. The study also noted that the presence of human *Escherichia coli* was localized within the watershed (Simmons et al., 2001). The EPA estimated that for watersheds of up to 20 square miles draining to small coastal bays, two to three days of droppings from a population of about 100 dogs could contribute enough bacteria and nutrients to temporarily close a bay to swimming and shellfishing (EPA, 1993). Other documented nonhuman sources of fecal coliform bacteria in urban watersheds include rats, raccoons, deer, and pigeons (Lim and Olivieri, 1982; Simmons et al., 2001).

Fecal indicator bacteria in first flush flows depend upon the accumulation of fecal material on impervious surfaces between rainfall events (Godfrey, 1993; Jeffferies et al., 1990) and indicator loads increase 0.5 to 2 times compared to dry weather flow (Ashley et al., 1993). Analysis of an urban stormwater database developed by the Center for Watershed Protection (2000) indicates that considerable variability can exist in storm-to-storm values for fecal coliforms, with concentrations spanning up to five orders of magnitude in a single sampling location. Discharge flows may resuspend sediments containing fecal indicator bacteria, with the result that receiving water may experience a tenfold increase in the number of fecal indicator bacteria, compared to low-flow periods (McDonald et al., 1982). Several studies provide evidence that fecal indicator bacteria can survive and even multiply in sediments in various parts of the urban-suburban drainage network (both surface and subsurface), suggesting that this network itself may perpetuate elevated levels of indicator bacteria and may be a major source of bacteria during storms if sediments are resuspended or scoured and flushed into adjacent waterbodies (Burton et al., 1987; Butler et al., 1995; Gannon and Busse, 1989; Marino, and Gannon, 1991; Olivieri et al., 1977; Simmons et al., 2001; Steuer et al., 1997).

Whereas it is assumed that potentially hazardous levels of pathogens are present when bacterial indicator levels exceed surface water quality standards, few studies have sought to identify pathogens or characterize public health risks attributable to surface water exposure following urban stormwater flows. In a study monitoring stormwater in agricultural and urban watersheds of New York City water supply reservoirs, Stern (1996) detected higher levels of *Giardia* and *Cryptosporidium*, as well as higher rates of confirmed viability of these protozoan parasites, in the urban watersheds. Jiang et al. (2001) applied polymerase chain reaction methods for detection of adenoviruses in rivers impacting coastal beaches and found them in 4 of 12 samples at levels of 5.3 to 3,332 plaque forming units (PFU) per liter. Bacterial indicators did not correlate with the presence of viruses, although F-specific coliphage (viruses that infect *E. coli*) were significantly correlated with adenovirus occurrence. Samples for this study were taken from rivers impacted by CSOs, so it was not possible to determine whether the viral pathogens or phages originated in sewage or stormwater. That study suggests that fecal indicator bacteria are not a sufficiently robust indicator to provide public health protection for exposures to recreational surface waters that receive CSOs. Recognizing the paucity of health information related to stormwater and CSO flows, the Water Environment Research Foundation (WERF)[15] recently issued a request for proposals for projects to assess the risk to public health posed by these flows.

Stormwater in Southwestern Pennsylvania

Nationally, urban runoff and storm sewers are the third leading source of impairment for lakes, ponds, and reservoirs and the second leading source of impairment for estuaries (EPA, 2002a). Runoff including urban runoff and storm sewers, road runoff, and small residential runoff is implicated in the impairment of 3,007 stream-miles and 6,797 lake-acres in the Commonwealth of Pennsylvania (PADEP, 2004a). Stormwater and urban runoff is listed as a problem that is expected to increase in many of the WRAS state water plans in the region.

ALCOSAN estimates that combined sewers collect sewage and stormwater from more than 60 square miles of its 204 square mile service area (TPRC, 2002). The balance of the area is served by separate sewers, but with the very high per capita flow rates in these sewers, it is clear that a large percentage of that flow is also stormwater. Furthermore, the breakdown of the service area into separate and combined sewer areas understates the magnitude of the problem. Large portions of several watersheds that contribute stormwater to the ALCOSAN service area

---

[15] WERF is a nonprofit organization that funds and manages water quality research for its subscribers through public-private partnerships between municipal utilities, corporations, academia, industry, and the federal government (see *http://www.werf.org for further information*).

lie outside the area. For example, only about one-third of the Turtle Creek watershed is within the ALCOSAN service area, but all of the stormwater from that 200 square mile drainage area flows through the ALCOSAN service area. Likewise, only about one-fourth of the Chartiers Creek watershed lies within the ALCOSAN service area, but all of the stormwater from that 300+ square mile watershed flows through the service area. During large rainfall events, these watersheds contribute huge volumes of water to the service area, a portion of which enters both the separate and the combined sewer systems. A rainfall that results in 1 inch of runoff delivers 232 million cubic feet of water for every 100 square miles of drainage area, and one-day rainfalls of this magnitude are common in the Pittsburgh region. Much of that drainage will be carried in stream channels, bypassing all sewer systems, but it is obvious that large quantities are entering the collection systems.

ACID MINE DRAINAGE

More than 12,000 miles of rivers and streams in the United States are adversely impacted by drainage from abandoned surface and underground mines. The majority of these streams are located in coal mining regions of the eastern United States, particularly Pennsylvania, West Virginia, and Maryland (Kleinmann et al., 2000). Figure 4-7 shows acid mine drainage streams in the Appalachian region.

Acid mine drainage (AMD) forms when sulfide minerals, particularly pyrite (FeS_2), have been exposed to oxidizing conditions from underground and surface mining and from other excavation activities such as highway construction. In the presence of oxygen and water, sulfide minerals oxidize to form sulfate-rich and often metal-laden drainage (Skousen, 1995). "Untreated AMD flowing into streams can severely degrade both habitat and water quality, often producing an environment devoid of most aquatic life and unfit for desired uses" (Kimmel, 1983). In addition, AMD can be toxic to vegetation and can reduce the potability of water supplies (Earle and Callaghan, 1998).

The PADEP estimates that cleanup of AMD-impacted watersheds in Pennsylvania will cost from $5 billion to $15 billion[16] (PADEP, 2004b).[17] Pennsylvania receives about $25 million annually from the federal Abandoned Mine Land (AML) Reclamation Fund to help address and abate abandoned mine problems, including AMD (OSM, 2003). In the first three years of the Growing Greener initiative,[18] PADEP distributed more than $130 million in more than 800 grants to local organizations and watershed associations to address problems of abandoned mines (see also Chapter 5). In 1982, the Western Pennsylvania Coalition for Abandoned Mine

[16] See *http://pa.water.usgs.gov/projects/amd/ for further information.*

[17] The U.S. Office of Surface Mining initiated a program called AMDTreat in February 2003 that is designed to allow more accurate predictions of the cost of AMD remediation; for further information, see *http://www.tips.osmre.gov/amdtreatPressLink.htm.*

[18] The Growing Greener Program, signed into law in 1999, was established to invest nearly $650 million over a five-year period to preserve farmland and protect open space; eliminate the maintenance backlog in State Parks; clean up abandoned mines and restore watersheds; and provide new and upgraded water and sewer systems. Four different agencies are involved in the program under the Environmental Stewardship and Watershed Protection Act and include PADEP, the Pennsylvania Department of Conservation and Natural Resources, PENNVEST, and the Pennsylvania Department of Agriculture. Further information on the program and the involvement of these agencies can be found at *http://www.dep.state.pa.us/growgreen/.*

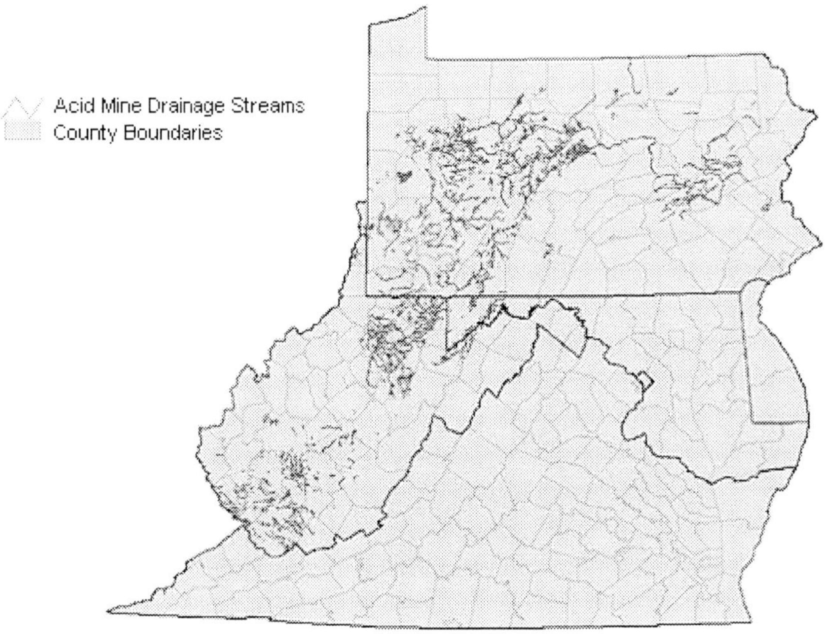

FIGURE 4-7 Acid mine drainage streams in Pennsylvania and West Virginia. SOURCE: PADEP, as cited at *http://www.leo.lehigh.edu/envirosci/enviroissue/amd/links/graphs.html*.

Reclamation (WPCAMR)[19] was formed with the goal of providing leadership for building local watershed-based support. In 1995, the U.S. Office of Surface Mining and EPA Region III signed an agreement establishing the Appalachian Clean Streams Initiative[20] to address AMD water quality problems resulting from abandoned coal mines in Maryland, Ohio, Pennsylvania, and West Virginia. As of 1999, Pennsylvania had received \$7.2 million in distributions under this program and it received just over \$2 million in 2003 (OSM, 2003). Continued funding for these ongoing programs is essential to future water quality improvement in southwestern Pennsylvania.

Characteristics of Acid Mine Drainage in Southwestern Pennsylvania

The contact of pyrite with large volumes of oxygenated water moving into mine voids yields dissolved ferrous iron (Fe^{2+}) and sulfuric acid ($2H^+$ and SO_4^{2-}). The ferrous iron subsequently oxidizes to ferric iron (Fe^{3+}), which precipitates as ferric hydroxide, $Fe(OH)_3$. This is the yellow-orange precipitate seen along AMD sluiceways and the bottoms of streams into which AMD discharges. AMD thus is typically a highly acidic solution bearing a large load of iron, either dissolved or precipitated as ferric hydroxide. AMD contact with rock and soil en route to surface discharge results in the dissolution of other metals such as aluminum, manganese, magnesium, and sodium. Mass concentrations of chemical constituents in AMD can

[19] Additional information on the WPCAMR can be found at *http://www.amrclearinghouse.org/WPCAMR/*.

[20] Additional information on the Clean Streams Program, formerly called the Appalachian Clean Stream Initiative, can be found at *http://www.osmre.gov/acsihome.htm#*. Since its inception, the program has expanded (as of 2003) to include cooperative agreements with 12 states (Alabama, Illinois, Indiana, Iowa, Kentucky, Maryland, Missouri, Ohio, Oklahoma, Pennsylvania, Virginia, and West Virginia).

range from tens to several thousands of milligrams per liter, depending on the local geology and hydrogeology, the flooded state of the mine, and the time since its abandonment (Lambert et al., 2004; Wood et al., 1999; Younger, 2000).

Stiles et al. (2004) characterized water quality types from mine discharges in the Monongahela basin. Data from 1,624 water samples collected from 84 mine discharge sites were compiled and grouped according to chemical, as well as other characteristics of the discharges. This analysis produced four basic clusters. The discharge groupings were governed primarily by three geochemical factors: total dissolved solids (TDS), degree of acid neutralization, and concentrations of metals. Most of the mine discharge sites (84 percent) were classified in a single cluster characterized by variable levels of pH, alkalinity, calcium, aluminum, and chloride and low levels of sodium, magnesium, iron, manganese, and sulfate. The smaller clusters were associated with waters that were high to very high in TDS content. Mine discharges with lower TDS content and positive alkalinity tend to be older discharges from mine voids that are flooded (Lambert et al., 2004; Stiles et al., 2004). Discharges with acidity and high TDS levels tend to be more recently initiated discharges from mine voids that are not completely flooded (Lambert et al., 2004). There are many old abandoned mine discharges in southwestern Pennsylvania, but newer discharges exist as well.

Effects of Acid Mine Drainage

The Monongahela and the Allegheny River basins have been influenced significantly by AMD for many decades. AMD inputs to streams have resulted in no-fishing designations for 1,071 stream-miles in the Monongahela River basin and 1,320 stream-miles in the Allegheny River basin (Sams and Beer, 2000). However, AMD loadings to many streams in the basins have decreased with time. As part of the USGS NAWQA program, the effects of coal mine drainage on stream water quality in the Allegheny and Monongahela River basins were evaluated (Anderson et al., 2000; Sams and Beer, 2000). Historical concentrations of sulfate (SO_4^{2-}), a relatively nonreactive tracer of AMD inputs, were used to evaluate the long-term trends in AMD impacts on streams in the two basins. Sulfate serves as an indicator of total AMD input. It is toxic itself only when present at very high concentrations. Sams and Beer (2000) found that AMD inputs to the Monongahela River have been much greater than the inputs to the Allegheny. In 1980, for example, the annual sulfate loads transported by the Allegheny and Monongahela Rivers to the Ohio River at Pittsburgh were 1.2 million and 1.35 million tons, respectively. The Monongahela River basin, although smaller in overall drainage area (7,340 square miles versus 11,700 square miles for the Allegheny), contributed 53 percent of the sulfate load. Further, the Monongahela River at Braddock exhibited a median sulfate concentration of 110 mg/L from 1965 to 1995. This is almost twice as high as the median sulfate concentration of 60 mg/L for samples collected over the same time period in the Allegheny River at New Kensington. The difference in the inputs to the Monongahela and Allegheny Rivers is explained by the magnitude of the mining operations that have occurred in each. According to Sams and Beer (2000), approximately 6,600 mines have operated in the Monongahela River basin, compared to 2,500 in the Allegheny River basin.

In areas of the Allegheny and Monongahela River basins where coal production is very low or has ceased altogether, AMD inputs to the streams have decreased, in some cases substantially (Sams and Beer, 2000). For example, in the Loyalhanna Creek, which eventually discharges to the Allegheny River, sulfate concentrations have decreased steadily since 1950, as shown in Figure 4-8. The decline in AMD chemical inputs to the Loyalhanna Creek and other surface waters in the basins is attributable to significantly reduced coal production, reclamation of abandoned mine lands since the late 1970s, implementation of AMD treatment at active mining operations and some abandoned mine sites, and decrease in the amount of readily available pyrite in the abandoned mine voids (Lambert et al., 2004; Sams and Beer, 2000). Although historical data indicate that the quality of discharges from many abandoned mines improves with time, this is not the case for all mine discharges. For example, discharges from mines that do not flood completely because of mine geometry and local hydrogeological conditions can remain highly acidic many decades after mine abandonment (Lambert et al., 2004). Moreover, as indicated by sulfate concentration data for Loyalhanna Creek in Figure 4-8, the rate of decrease in AMD constituent concentrations slows with time, such that long periods of time will be required for reduction to levels approaching natural background concentrations (Sams and Beer, 2000).

Regional groundwater quality is also affected by AMD. Another NAWQA study (Anderson et al., 2000; see also Appendix B) sampled 45 domestic water supply wells in the high-sulfur coal region of the Appalachian coal fields. Compared to groundwater in unmined areas of the coal-bearing rocks, water in shallow private domestic wells near reclaimed surface coal mines had higher concentrations of sulfate, iron, and manganese. Table 4-13 summarizes

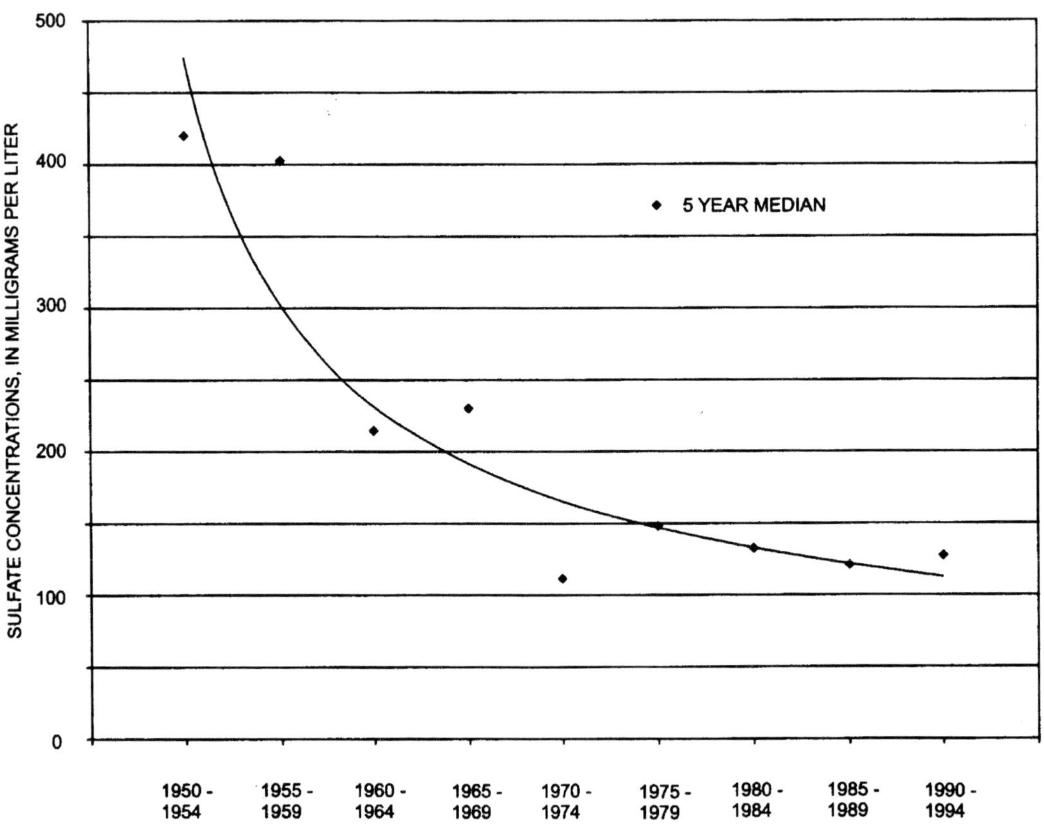

FIGURE 4-8 Median 5-year sulfate concentrations for the Loyalhanna Creek at Loyalhanna Dam, 1950-1995. SOURCE: Sams and Beer, 2000.

TABLE 4-13 Domestic Wells in the Allegheny and Monongahela Watersheds Exceeding Secondary Maximum Contaminant Levels (Data From 1996-1998)

Contaminant	Mined Area (%)	Unmined Area (%)
Sulfate	20	0
Fe	60	20
Mn	70	47
TDS	20	7

SOURCE: Adapted from Anderson et al., 2000.

the percentage of tested wells not meeting secondary maximum contaminant levels (MCLs) in both mined and unmined area wells.[21]

In addition to direct water quality effects due to AMD, mining activities can also affect water quality through changes to the subsurface during mine subsidence. These effects are described in Box 4-5.

AGRICULTURE

Conventional crop production activities entail the application of pesticides and fertilizers to boost crop production. As noted in Chapter 3, these inputs can harm water quality when carried by runoff into surface waters or by leaching into groundwaters. Further, disturbance of soils through tillage and poor livestock management practices can increase erosion, resulting in excess sedimentation of surface waterbodies. Livestock manure applied to crop and pasture lands can be a source of nutrients and pathogens, as can more direct input from livestock with access to streams or from feedlots and loafing lots directly adjacent to streams. In the latter uses, livestock also frequently trample and degrade streamside vegetation, destroying buffers that might otherwise intercept contaminants from the manure. Land application of biosolids on agricultural land is an additional concern, because of potential health risks and water quality. In the steep terrain of the region, erosion and sedimentation from agricultural activities can be considerable, and near-stream livestock activity can generate a combination of sediment and manure entering a stream, with potential for resuspension and possible regrowth of bacteria in the sediments. Accordingly, it is significant that agriculture is the largest land use in the region after forest land (see Figure 4-9).

Land in farms in the study region is presented in Table 4-14. Washington and Somerset Counties are the dominant agricultural counties with 261,139 and 223,323 acres, respectively (see Table 4-14). About 60 percent of the land in farms is cropland, with the remainder allocated to pasture, forest, and other uses. Primary agricultural production in the study region consists of sheep, dairy, hay, and truck farm crops. According to the 1997 U.S. Census of Agriculture, four counties in the region (Beaver, Indiana, Washington, and Westmoreland) rank among the top 100 in the United States in direct sales of farm products to consumers (USDA, 1997).

[21] MCLs are the maximum allowable amounts of contaminants in drinking water and are set by the EPA through the National Primary Drinking Water Regulations. Nonmandatory secondary MCLs are also set for 15 contaminants that affect the aesthetics (taste, smell) of drinking water; see *http://www.epa.gov/safewater/mcl.html* for further information.

BOX 4-5
Longwall Mining in Southwestern Pennsylvania

Coal mining has helped to shape both southwestern Pennsylvania's history and its current landscape. For many years, underground mining in this area used a method known as room-and-pillar, in which about 50 percent of the coal seam is left in place as "pillars" for roof support. In Washington and Greene Counties and the adjacent parts of Ohio and West Virginia, modern, active, high-extraction mining removes coal by the longwall method, in which 100 percent of the coal is removed within a large block, or "panel" of coal, using a longwall mining machine. Panels are typically 800 to 1,500 feet wide and several thousand feet to several miles long. Special hydraulic devices that support the roof are advanced as the mining machine progresses. As the coal is mined and the machine moves forward, supports are removed, and the overlying rocks and ground surface subside into the void in the "wake" of the mining. Longwall mining, which became common in Pennsylvania in the early 1980s, operates almost exclusively in the Pittsburgh coal seam, at depths of 300 to 800 feet below the surface. Because this method of mining is faster, requires fewer employees, and extracts virtually all the coal, it is more profitable than the older room-and-pillar technique. In the area's two coalfield counties, about 20 percent of the land lies above longwall mines, and 60 percent or more may eventually be undermined in the next half century (Hopey, 2003a,b; PADEP, 1999).

Although longwall mining subsidence received considerable attention recently in a series of articles in the *Pittsburgh Post-Gazette* relating the impacts on the area's historic structures, subsidence has affected local structures and natural features since the 1980s and has been the focus of numerous studies (Kern et al., 2002; Kohli, 2002). Subsidence can cause damage to structures, ranging from small cracks in plaster to sinking, rending, or buckling of foundation support walls and footings. Hydrologic effects on both surface and groundwater are also common and may range from small changes in streamflow or water well levels to profound impacts such as stream diversion and dewatering of aquifers and well supplies (Booth, 1984, 1990; Carver and Rauch, 1994). In some cases, changes in surface slope due to subsidence can disrupt flow in streams, ditches, canals, and water and sewer lines. While some impacts are short term and some may even be positive (increased well yield in some cases), damage from subsidence is commonly long-term or permanent (Rauch, 1989).

Final subsidence troughs, at the completion of mining, are roughly elliptical in shape and have a surface extent larger than the area of coal extraction. Ground movements have both horizontal and vertical components, with relative magnitude depending on the location within the subsidence trough. An area of compressive strain develops over the central part of the collapse, and zones of tensile strain make up the surrounding collapse structure and usually extend beyond the collapsed area within the mine. Characteristics and effects of the subsidence in any single longwall operation depend on many factors: topographic and hydrologic setting, hydrologic characteristics of the rock units, presence of existing rock fractures, depth of the mine, relative width and length of the panel, thickness and type of the overlying rock and soil, and thickness of the mined coal.

Although subsidence is also found in areas of room-and-pillar mining, the effects of longwall mine subsidence are more immediate, and they follow a dynamic succession as the longwall panel advances underground. As the active face of the underground mining approaches the subsurface beneath a structure or natural feature, the object may at first lie in the zone of tensile strain, As mining advances beneath the object, it may then experience compression forces, and as the longwall

As noted previously, agriculture is listed as a cause of impairment for a significant portion of waters in southwestern Pennsylvania (PADEP, 2004a). The listings reflect the aforementioned multiple effects of agriculture on receiving waters.

The impact of agricultural activities on water quality depends on the types of agricultural land use (e.g., specific crop types) and the specific agricultural practices utilized. In some cases, face moves beyond the structure, tensile forces may again be in effect. This "dynamic subsidence" is the changing of the ground surface as the longwall passes through the area. As a result of these changes, cracks in structures, pavement, or aquifers may open, close, and even open again (PADEP, 1999). Following the dynamic process, there may be slow, long-term surface movement (Luo and Peng, 2000).

Observed effects on structures include buckling of foundations, tilting of walls and supports out of plumb, deformation of door and window frames, cracks in walls and floors, separation of building components, and shifting of foundation walls and floors. On a field tour of Greene County in December 2002, the committee observed cracking and shifting of the foundation of the County Animal Shelter, which had been evacuated and abandoned, and the uneven surface and numerous repairs in the pavement of an immediately adjacent highway.

According to Rauch (1989), deep aquifers and subsurface water supplies are partially to totally dewatered in the zones above subsided deep mines and typically have no short-term recovery. Water levels in deep wells following mining are generally below pre-mining levels. Wells that are shallower in the subsidence profile may suffer only partial and temporary dewatering.

Subsidence can produce dramatic changes in surface waters, causing shifting of course and ponding (Peng et al., 1994). Carver and Rauch (1994) observed dewatering of streams, reduced discharges, and changes in baseflow conditions, with altered baseflow in recovered streams compared to unaltered streams. In a study in north central West Virginia, Cifelli and Rauch (1986) determined that baseflow streams were significantly affected where at least 10 percent of the watershed had subsided and dried up entirely where at least 25 to 30 percent of the watershed was undermined and subsided.

Stout (2003) compared diversity, longevity, and functional organization of benthic communities in first- and second-order headwater streams in longwall undermined areas and in reference streams in southwestern Pennsylvania and northern West Virginia. Of four undermined streams in southwestern Pennsylvania, one stream was completely dry, two were dewatered in mid-reaches, and one stream was apparently unaffected; however, detailed examination of the latter stream revealed that 52 percent of the total length of the headwater stream network in the associated watershed had been lost due to subsidence. Benthic communities in longwall-undermined streams were significantly different from benthic communities of reference streams and retained approximately 50 percent taxa richness.

Pennsylvania's Act 54, the Bituminous Mine Subsidence and Land Conservation Act, requires mine operators to provide water supply replacement and subsidence damage repair to affected properties or to provide compensation as stated in the act. Recent controversies surrounding Act 54 and historic structures have involved a 1994 amendment allowing mining beneath structures built before 1966, as long as the property owner is compensated for structural damage and water loss. Prior to this amendment, coal operators were required to leave pillars of coal in place to support the structures. Differences exist between state law and federal historic preservation requirements. Several state legislators had proposed amendments tightening protections for historic buildings, water supplies, streams, and farms (Hopey, 2003a). A U.S. Office of Surface Mining review conducted in 2001 found that state law and regulations failed to adequately protect water supplies, homes, and surface properties damaged by longwall mining. In response, the state proposed regulatory changes and presented them for public hearings and comments in October 2003. Final adoption of the changes is expected in fall 2004 (Hopey, 2003c).

such as orchard and tobacco production, former agricultural activities may also be of concern for potential present-day soil and water quality impacts due to intensive use of pesticides having persistent residues. Other important factors include the locations of agricultural activities in relation to streams and recharge areas and the soil features and topography that affect runoff, leaching, and run-in. Box 4-6 describes agricultural water quality problems in the Dunkard Creek watershed on the Pennsylvania-West Virginia border in Greene County. These water quality issues may be of concern in other areas of the study region where livestock are a major

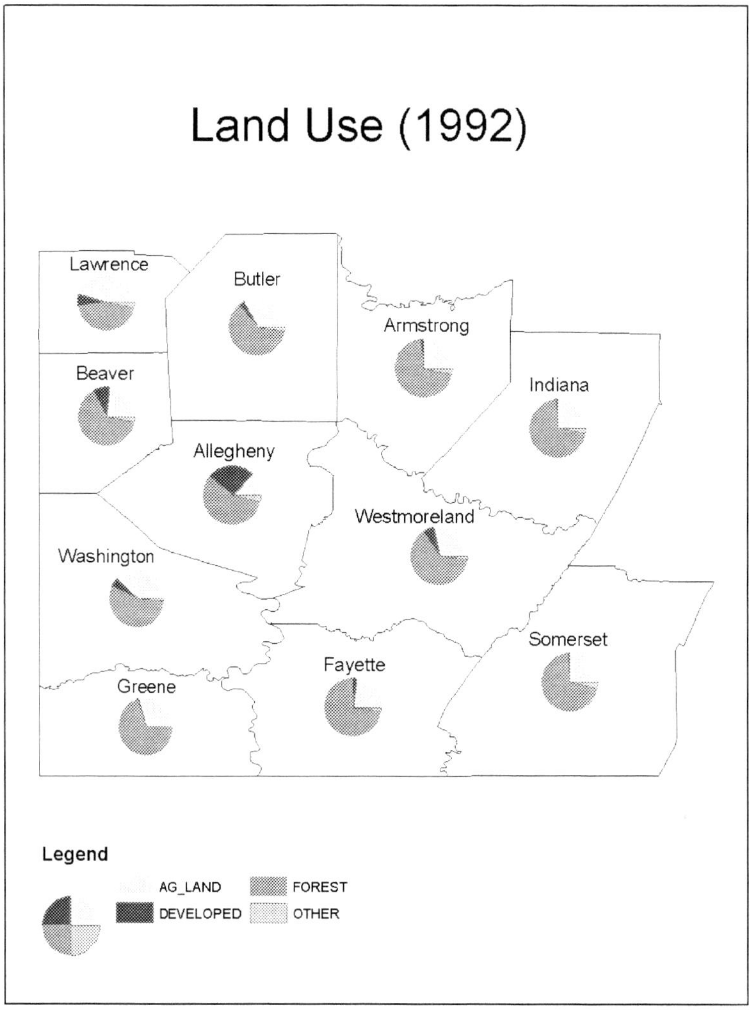

FIGURE 4-9 Major land uses in southwestern Pennsylvania as of 1992. SOURCE: Generated from 1992 USGS National Land Cover Data. Available on-line at *http://landcover.usgs.gov/natllandcover.asp.*

TABLE 4-14 Farm Acreage in Southwestern Pennsylvania and Percentage Used as Cropland

County	Land in Farms (acres)	% Cropland
Allegheny	33,788	56.79
Armstrong	130,637	63.23
Beaver	62,801	61.52
Butler	143,985	66.91
Fayette	125,034	56.63
Greene	141,684	52.88
Indiana	157,286	57.88
Lawrence	88,987	66.34
Somerset	223,323	60.42
Washington	261,139	54.89
Westmoreland	150,967	68.60

SOURCE: Adapted from USDA, 2002.

part of agricultural production. Depending on the water quality problem (e.g., sediment, pathogens, pesticides), changes in agricultural land use and/or adoption of best management practices (BMPs) for the specific problem can reduce pollution loads from agricultural sources.

Few studies have characterized the public health impact of agricultural runoff. An outbreak of *Escherichia coli* O157:H7 and *Campylobacter jejeuni* in May and June 2000 in the rural town of Walkerton, Ontario, resulted in six deaths among the 1,346 reported cases and 65 hospitalizations. Laboratory and field investigations by the Bruce-Grey-Owen Sound Health Unit, Health Canada, and the Ontario Ministry of Health and Long-Term Care determined that the likely source of the outbreak was manure from adjacent farms and that surface water contamination from the farms entered a municipal drinking water well during a period of heavy rainfall and flooding (Clark et al., 2003; Health Canada, 2000). Notably, extreme precipitation events preceded 51 percent of waterborne disease outbreaks occurring between 1948 and 1994 (see discussion of Curriero et al., 2001, in Box 4-2). The potential for *Cryptosporidium* to contaminate surface water supplies is illustrated by the fact that agricultural runoff may contain oocysts of this parasite (Madore et al., 1987; Ongerth and Stibbs, 1987; Rose et al., 1988), and aperiod of wet weather preceded the well-known 1993 cryptosporidiosis outbreak in Milwaukee, Wisconsin.

Land Application of Biosolids

Biosolids—wastewater solids that have been specially treated to produce fertilizers or soil amendments—are used in agriculture, landscaping, and mine reclamation to promote plant growth and soil regeneration. In Pennsylvania, municipal wastewater treatment plants produce an estimated 400,000 dry tons of solids per year, and OSTDSs an estimated 605 million gallons of residential septage. Biosolids application is a means of recycling this material beneficially, to supply nutrients (N and P), micronutrients (trace metals), and organic amendments to soil and plants.

Biosolids contain several pollutants that can potentially affect water resources, including pathogens, nutrients, and metals. For this reason, federal and state regulations address treatment and quality of biosolids, site criteria, application procedures and rates, and cumulative loading rates (see NRC, 2002). Although the committee is unaware of any data demonstrating that land application of biosolids is adversely affecting regional water quality, this is a relatively common practice on southwest Pennsylvania's agricultural and reclamation lands. Although the Region V office of PADEP maintains records of biosolids generation and application, these records were not readily available.

The PADEP regulates the treatment facilities and the land application sites, the latter with assistance from many of the county soil conservation districts. Regulations established in 25 PA Code § 271 follow the federal 503 biosolids regulations, with more stringent site requirements for application sites. All land-applied biosolids must meet criteria for "exceptional quality" or "non-exceptional quality." These quality criteria address metals concentrations, pathogen reduction, and vector attraction reduction, with non-exceptional quality criteria being less stringent than exceptional quality. Biosolids that do not meet the regulatory criteria must be incinerated or landfilled (PADEP, 2001).

Treatment processes for biosolids can include digestion, lime stabilization, pasteurization, and composting to reduce odors, pathogens, and vector attraction. Facilities that produce

BOX 4-6
Dunkard Creek Watershed: A Snapshot of Agricultural and Rural Land Practices

Dunkard Creek watershed, a tributary of the Monongahela River, straddles the Pennsylvania-West Virginia state boundary and lies within Greene County, Pennsylvania, and Monongalia County, West Virginia. The main stem of the creek originates approximately 2.5 miles west of Blacksville, West Virginia, at the confluence of the Pennsylvania Fork and the West Virginia Fork of Dunkard Creek. From there, it flows 17 channel-miles eastward, crossing the state boundary several times and discharging to the Monongahela north of Point Marion, Pennsylvania. The watershed acreage is almost evenly divided between the two states, with 71,350 acres in West Virginia and 75,000 acres in Pennsylvania.

Population in the watershed is estimated at 18,000. Blacksville, West Virginia, and Mt. Morris and Bobtown, Pennsylvania, are the major towns, although none have more than about 200 residents. Coal mining has dominated the watershed's economy, and is responsible for AMD-impaired stream listings in both states, as well as other water quality problems in the watershed. Although diverse and stable employment opportunities in nearby Morgantown, West Virginia, have tempered the effects of downturns in mining and other rural employment, U.S. Census statistics and U.S. Department of Agriculture programs for limited resource farmers indicate that limited incomes and poverty are common in the watershed area. Water quality efforts in the watershed have involved formal interstate cooperation, and collaborative citizens' watershed groups, the Dunkard Creek Watershed Association (West Virginia) and the East Dunkard Watershed Association (Pennsylvania), are very active in education, remediation, and grant-seeking. During the past several years, these groups have won several awards and more than $1 million in grants, including more than $900,000 for treating AMD and a large EPA Watershed Initiative Grant.

A 1999 watershed assessment of the upper watershed by the West Virginia Conservation Partnership (1999) and a Rivers Conservation Plan (Greene County Conservation District, 2000) by the Greene County, Pennsylvania, Soil Conservation District reviewed water quality problems affecting the watershed. Woodlands (60 percent) and grassland (20 percent) are dominant land uses, particularly in the upper West Virginia watershed. Although agricultural land use and water quality problems are found predominantly in this part of the watershed, similar problems have been observed where agriculture is prevalent in the lower watershed. These problems may also be typical of some of the other agricultural lands in the 11-county study area, especially those where livestock agriculture is predominant.

Many of the steeply sloping farmlands in the watershed are used as pasture. Pasture and grasslands within the watershed have an average slope of 25-30 percent, even though 20 percent of the floodplain areas exist as grassland. Approximately 75 percent of the grassland is eroding at a rate of T or greater, where T is the soil erosion tolerance factor, or the maximum amount of erosion the soil can sustain and still serve as a medium for plant growth, retain water and plant nutrients, and allow the entry of air and water while protecting the underlying soil from erosion (NRCS, 1999). The soil erosion factor in this area of West Virginia and Pennsylvania is approximately 3 tons per acre (Jeffrey Skousen, West Virginia University, personal communication, 2004). Common management of pasture and grassland in the watershed is continuous grazing and mechanical brush control, with application of lime and fertilizer every 6-10 years. Grassland areas with horses exhibit a lower

biosolids are required to obtain a permit that specifies quality criteria, management practices, site restrictions, monitoring, and reporting. Biosolids generators and land appliers must also receive approved training.

Land application sites are regulated under a general permitting system based on biosolids quality, with more stringent site restrictions for application of non-exceptional quality biosolids. Regulations mandate management practices and restrictions for application sites, depending on percentage of ground cover due to equine habits of "spot grazing." These areas are very vulnerable to higher erosion rates.

More than half of the area experiences problems with increased sediment and nutrient loads in streams due to winter feeding locations and lack of improved animal watering facilities. A common practice of many agricultural land users is to feed and/or confine cattle to low-lying areas in winter, in close proximity to the farmsteads, which commonly exist near small streams. Most animal waste is stacked nearby or spread on fields throughout the year. There are no large confined animal feeding operations in the watershed, and approximately 90 percent of the cattle operations are part-time farmers with small beef cattle farms. Due to lack of improved watering facilities, cattle commonly have direct access to streams, and in many areas, livestock have denuded streamside vegetation. Cropland is a small portion of the total agricultural acres, with corn as the major crop and small acreages of oats, soybeans, and tobacco. Most cropland is 8-15 percent slope with soils of moderate fertility, and manures and fertilizers are applied at estimated rates. Management is generally crop rotation, both conventional and no-till.

Limited water quality sampling by both states during 1996-1997 earned surface water quality designations of "moderate" by West Virginia and "good" by Pennsylvania. Aside from mining impacts, sampling results indicated that the main stem of Dunkard Creek in West Virginia periodically failed to meet the state's fecal coliform standards. Recent upgrades and installations of sewage treatment plants in Bobtown, Mt. Morris, and Blacksville have eliminated many raw sewage problems. However, there have not been subsequent detailed investigations to determine the relative contributions of problematic livestock practices and faulty septic systems or straight piping of household wastewater.

Along with acid mine drainage, stream bank erosion and sedimentation are major concerns among local watershed association members and other citizens. Erosion and sedimentation result from a combination of natural factors (meandering streams, highly erodable soils, steep terrain), agricultural practices, timber management practices on small-scale logging operations, mining, and dirt and gravel roads. Average annual stream bank erosion was estimated to be 24,000 tons per year on the main stem in West Virginia and an additional 20,000 tons on the tributaries.

Since 1976, the Pennsylvania Fish and Boat Commission has extensively sampled the fish populations of Dunkard Creek and had documented the stream as one of the highest-density smallmouth bass populations in a warm-water stream in southwestern Pennsylvania. In the stream's lower reaches, acid mine drainage and metals have severe adverse impacts on the fish and macroinvertebrate populations (including a widespread fish kill in 1998), but even in areas not severely effected by AMD, the local fishery has experienced problems. Declines in the quality of this fishery in the late 1980s to early 1990s were attributed to overfishing, and catch-and-release regulations were implemented in 1995 on a portion of the creek. Studies at a catch-and-release site and a control site with standard regulations indicated trends of continued decline from 1984-2000 at both sites. Researchers suspected habitat and/or water quality deterioration and determined that erosion and sedimentation have deleterious impacts on the fish population (58 PA Code § 65.24).

As part of the watershed corrective measures in the Pennsylvania portion of the watershed, the Greene County Conservation District provides technical assistance and cost share funds for agricultural BMPs such as stream bank fencing, stabilized crossings, riparian buffers, rotational grazing, off-stream livestock watering facilities, and barnyard or feedlot area improvements (see *http://www.county. greenepa.net/secured/gc/depts/pd/conserv/prog.htm* for further information).

the particular biosolids being applied. Application sites must have an implemented farm conservation plan or erosion-sedimentation control plan. Application rates are established by agronomic rates of crops, and cumulative metals loading at the application site is subject to regulatory limits. Soils must be sampled and analyzed for metals concentrations prior to application of non-exceptional quality biosolids at a site (exceptional quality biosolids have sufficiently low metals concentrations that cumulative loading rates do not apply). Regulations specify injection and incorporation requirements and site restrictions for food, feed, and fiber crops, as well as turf, animals, and public access (PADEP, 2001; Stehouwer, 1999a).

Research conducted by Pennsylvania State University agronomists found that between 1978 and 1999, biosolids quality, in terms of metals concentrations, greatly improved and that more than 95 percent of Pennsylvania biosolids have metals concentrations well below the state's most stringent regulatory limits (Stehouwer, 1999b). Because the composition of individual biosolids will vary from values examined in their assessment, that report recommended continuing to monitor cumulative loading whenever biosolids are applied to soil. According to the PADEP Region V biosolids coordinator, some sites in the region have had 20 years of biosolids application with no increases in regulated metals (William Graham, PADEP Region V, personal communication, 2004).

Pennsylvania State University agronomists conducted a three-year assessment of the effects of biosolids utilization on soil and crop quality. The assessment, involving 20 farms in 18 counties, compared land application agricultural sites with control sites having similar soils, crops, and management practices. A site in Beaver County, which received a total biosolids application of 45.3 tons per acre between 1984 and 2001, was the only southwestern Pennsylvania location. Statewide results showed that crop yields, nutrient contents, and trace element contents were similar in the biosolids and control plots, indicating no adverse effects of biosolids on crop quality and no increase of trace elements entering the food chain. At the end of the growing season, soils in the biosolids fields had higher nitrate levels, leading researchers to suggest that winter cover crops be planted to take up excess nitrogen and reduce the potential for nitrate leaching or runoff. Similar to animal manures, repeated application of biosolids led to increased soil phosphorus levels, and researchers recommended that the environmental significance of phosphorus be examined as the state moves to phosphorus-based nutrient management (Stehouwer, 2003).

For reclamation use at active permitted mining sites, the PADEP District Mining Office approves actual land application of biosolids. Biosolids used must come from a publicly owned wastewater treatment facility meeting regulatory biosolids criteria and having a permit issued by the Biosolids Division of the regional office.

SUMMARY

The relative contribution of different sources to microbial loading to surface and groundwaters in southwestern Pennsylvania cannot be determined with available information. However, sufficient information is available to determine that improperly managed wastewaters resulting from human activities are degrading the microbiological water quality in the region. Wet weather biological water quality in the main stem rivers is demonstrably worse than dry weather biological water quality, suggesting that stormwater and sewer overflows may be important contributors. Biological water quality in many tributaries does not meet standards in either wet or dry weather, suggesting that failing on-site treatment and disposal units may be important contributors. The contribution of agriculture to pathogen loading in rural areas of southwestern Pennsylvania could not be determined, but this is a well-known pathogen source in other regions, and many livestock management practices in the study region are likely to contribute pathogens to streams. Relative nonpoint contributions of human and nonhuman pathogen sources in both urban and rural watershed are not known.

Acid mine drainage is a significant cause of water quality impairment in the region, predominately affecting streams and tributaries. This regional water quality issue extends beyond Pennsylvania to encompass much of the Appalachian Range. Presently, this problem is being addressed by multiple jurisdictions including federal and state programs. The continuing flow of financial support to combat this water pollution problem is essential to future environmental water quality improvement.

REFERENCES

ACHD (Allegheny County Health Department). 2004. ACHD/GSPH Study of River Water Quality and Human Health; Ongoing study of rowers in southwestern Pennsylvania. Pittsburgh, PA: ACHD.

Alderiso, K., D. Wait, and M. Sobsey. 1996. Detection and characterization of male-specific RNA coliphages in a New York City reservoir to distinguish between human and non-human sources of contamination. In Proceedings of a Symposium on New York City Water Supply Studies, TPS-96-2. Herndon, VA: American Water Resources Association.

Anderson, D., R. Otis, J. McNeillie, and R. Apfel. 1994. In-situ lysimeter investigation of pollutant attenuation in the vadose zone of a fine sand. In On-Site Wastewater Treatment: Proceedings of the Seventh International Symposium on Individual and Small Community Sewage Systems. St. Joseph, MI: American Society of Agricultural Engineers.

Anderson, R., K. Beer, T. Buckwalter, M. Clark, S. McAuley, J. Sams, and D. Williams. 2000. Water Quality in the Allegheny and Monongahela River Basins: Pennsylvania, West Virginia, New York, and Maryland (1996-98). Denver, CO: U.S. Geological Survey.

Ashley, R., D. Wotherspoon, B. Coghlan, and E. Ristenpart. 1993. Cohesive sediment erosion in combined sewers. In Procedures of the 6th International Conference of Urban Storm Drainage, J. Marsalek and H. Torno (eds.). Victoria, Canada: Seapoint Publishing.

Booth, C. 1984. The hydrogeological impact of deep longwall mining: Appalachian Plateau, Pennsylvania. In Proceedings of the National Water Well Association Conference on the Impact of Mining on Ground Water. Dublin, OH: National Water Well Association.

Booth, C. 1990. Hydrogeological significance of subsurface coal mining. In Water Resources in Pennsylvania: Availability, Quality, and Management, S. Majumdar, E. Miller, and R. Parizek (eds.). Harrisburg, PA: Pennsylvania Academy of Science.

Borchardt, M., P. Bertz, S. Spencer, and D. Battigelli. 2003. Incidence of enteric viruses in groundwater from household wells in Wisconsin. Applied and Environmental Microbiology 69(2):1172-1180.

Burm, R., and R. Vaughan. 1966. Bacteriological comparison between combined and separate sewer discharges in southeastern Michigan. Journal of the Water Pollution Control Federation. 38:400-409.

Burton, A., D. Gunnison, and G. Lanza. 1987. Survival of pathogenic bacteria in various freshwater sediments. Applied and Environmental Microbiology 53(4):633-638.

Butler, D., Y. Xiao, S. Karunaratne, and S. Thedchanamoorthy. 1995. The gully pot as a physical, chemical, and biological reactor. Water Science and Technology 31(7):219-228.

Caccio, S., M. Giocomo, F. Aulicino, and E. Pozio. 2003. *Giardia* cysts in water treatment plants in Italy. Applied and Environmental Microbiology 69(6):3393-3398.

Carver, L., and H. Rauch. 1994. Hydrogeologic effects of subsidence at a longwall mine in the Pittsburgh Coal Seam. In Proceedings of the 13th Conference on Ground Control in Mining. Morgantown, WV: West Virginia University, Department of Mining Engineering.

Center for Watershed Protection. 2000. Microbes and urban watersheds: Concentrations, sources, and pathways. The Practice of Watershed Protection 3(1):554-565.

Chambers, P., M. Allard, S. Walker, J. Marsalek, J. Lawrence, M. Servos, J. Busnarda, K. Munger, K. Adare, C. Jefferson, R. Kent, and M. Wong. 1997. Impacts of municipal wastewater effluents on Canadian waters: A review. Water Quality Research Journal of Canada 32(4):659-713.

Chauret, C., N. Armstrong, J. Fisher, R. Sharma, S. Springthorpe, and S. Sattar. 1995. Correlating *Cryptosporidium* and *Giardia* with microbial indicators. Journal of the American Water Works Association 87(11):76-84.

Cifelli, R., and H. Rauch. 1986. Dewatering effects from selected underground coal mines in north-central West Virginia. In Proceedings of the 2nd Workshop on Surface Subsidence Due to Underground Mining. Morgantown, WV: West Virginia University.

Clark C., L. Price, R. Ahmed, D. Woodward, P. Melito, F. Rodgers, F. Jamieson, B. Ciebin, A. Li, and A. Ellis. 2003. Characterization of waterborne outbreak-associated *Campylobacter jejuni*. Emerging Infectious Diseases. Available on-line at *http://www.cdc.gov/ncidod/EID/vol9no10/02-0584.htm*. Accessed June 21, 2004.

Curriero, F., J. Patz, J. Rose, and S. Lele. 2001. The association between extreme precipitation and waterborne disease outbreaks in the United States, 1948-1994. American Journal of Public Health 91(8):1194-1199.

Desvousges, W., V. Smith, and A. Fisher. 1987. Option price estimates for water quality improvements: A contingent valuation study for the Monongahela River. Journal of Environmental Economics and Management 14:248-67.

DOC (United States Department of Commerce) and HUD (U.S. Department of Housing and Urban Development). 1997. American Housing Survey for the Pittsburgh Metropolitan Area in 1995. Current Housing Reports H170/95-13. Washington, DC: DOC Bureau of the Census and HUD Office of Policy Development and Research.

Earle, J., and T. Callaghan. 1998. Impacts of mine drainage on aquatic life, water uses and man-made structures. In Coal Mine Drainage Prediction and Pollution Prevention in Pennsylvania. Harrisburg, PA: Pennsylvania Department of Environmental Protection.

Edwards, D. 1993. Troubled waters in Milwaukee. ASM News 59(7):342-345.

Ellis, J. and Y. Wang. 1995. Bacteriology of urban runoff: The combined sewer as a bacterial reactor and generator. Water Science and Technology 31(7):303-310.

Environment Canada. 1999. Summary and Update of the 1997 Science Assessment of the Impacts of Municipal Wastewater Effluents (MWWE) on Canadian Waters and Human Health. Available on-line at *http://www.ec.gc.ca/etad/default.asp?lang=En&n=36A7F162-11*. Accessed April 27, 2004.

EPA (U.S. Environmental Protection Agency). 1980. Design Manual: Onsite Wastewater Treatment and Disposal System. EPA/625/1-80/012. Cincinnati, OH: Office of Research and Development and Office of Water.

EPA. 1993. Guidance Specifying Management Measures for Sources of Nonpoint Pollution in Coastal Waters. Washington, DC: Office of Water.

EPA 1997. Response to Congress on Use of a Decentralized Wastewater Treatment System. EPA 832-R-97-001b. Available on-line at *http://www.epa.gov/owm/mtb/decent/response/*. Accessed June 21, 2004.

EPA. 2000. Draft EPA Guidelines for Management of Onsite/Decentralized Wastewater Systems. Federal Register 65(195): 59840-59841. Washington, DC: Office of Wastewater Management.

EPA. 2001a. EPA's Report to Congress on Implementation and Enforcement of the CSO Control Policy. EPA 833-R-01-003. Available on-line at *http://cfpub1.epa.gov/npdes/cso/cpolicy_report.cfm?program_id=5*. Accessed June 21, 2004.

EPA. 2001b. Proposed Rule to Protect Communities from Overflowing Sewers. EPA-833-01-F-001. Washington, DC: Office of Water.

EPA. 2002a. Report to Congress on the Impacts and Control of Combined Sewer Overflows and Sanitary Sewer Overflows; Availability of Public Health Experts Workshop Summary. EPA 833-R-02-002. Federal Register 67(248):78802. Available on-line at *http://www.epa.gov/fedrgstr/EPA-WATER/2002/December/Day-26/w32566.htm*. Accessed June 21, 2004.

EPA. 2002b. Onsite Wastewater Treatment Systems Manual. EPA-625-R-00-008. Washington, DC: EPA Office of Water and Office of Research and Development.

EPA. 2002c. National Water Quality Inventory: 2000 Report. EPA-841-R-02-001. Washington, DC: Office of Water.

EPA. 2003a. Fact Sheet: National Listing of Fish and Wildlife Advisories. EPA 823-F-03-003. Available on-line at *http://www.epa.gov/waterscience/fish/advisories/factsheet.pdf*. Accessed April 26, 2004.

EPA. 2003b. Voluntary National Guidelines for Management of Onsite and Clustered Wastewater Treatment Systems. EPA-832-B-03-001. Washington, DC: Office of Water.

Fleeger, G. 1999. The Geology of Pennsylvania's Groundwater, 3rd Edition. Pennsylvania Geological Survey Educational Series 3. Available on-line at *http://www.dcnr.state.pa.us/topogeo/education/es3.pdf*. Accessed March 31, 2004.

Gannon, J., and M. Busse. 1989. *E. coli* and enterococci levels in urban stormwater, river water, and chlorinated treatment plant effluent. Water Resources 23(9):1167-1176.

Gibson, C., K. Stadterman, S. States, and J. Sykora. 1998. Combined sewer overflows: A source of *Cryptosporidium* and *Giardia*. Water Science and Technology 38(12):67-72.

Godfrey, A. 1993. Sources and fate of microbial contaminants. In Recreational Water Quality Management, D. Kay and R. Hanbury (eds.). London: Ellis Horwood.

Greene County Conservation District. 2000. Rivers Conservation Plan for the Dunkard Creek Watershed. Available on-line at *http://www.dcnr.state.pa.us/brc/rivers/riversconservation/registry/18fullplan.pdf*. Accessed November 30, 2004.

Hamilton, J. 2001. Man uses GPS to map dog doo. Associated Press, Boulder, CO. November 29.

Health Canada. 2000. Waterborne outbreak of gastroenteritis associated with a contaminated municipal water supply, Walkerton, Ontario, May-June 2000. Canada Communicable Disease Report. Available on-line *at http://www.hc-sc.gc.ca/pphb-dgspsp/publicat/ccdr-rmtc/00vol26/dr2620eb.html*. Accessed June 21, 2004.

Herwaldt, B., G. Craun, S. Stokes, and D. Juranek. 1991. Surveillance summaries: Waterborne-disease outbreaks, 1989-1990. Morbidity and Mortality Weekly Report 40(SS-3):1-21.

Hopey, D. 2003a. How longwall mining works. Pittsburgh Post-Gazette, November 23.

Hopey, D. 2003b. Sinking history: Longwall mines put holes in the past. Pittsburgh Post-Gazette, November 23.

Hopey, D. 2003c. Some legislators hope to toughen mining law. Pittsburgh Post-Gazette, November 25.

Jefferies, C., K. Young, and I. McGregor. 1990. Microbial aspects of sewage and sewage sludge in Dundee. Water Science and Technology 22(10-11):47-52.

Jiang, S., R. Noble, and C. Weiping. 2001. Human adenoviruses and coliphages in urban runoff-impacted coastal waters of Southern California. Applied and Environmental Microbiology 67(1):179-184.

Kern, J., D. Falkenstern, and R. Stingelin. 2002. Effects of Longwall Mining on Real Property Value and the Tax Base of Greene and Washington Counties, Pennsylvania. Harrisburg, PA: PADEP, Bureau of Mining and Reclamation.

Kimmel, W. 1983. The impact of acid mine drainage on the stream ecosystem. In Pennsylvania Coal: Resources, Technology and Utilization, S. Majundar and W. Miller, (eds.). Harrisburg, PA: Pennsylvania Academy of Science.

Kleinman, R., R. Hornberger, B. Leavitt, and D. Hyman. 2000. Introduction and recommendations. In Prediction of Water Quality at Surface Coal Mines. Morgantown, WV: West Virginia University, National Mine Land Reclamation Center.

Knowles, G. 1998. SepticStats: An Overview. Available on-line at *http://www.nesc.wvu.edu/images/SepticStat.pdf*. Accessed March 23, 2004.

Kohli, K. 2002. Mitigation measures to minimize subsidence damages caused by longwall mining to historic structures and their effectiveness. SME Paper No 02-052. Littleton, CO: Society for Mining, Metallurgy, and Exploration.

Lambert, D., K. McDonough, and D. Dzombak. 2004. Long-term changes in quality of discharge water from abandoned underground coal mines in Uniontown Syncline, Fayette County, PA, USA. Water Research 38:277-288.

Larsen, T., K. Broch, and M. Anderson. 1998. First flush effects in an urban catchment area in Aalborg. Water Science and Technology 37(1):251-257.

Lee, S., D. Levy, G. Craun, M. Beach, and R. Calderon. 2002. Surveillance for waterborne-disease outbreaks—United States, 1999-2000. Morbidity and Mortality Weekly Report 51(SS-8):1-47.

Lim, S., and V. Olivieri. 1982. Jones Falls Urban Runoff Project: Sources of Microorganisms in Urban Runoff. Baltimore, MD: Johns Hopkins School of Public Health & Hygiene.

Luo, Y., and S. Peng. 2000. Long-term subsidence associated with longwall mining—Its causes, development, and magnitude. Presented at Society of Mining Engineering Annual Meeting, Denver, CO, 1999.

MacKenzie, W., N. Hoxie, M. Proctor, M. Gradus, K. Blair, D. Peterson, J. Kazmierczak, D. Addiss, K. Fox, J. Rose, and J. Davis. 1994. A massive outbreak in Milwaukee of *Cryptosporidium* infection transmitted through the public water supply. New England Journal of Medicine 331(3):161-167.

Madore, M., J. Rose, C. Gerba, M. Arrowood, and C. Sterling. 1987. Occurrence of *Cryptosporidium* oocysts in sewage effluents and select surface waters. Journal Parasitology 73:702-705.

Makepeace, D., D. Smith, and S. Stanley. 1995. Urban stormwater quality: Summary of contaminant data. Critical Reviews in Environmental Science and Technology 25:93-139.

Marino, R., and J. Gannon. 1991. Survival of fecal coliforms and fecal Streptococci in storm drain sediments. Water Resources 9:1089-1098.

Marsalek, J., and B. Chocat. 2002. International report: Stormwater management. Water Science and Technology 46(6-7):1-17.

McDonald, A., D. Kay, and A. Jenkins. 1982. Generation of faecal and total coliform surges by storm flow manipulation in the absence of normal hydrometerological stimuli. Applied and Environmental Microbiology 44:292-300.

Metcalf & Eddy, Inc. 1991. Wastewater Engineering: Treatment, Disposal, Reuse, 3rd Edition, G. Tchobanoglous and F. Burton (eds.). New York: McGraw Hill.

MTAS (University of Tennessee Municipal Technical Assistance Service). 1997. Wastewater justice: Its complexion in small places. Doc. R14-1050-36-001-97. Knoxville, TN: University of Tennessee, Municipal Technical Assistance Service.

Nelson, V., S. Dix, and F. Shepard. 1999. Advanced On-Site Wastewater Treatment and Management Scoping Study: Assessment of Short Term Opportunities and Long Run Potential. Prepared for the Electric Power Research Institute, the National Rural Electric Cooperative Association and the Water Environment Research Foundation.

Novotny, V., H. Sung, R. Bannermann, and K. Baum. 1985. Estimating nonpoint pollution from small urban watersheds. Journal of the Water Pollution Control Federation 57(4):339-348.

NRC (National Research Council). 2002. Biosolids Applied to Land: Advancing Standards and Practices. Washington, DC: National Academy Press.

NRC. 2004. Indicators for Waterborne Pathogens. Washington, DC: National Academies Press.

NRCS (Natural Resources Conservation Service). 1999. National Soil Survey Handbook, Title 430-VI. Washington, DC: U.S. Government Printing Office.

Olivieri, V., C. Kruse, K. Kawata, and J. Smith. 1977. Microorganisms in urban stormwater. EPA/600/2/77/087. Washington, DC: Office of Research and Development.

Ongerth, J., and H. Stibbs. 1987. Identification of *Cryptosporidium* oocysts in river water. Applied and Environmental Microbiology 53:672-676.

OSM (Office of Surface Mining). 2003. Abandoned Mine Land Reclamation: Reclamation of Abandoned Mine Land That Took Place Before the Surface Mining Law was Passed in 1977. Available on-line at *http://www.osmre.gov/annualreports/03aml.pdf.*

PADEP (Pennsylvania Department of Environmental Protection). 1999. The Effects of Subsidence Resulting From Underground Bituminous Coal Mining on Surface Structures and Features and Water Resources. Harrisburg, PA: Office of Mineral Resources Management.

PADEP. 2001. Understanding Biosolids Land Application in Your Community. Fact Sheet 3800-FS-DEP 2649. Harrisburg, PA: PADEP.

PADEP. 2002. Clean Water Act, Section 303(d) List of Impaired Water Bodies, Draft, Version 5. Harrisburg, PA: PADEP.

PADEP. 2003. Frequently Asked Questions About Private Water Wells in Pennsylvania. Fact Sheet 3800-FS-DEP2657. Available on-line at *http://www.dep.state.pa.us/dep/*

deputate/watermgt/wc/Subjects/SrceProt/well/questions/default.htm. Accessed March 18, 2004.

PADEP. 2004a. 2004 Pennsylvania Integrated Water Quality Monitoring and Assessment Report. Available on-line at *http://www.dep.state.pa.us/dep/deputate/watermgt/Wqp/WQStandards/303d-Report.htm.* Accessed June 21, 2004.

PADEP. 2004b. Reclaim PA: Pennsylvania's Abandoned Mine Reclamation and Well Plugging Program. Available on-line at *http://www.dep.state.pa.us/dep/deputate/reclaimpa/reclaimpahome.htm.* Accessed June 12, 2004.

Peng, F., Z. Sun, and S. Peng. 1994. Disturbance of surface stream due to longwall mining. In Proceedings of the International Land Reclamation and Mine Drainage Conference, U.S. Bureau of Mines Special Publication SP 06D-94. Available on-line at *http://www.ott.wrcc.osmre.gov/library/proceed/intl1994/volume4.pdf.*

Puchalsky, R. and A. LaPlante. 1998. Troubled Waters: A Report on Toxic Releases into America's Waterways. Available on-line at *http://uspirg.org/uspirg.asp?id2=7007&id3=USPIRG&.* Accessed June 6, 2004.

Rauch, H. 1989. Ground water impacts from surface and underground coal mining. In Proceedings of the Conference on West Virginia Ground Water, 1987, Status and Future Directions. Morgantown, WV: West Virginia University, West Virginia Water Research Institute.

Rodricks, D. 2003. Unleashed dogs spoil a walk—and topsoil. Baltimore Sun, December 7.

Rose, J., H. Darbin, C. Gerba. 1988. Correlations of the protozoa, *Cryptosporidium* and *Giardia* with water quality variables in a watershed. Water Science and Technology 20:271-276.

Rose, J., S. Daeschner, D. Easterling, F. Curriero, S. Lele, and J.A. Patz. 2000. Climate and waterborne disease oubreaks. Journal of the American Water Works Association 92(9):77-87.

Sams, J., and K. Beer. 2000. Effects of coal-mine drainage on stream water quality in the Allegheny and Monongahela River basins—Sulfate transport and trends. Water Resources Investigations Report 99-4208. Lemoyne, PA: USGS, National Water Quality Assessment Program.

Scandura, J., and M. Sobsey. 1997. Viral and bacterial contamination of groundwater from on-site sewage treatment systems. Water Science and Technology 35(11-12):141-146.

Schiff, K., and P. Kinney. 2001. Tracking sources of bacterial contamination in stormwater discharges to Mission Bay, California. Water Environment Research 73(5):534-542.

Simmons, G., Jr., D. Waye, S. Herbein, S. Myers, and E. Walker. 2001. Estimating nonpoint fecal coliform sources in Northern Virginia's Four Mile Run watershed. In Proceedings of the 2000 Virginia Water Resources Research Symposium. Virginia Water Resources Research Center Special Report SR-19-2000. Blacksburg, VA: VWRRC.

Skousen, J. 1995. Acid mine drainage. Green Lands 25(2):52-55.

Smullen J., A. Shallcross, and K. Cave. 1999. Updating the U.S. nationwide urban runoff quality database. Water Science and Technology 39(12):9-12.

States, S., K. Stadterman, L. Ammon, P. Vogel, J. Baldizar, D. Wright, L. Conley, and J. Sykora. 1997. Protozoa in river water: Sources, occurrence, and treatment. Journal of the American Water Works Association 89(9):74-83.

Stehouwer, R. 1999a. Land application of sewage sludge in Pennsylvania: A plain English tour of the regulations. University Park, PA: Penn State College of Agricultural Sciences Cooperative Extension.

Stehouwer, R. 1999b. Land application of sewage sludge in Pennsylvania: Biosolids quality. University Park, PA: Penn State College of Agricultural Sciences Cooperative Extension.

Stehouwer, R. 2003. Land application of sewage sludge in Pennsylvania: Effects of biosolids on soil and crop quality. University Park, PA: Penn State College of Agricultural Sciences Cooperative Extension.

Stern, D. 1996. Initial investigation of the sources and sinks of *Cryptosporidium* and *Giardia* within the watersheds of the New York City water supply system. In Proceedings of a Symposium on New York City Water Supply Studies. TPS-96-2. Herndon, VA: American Water Resources Association.

Steuer, J., W. Selbig, N. Hornewer, and J. Prey. 1997. Sources of contamination in an urban basin in Marquette, Michigan, and an analysis of concentrations, loads, and data quality. USGS Water Resources Investigation Report 97-4242. Reston, VA: USGS.

Stiles, J., J. Donovan, D. Dzombak, R. Capo, L. Cook. 2004. Geochemical cluster analysis of mine water quality within the Monongahela basin. Presented at the 2004 National Meeting of the American Society of Mining and Reclamation and the 25th West Virginia Surface Mine Drainage Task Force, Morgantown, WV, April 18-24.

Stout, B. 2003. Impacts of longwall mining on the diversity, longevity, and functionality of benthic macroinvertebrate communities in central Appalachian headwater streams. Presented at North American Benthological Society Annual Meeting, Athens, GA.

Strawley, G. 2002. Pennsylvania Development: Community's Growth Troubles Start in Its Sewers. Water Infrastructure Network News. Available on-line at *www.win-water.org/win_news/ 053002article.html.* Accessed June 21, 2004.

Sykora, J., S. States, W. Bancroft, S. Boutros, M. Shapiro, and L. Conley. 1986. Monitoring of water and wastewater for *Giardia.* In Proceedings of the American Water Works Association (AWWA) Water Quality Technology Conference. Denver, CO: AWWA.

TPRC (Third Party Review Committee). 2002. Third Party Review of the ALCOSAN Regional Long Term Wet Weather Control Concept Plan. Pittsburgh, PA: ALCOSAN.

USACE (U.S. Army Corps of Engineers). 1997. Montour Run Watershed, Allegheny County, Pennsylvania: Water Quality and Aquatic Life Resources. Pittsburgh, PA: USACE.

USACE. 2000. Nine Mile Run Allegheny County, PA: Aquatic Ecosystem Restoration Water Quality and Aquatic Life Report. Pittsburgh PA: USACE.

USDA (U.S. Department of Agriculture). 1997. U.S. Census of Agriculture. Washington, DC: National Agricultural Statistics Service.

USDA. 2002. U.S. Census of Agriculture. Pennsylvania State and Country Data. Available on-line at *http://www.nass.usda.gov/census/census02/volume1/PAVolume104.pdf.* Accessed November 16, 2004.

Van Metre, P., and B. Mahler. 2003. The contribution of particles washed from rooftops to contaminant loading to urban streams. Chemosphere 52(10):1727-1741.

Walker, W., R. NcNutt, and C. Maslanka. 1999. The potential contribution of urban runoff to surface sediments of the Passaic River: Source and chemical characteristics. Chemosphere 38(2):363-377.

WEF (Water Environment Federation). 1999. Prevention and Control of Sewer System Overflows. MOP FD-17, 2nd edition. Alexandria, VA: WEF.

West Virginia Conservation Partnership. 1999. Resource Assessment for the Dunkard Creek Watershed. Morgantown, WV: West Virginia Conservation Partnership.

Wood, S., P. Younger, and N. Robins. 1999. Long-term changes in the quality of polluted minewater discharges from abandoned underground coal workings in Scotland. Quarterly Journal of Engineering Geology and Hydrogeology 32(1):69-79.

WRAS. 2001. Watershed Restoration Action Strategy: State Water Plan Subbasin 19D: Lower Youghiogheny River Watershed: Westmoreland and Fayette Counties. Available on-line at *http://www.dep.state.pa.us/dep/deputate/watermgt/wc/Subjects/Nonpointsourcepollution/Initiatives/WRASLISTINFO/WrasPlans/WRAS-19D.pdf.* Accessed June 21, 2004.

WRAS. 2003a. Watershed Restoration Action Strategy: State Water Plan Subbasin 17C: Redbank Creek Watershed (Allegheny River): Jefferson, Armstrong, Clarion, and Clearfield Counties. Available on-line at *http://www.dep.state.pa.us/dep/deputate/watermgt/wc/Subjects/Nonpointsourcepollution/Initiatives/WRASLISTINFO/WrasPlans/WRAS-17C.pdf.* Accessed June 21, 2004.

WRAS. 2003b. Watershed Restoration Action Strategy: State Water Plan Subbasin 19E: Upper Youghiogheny River Watershed (Laurel Creek and Indian Creek): Fayette, Somerset, and Wetmoreland Counties. Available on-line at *http://www.dep.state.pa.us/dep/deputate/watermgt/wc/Subjects/Nonpointsourcepollution/Initiatives/WRASLISTINFO/WrasPlans/WRAS-19E.pdf.* Accessed June 21, 2004.

WRAS. 2003c. Watershed Restoration Action Strategy: State Water Plan Subbasin 20F: Chartiers Creek Watershed (Ohio River): Washington and Allegheny Counties. Available on-line at *http://www.dep.state.pa.us/dep/deputate/watermgt/wc/Subjects/Nonpointsourcepollution/Initiatives/WRASLISTINFO/WrasPlans/WRAS-20F.pdf.* Accessed June 21, 2004.

WRAS. 2003d. Watershed Restoration Action Strategy: State Water Plan Subbasin 18E: Stonycreek River and Little Conemaugh River Watersheds: Somerset and Cambria Counties. Available on-line at *http://www.dep.state.pa.us/dep/deputate/watermgt/wc/Subjects/Nonpointsourcepollution/Initiatives/WRASLISTINFO/WrasPlans/WRAS-18E.pdf.* Accessed June 21, 2004.

WSIP (Southwestern Pennsylvania Water and Sewer Infrastructure Project Steering Committee). 2002. Investing in Clean Water: A Report from the Southwestern Pennsylvania Water and Sewer Infrastructure Project Steering Committee. Pittsburgh, PA: Campaign for Clean Water.

Younger, P. 2000. Predicting temporal changes in total iron concentrations in groundwaters flowing from abandoned deep mines: A first approximation. Journal of Contaminant Hydrology 44:47-69.

5

Water Quality Improvement: Decision-Making Strategies and Technical Solutions

INTRODUCTION

This chapter explores various wastewater management techniques and recommends actions to address water quality problems—especially wet weather-related problems—in southwestern Pennsylvania. As discussed in Chapters 3 and 4, the aquatic environment of southwestern Pennsylvania is impaired for a variety of designated beneficial uses including recreational use due to the likelihood of waterborne pathogens in surface waters and aquatic life use due to acid mine drainage (AMD). Although AMD is a significant cause of water quality impairment in the region, especially in the predominantly rural counties of southwestern Pennsylvania, this problem is most appropriately addressed at the state and federal levels. Continued funding for these ongoing efforts (see Chapter 4 for further information) is essential to improve water quality in southwestern Pennsylvania.

A fundamental prerequisite to the formulation of cost-effective plans for reducing water quality impairments in southwestern Pennsylvania is a systematic and extensive set of water quality data covering both sources of impairments and in-stream responses. The data should be sufficient to accurately assess the different sources of contamination and their impacts on receiving streams. Toward this end, Chapter 3 summarizes data available for the region. Increased monitoring by different groups and agencies has taken place over the past several years, and the available data are sufficient to conclude that serious water quality problems exist in southwestern Pennsylvania. However, there are not sufficient data to determine the relative seriousness of the related environmental and human health problems, the relative importance of the potential sources of contamination, and the improvements that are likely to result from alternative pollution control measures.

Despite these limitations, the Allegheny County Sanitary Authority (ALCOSAN) and its member communities are now facing enforcement action by the U.S. Environmental Protection Agency (EPA) through the Pennsylvania Department of Environmental Protection (PADEP) and the Allegheny County Health Department (ACHD) (see more below) for violations of the Clean Water Act related to combined sewer overflows (CSOs) and sanitary sewer overflows (SSOs). Thus, remedial actions are anticipated that will alter the relative contribution of different sources to the water quality problems in the region. It will be critical that evaluation of water quality improvements related to these activities be undertaken. Further, the implementation of solutions for identified sources of impairment does *not* preclude the need for additional information related to other sources and their contributions to water quality impairment. Monitoring and modeling efforts should proceed in conjunction with, and inform decisions with respect to, a variety of mandated water quality improvements currently being pursued (e.g., those listed in Table 5-7).

Several entities (ALCOSAN, 1999; TPRC, 2002; WSIP, 2002) have estimated recently that addressing the region's CSO and SSO problems by conventional means, using a combination of storage, conveyance, and treatment improvements could cost several billion dollars. Although the problem of excessive discharge of untreated wastewater from CSOs and

SSOs is well documented, the data presented to the committee and those uncovered by its own research as summarized in Chapter 3 are inadequate to arrive at a definitive conclusion as to (1) the impact of these discharges on water quality in receiving streams and (2) what should be done to address the issue in the context of federal CSO policy.

Investing large sums of capital based only on currently available data may not ultimately solve the most important problems or provide appropriate solutions. Although it is true that no amount of additional data and analyses would remove all uncertainty about water quality investments, it is clear that currently available information is lacking in several critical areas, including the following:

- the nature and magnitude of CSO effects on receiving streams during wet weather events;
- whether effects are limited to indicator microorganisms (i.e., bacterial indicators of fecal contamination and, indirectly, the presence and quantity of fecal pathogens) and the extent to which they include floatable and settleable solids;
- how much surface water runoff from separate stormwater conveyances affects water quality in receiving streams during wet weather events;
- whether present discharges constitute a threat to the public as evidenced by health data; and
- the extent of the effects of present and potential small community and on-site systems.

The causes and nature of water quality impairments, the parties responsible, and the individuals and waterways affected differ for each of the problem contaminants in the region. A comprehensive watershed-based approach is needed to address the spectrum of water quality problems; such a systematic approach should recognize interrelationships among problems and the need for the parties responsible for each water quality problem to share in its solution. Responsible groups may be the public at large, a segment of the population, individuals, or a particular industry or group of industries. Recognition of payment capacity of individuals and the region as a whole should also be considered in reaching equitable solutions (see Chapter 6 for further information).

EXISTING INFRASTRUCTURE AND INSTITUTIONS FOR WATER QUALITY MANAGEMENT

The Commonwealth of Pennsylvania and local governments in the Pittsburgh region have a long history of planning, regulations, capital investments, and development of managerial expertise to control water pollution. It is evident that more is needed, particularly in the management of CSOs, SSOs, separate storm sewers, and other sources of pollution. Future actions will build on or modify existing infrastructure and managerial institutions. Some of those facilities and arrangements are discussed in the sections that follow.

Sewer systems that convey wastewater or combined wastewater and stormwater to sewage treatment plants generally have multiple components and multiple owners. First, pipes within a residence collect wastewater and carry it to a house lateral pipe (see Figure 4-5). House lateral pipes are underground and owned by the homeowners. Laterals typically comprise 50 percent of the total length of pipe in a sewer system and are connected to a street sewer pipe; as a general rule, they may account for a substantial portion of the total infiltration and inflow into the sewer system. Because a large portion of them are located on private property, and measuring flows in laterals is not common practice, identifying sources and fixing them requires special detective work and authority to order corrective measures. Sewer pipes are generally owned by the municipality that owns the street above it. This street sewer pipe joins with others and enters an interceptor. This larger pipe might be owned by a municipality, a sewer authority, or the same authority or organization that owns a wastewater treatment plant (WWTP). Finally, major interceptors carry the flow from all the small interceptors and pipes to the treatment plant. Major interceptors are generally owned by the same organization that owns and operates the treatment plant. These complexities are compounded in southwestern Pennsylvania because of fragmented ownership and responsibilities.

The Pennsylvania Sewage Facilities Act, commonly referred to as Act 537, was passed in 1966. It requires all municipalities to formulate and implement plans for management of current and future sewage. Individual municipalities may choose to administer their own plans, or they may choose to participate in a joint local agency (JLA) with other municipalities or county health department. These plans, subject to review and approval by PADEP, must be modified when new land development is proposed or other changes occur.

By far the largest JLA in the region is ALCOSAN (see Chapter 2 for further information). ALCOSAN owns and operates 83 miles of major interceptors and a wastewater treatment plant that provides primary and secondary treatment of up to 225 million gallons per day. Eighty-three communities are within the ALCOSAN service area (see Figure 1-1), and a total of 12 sewer authorities serve many of these communities. Although these "partner communities" own and operate their own sewer collection infrastructure (street sewer pipes and smaller interceptors), they do not operate WWTPs. All collected sewage (and stormwater flow where combined systems are used, unless released during a CSO event) is sent to the ALCOSAN interceptors and eventually to the central wastewater treatment plant. Thus, ownership and management of wastewater collection and treatment facilities for the most populous county in the region is extremely fragmented. The institutional complexities of the region are discussed more fully in Chapter 6.

This complexity is illustrated by the fact that as of March 2004, there were 591 Act 537 plans in the 11-county area. They are summarized by county in Table 5-1. Any development requiring the extension of sewer systems not included in an existing plan will trigger a plan revision. Twenty-eight percent of the plans have been revised in the past five years. Many of the plans cover rural areas in which no sewer systems are located or very little development is occurring. Indeed, more than half of the municipalities have had no reason to revise their plans for 20 years or more. A map of the sewered areas is provided in Figure 5-1. Act 537 also required municipalities to establish a permitting program for on-lot treatment and disposal systems (OLDS; referred to as on-site sewage treatment and disposal systems [OSTDS] in this report) for individual lots and community OSTDSs with design flows of up to 10,000 gallons per day. Like the sewage disposal plans, these programs are to be operated by individual municipalities or through JLAs. Each program is administered by a sewage enforcement officer (SEO) who is trained by PADEP. SEOs are responsible for determining the adequacy of sites to support OSTDSs and ensuring that system designs comply with Chapter 73 of PADEP's regulations. The PADEP has developed a home buyers' guide, a homeowners' guide to maintenance, a manual to SEO's decisions about repairs, and other educational material that is

TABLE 5-1 Act 537 Plans by County by Age of Plan

County	Age of Plan (years)				
	<5	5-9	10-19	>20	All
Allegheny	75	9	5	41	130
Armstrong	5	3	0	36	44
Beaver	7	3	4	40	54
Butler	8	7	4	28	47
Fayette	13	2	4	24	43
Greene	6	1	0	19	26
Indiana	15	8	4	11	38
Lawrence	7	2	2	16	27
Somerset	5	1	6	38	50
Washington	14	7	8	38	67
Westmoreland	9	7	6	43	65
Total	164	50	43	334	591

SOURCE: Data from *www.dep.state.pa.us/dep/deputate/watermgt/ Wqp/WQP_WM/537Map/.*

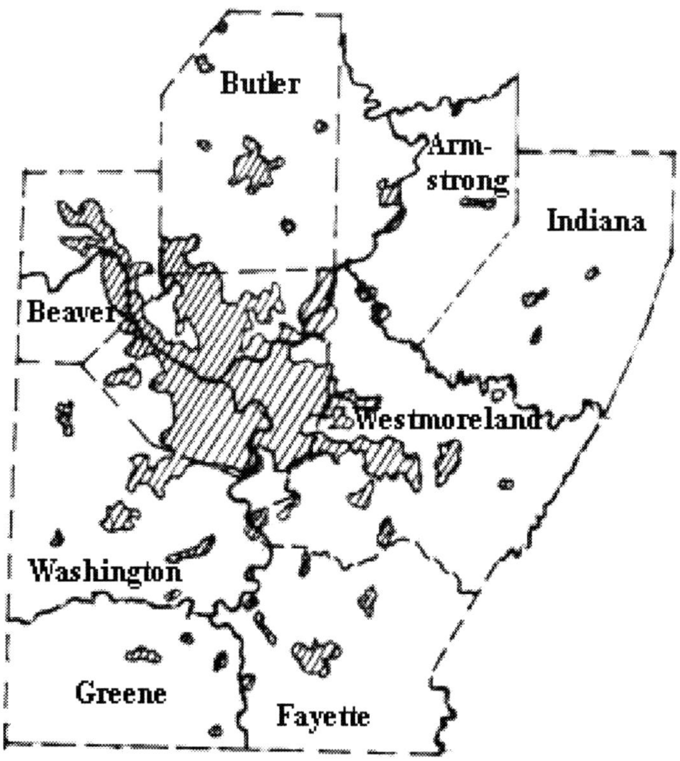

FIGURE 5-1 Approximate boundaries of sewered areas in Allegheny and surrounding counties.
NOTE: Lawrence and Somerset Counties of southwestern Pennsylvania (see also Box 1-2) are not included.
SOURCE: Adapted from SPC map, “Sewer Service Areas in the SPC Nine County Region.”

available on-line,[1] but once a system is permitted, there is no program for regular inspection and maintenance. Pennsylvania's stormwater program operates under authority of the Storm Water Management Act of 1978 (Act 167) and EPA regulations under the 1987 amendments to the Clean Water Act. Act 167 provides grants to counties to prepare stormwater regulations for designated watersheds, although initiation of planning is left to the counties. A county desiring to develop a plan then makes a proposal to PADEP for financial assistance. A requirement of the plan is preparation of a model stormwater ordinance, which must be adopted by any municipality within the watershed that does not have equivalent language in its building subdivision or land development codes. The ordinances address control of stormwater runoff from new development; they do not require retrofitting of existing development.

A recent map of Pennsylvania watersheds for which stormwater plans have been prepared[2] shows that only six plans have been prepared for the approximately 60 watersheds within the 11-county area of southwestern Pennsylvania. Approved plans and dates of approval include the following:

- Pine Creek, Girty Run, and Deer Creek in Allegheny County (1986);
- Turtle Creek in Allegheny and Westmoreland Counties (1991 and 1992);
- Montour Run in Allegheny County (1990); and
- Monongahela River in Allegheny County (1990).

Two other plans are in preparation, one for Little Sewickley Creek in Allegheny County and one for Cokes Creek in Somerset County. The original Turtle Creek plan was prepared in 1986; its purpose was to reduce the impact of new development on peak discharges and related downstream effects on property and traffic. The plan established standards for post-development peak discharge rates from new development. Standards ranged from pre-development rates to 50 percent of post-development rates, depending on location of development within the watershed. The model ordinance affected about 30 municipalities in the watershed (John Maslanik, ATS-Chester Engineers, personal communication, 2004).

Amendments to the Clean Water Act in 1987 directed the EPA to expand the National Pollutant Discharge Elimination System (NPDES) permit system to include stormwater runoff. The EPA implemented that directive in two phases. The Phase I Rule was published on November 16, 1990 (55 Federal Register 47990) and required all operators of medium and large municipal separate storm sewer systems to obtain an NPDES permit and develop a management program to reduce the discharge of harmful pollutants. These regulations covered certain categories of stormwater associated with industrial activity and discharges of stormwater from urban areas with a population of 100,000 or more (EPA, 1996). Notably, Phase I included Allegheny County.

The EPA promulgated rules specifying who must apply for Phase II permits in August 1995; final Phase II rules were published in 1999 (EPA, 1999). That program automatically covers all municipal separate storm sewer systems located in an urbanized area as defined by the U.S. Bureau of the Census. A few special waivers apply, and small municipal separate storm

---

[1] See *http:// www.dep.state.pa.us*, "Subjects," "Wastewater" for further information.

[2] Map available on-line at *http://www.dep.state.pa.us/dep/deputate/watermgt/wc/subjects/stormwatermanagement/ Stormwater_11_18_02_web.jpg.*

sewer systems outside urban areas may be waived by a permitting agency on a case-by-case basis.

In December 2002, PADEP updated its stormwater policy to include protection of water quality as mandated by the Phase I and Phase II rules (PADEP, 2002). The federal regulations for Phase II require six minimum measures:

1. public involvement in planning and decision processes;
2. detection and elimination of illicit discharges to storm sewers;
3. reduction of pollutants in stormwater runoff from construction sites;
4. management of post-construction runoff from new development and redevelopment;
5. pollution prevention and good housekeeping; and
6. public education and outreach.

The PADEP has developed several guidance documents to assist communities in achieving compliance with federal regulations, including a handbook of best management practices (BMPs) for developing areas[3] and a model stormwater ordinance for municipalities (PADEP, 2002). The revised policy establishes post-construction management requirements that emphasize groundwater infiltration and BMPs to control volumes and rates of discharge. The policy sought to integrate those permits with its Act 167 authorities. Acceptable BMPs to promote groundwater infiltration, water quality, and rate and volume control are listed in Table 5-2.

The primary limitation of all of these stormwater programs is that they are focused mostly on controlling runoff from new development. They do cover illicit connections to storm sewers and redevelopment projects, but for the most part they have little effect on high density development in the urban core of Pittsburgh and surrounding communities that were developed prior to enactment of these programs.

DRIVERS FOR WATER QUALITY IMPROVEMENT UNDER WET WEATHER CONDITIONS

As discussed in Chapter 4, although the relative contribution of different sources of microbial loading to surface and groundwaters in southwestern Pennsylvania cannot be determined with available information, sufficient information is available to determine that improperly managed wastewaters resulting from human activities are degrading the microbiological water quality in the region. Furthermore, available water quality and human health data are insufficient to reach sound conclusions about the seriousness of this problem. Wet weather microbiological water quality in the main stem rivers is demonstrably worse than dry weather quality, suggesting that stormwater and sewer overflows may be important contributors. In this regard, bacterial (predominantly coliform) indicator levels greatly exceed acceptable standards for body contact recreation for a significant portion of the year in the three main stem rivers and many of their tributaries, especially during and immediately after precipitation events. This is a particular issue in and around the City of Pittsburgh that is probably exacerbated by upstream and downstream sources of microbial loading, including agricultural runoff. Microbiological water quality in many tributaries does not meet standards in

[3] Available on-line at http://www.pacd.org/products/bmp/bmp_handbook.htm.

TABLE 5-2 Best Management Practices Deemed Acceptable to PADEP for Stated Purposes

Groundwater Infiltration	Water Quality	Rate and Volume Control
Permeable paving	Permeable paving	Permeable paving
Grass swale	Bioretention	Stormwater infiltration
Bioretention	Grass swale	Dry ponds
	Filter strip	Stormwater wetlands
	Wet pond (extended detention pond)	Wet ponds (extended detention pond)
	Rooftop runoff management	Rooftop runoff management
		Bioretention

SOURCE: Adapted from www.dep.state.pa.us ("Stormwater").

either wet or dry weather, suggesting that failing OSTDSs and sewers may be important contributors.

Notwithstanding the data limitations, the most pressing water quality problem in the region from a regulatory perspective is caused by CSOs, SSOs, and stormwater drainage resulting from wet weather conditions. In fact, the EPA views CSO and SSO problems as sufficiently serious that it has promulgated regulations requiring CSO, SSO, and stormwater controls. Because southwestern Pennsylvania has one of the nation's most extensive CSO and SSO control problems, EPA, acting through PADEP and ACHD, has issued a series of administrative consent orders to many of the communities served by ALCOSAN (see also footnote 1, Chapter 1) to address this problem. It is important to state that the ALCOSAN draft consent decree with EPA remained attorney-client privileged[4] as this report neared completion in December 2004 so the committee's conclusions and recommendations may not be consistent with whatever actions result from the final legal agreements that may be reached. Thus, the following discussion concerning CSO and SSO problems should be viewed with this caution in mind. However, the consent orders for controlling CSOs and SSOs in the ALCOSAN partner communities have become publicly available during the study period, and many of the committee's recommendations are aligned with these activities and are discussed later in this chapter.

EPA's Regulatory Approach to CSO Remediation

EPA has produced a variety of guidance for municipalities to manage CSOs, including the following:

- *Combined Sewer Overflow Policy* (EPA, 1994)
- *Combined Sewer Overflows Guidance for Long-Term Control Plan* (EPA, 1995a)
- *Combined Sewer Overflows Guidance for Nine Minimum Control Measures* (EPA, 1995b)
- *Combined Sewer Overflows Guidance for Funding Options* (EPA, 1995c)
- *Combined Sewer Overflows Screening and Ranking Guidance* (EPA, 1995d)

[4] The availability of information and of the parties involved in this litigation to fully cooperate with the committee constrained this study at times and will likely continue to impede any process that seeks all available information and the candid input of knowledgeable experts.

- *Combined Sewer Overflows Guidance for Financial Capability Assessment and Schedule Development* (EPA, 1997)
- *Combined Sewer Overflows Guidance for Monitoring and Modeling* (EPA, 1999)
- *Guidance: Coordinating Combined Sewer Overflows (CSO) Long-Term Planning with Water Quality Standards Reviews* (EPA, 2001b)

Owners of CSOs are required to obtain discharge permits and establish programs that would implement nine minimum control measures (discussed below), and each is required to develop and implement a long-term control plan (LTCP) for controlling CSOs. The 1994 EPA CSO policy states:

> EPA expects a permittee's long-term CSO control plan to give the highest priority to controlling overflows to sensitive areas. Sensitive areas, as determined by the NPDES authority in coordination with state and federal agencies, as appropriate, include designated Outstanding National Resource Waters, National Marine Sanctuaries, waters with threatened or endangered species and their habitat, waters with primary contact recreation, public drinking water intakes or their designated protection areas, and shellfish beds.

In Pittsburgh's case, the sensitive areas are those waters below CSOs with primary contact recreation as their designated use and those that serve as public drinking water intakes (see Figure 4-4).

There are two basic remedial approaches for controlling CSOs: (1) the demonstration approach and (2) the presumption approach. In brief, the "demonstration approach" relies on data collection and simulation to demonstrate that a proposed management strategy will result in meeting water quality standards and considers all factors that are likely to influence success; there is no reliance on criteria governing by how much CSOs may be reduced. The demonstration approach seems inherently advantageous because it relies on actual data collection and analysis and also has the benefit of lending itself to adaptive implementation (described later) by determining the progressive performance, in the watershed context, of each measure undertaken. Although the demonstration approach has the advantage of focusing investment on measures likely to achieve water quality standards, because it relies on time-consuming data collection and analysis, it could result in delaying the reduction of pollutants to receiving streams.

In contrast, the "presumption approach" presumes that meeting certain criteria, including a statistical reduction (85 percent) of the annual volume of wet weather overflows, is likely to result in meeting water quality standards as reasonably determined by the regulatory agency. Under this approach, if the owner of CSOs has satisfied given criteria for reducing CSOs and some uncertainty remains about satisfaction of water quality standards, the owner is given the presumption that CSOs are no longer contributing to noncompliance with water quality standards. Satisfaction of the criteria for reducing CSOs usually involves large capital investments. The presumption approach was included as an alternative by EPA in the 1994 CSO policy because data and modeling of wet weather events do not always provide a clear picture of the level of CSO controls that are necessary to meet water quality standards.

More specifically, EPA (1994) noted that "because data and modeling of wet weather events often do not give a clear understanding of the level of CSO controls necessary to protect water quality standards, one of three technology and performance standards could be used to satisfy a presumption that water quality standards would be met." The criteria include the following:

- a limit on the number of overflow events per year;
- elimination or the capture for treatment of no less than 85 percent by volume of the CSO discharge; or
- elimination or removal of no less than the mass of the pollutants identified as causing water quality impairment through the sewer system.

In later guidance, EPA (1999) added some clarification to its CSO control policy:

> Because CSOs are subject to the technology-based requirements of the Clean Water Act (CWA), permitting authorities must specifically determine best available technology economically achievable (BAT)/best conventional pollutant control technology (BCT) **on a case-by-case basis using best professional judgment (BPJ) during the permitting process** [emphasis added]. . .Therefore, evaluation of CSO controls beyond the nine minimum controls may appropriately focus primarily on water quality issues. . .State and Federal NPDES authorities must coordinate throughout the planning process to ensure that after implementation of the controls in the proposed LTCP, CSOs will not cause or contribute to nonattainment of WQS.

The CSO policy is clear—attainment of water quality standards is a requirement. However, the policy also recognizes that an unambiguous determination of what constitutes compliance with water quality standards may not be possible. The policy appears to indicate that when there is significant uncertainty about whether a plan will lead to compliance with water quality standards, the permittee is entitled to the presumption approach if the plan satisfies one of the three aforementioned criteria.

As discussed in the next section, ALCOSAN proposed to rely primarily on the 85 percent reduction criterion as described in its March 1999 report *Draft Combined Sewer Overflow Program Phase I Activity Report: Regional Long Term Wet Weather Control Concept Plan* (ALCOSAN, 1999, pp. 1-2, 3-1). An independent third party review (TPR) of ALCOSAN's draft LTCP has cast serious doubts on whether the 85 percent reduction criterion would satisfy water quality standards. That analysis itself was based on several reasonable but unverifiable assumptions that are discussed in the TPR report (TPRC, 2002). Until the fundamental gaps in knowledge of regional water quality are filled (see Chapters 3 and 4 for further information), it remains unclear whether achieving an 85 percent reduction in CSO volume would satisfy water quality standards.

Use of the demonstration approach for controlling CSOs places a heavy burden of proof on the region to demonstrate that a particular control plan will satisfy water quality standards. As discussed later in this chapter, it is difficult with complex models to get an adequate estimate of uncertainty or to know precisely when a satisfactory "demonstration" has been achieved. Results of water quality modeling must be combined with substantial professional judgment in making a determination about compliance.

The demonstration approach for controlling CSOs can be used in southwestern Pennsylvania by incorporating a strategy of *adaptive implementation*, which is discussed in detail later in this chapter. In brief, it begins with monitoring actual CSO discharges and their water quality impacts. Field monitoring should be coupled with water quality models that enable planners to estimate the extent to which reductions in discharges will be necessary to meet water quality standards. Because effects of CSOs and discharges from separate storm sewers are intermingled in the region's primary receiving streams, monitoring and modeling of CSOs and their impacts on streams during wet weather events should occur simultaneously with monitoring and modeling of separate stormwater sewer systems during wet weather events. Conventional control strategies for reducing pollutant loading from CSOs should be conducted in parallel with experiments on innovative but unproven technology such as vortex separators (see more below). All of these investigations can be conducted over a relatively short period (e.g., three to five years). Upon completion, information available at that time should be used to help judge which CSO control strategies are cost-effective and subject to acceptable levels of uncertainty.

ALCOSAN's Long Term Wet Weather Control Concept Plan

To address the EPA and Commonwealth of Pennsylvania wet weather regulatory requirements, in March 1999 ALCOSAN produced a draft LTCP; its fundamental goal "is to improve and preserve the water environment in the ALCOSAN service area and to fulfill ALCOSAN's obligations under the Clean Water Act and the Pennsylvania Clean Streams Law."

More specifically, the draft LTCP (ALCOSAN, 1999, p. 1-1) has three primary phases to attain wet weather water quality standards that are summarized below:

1. implement a program for nine minimum CSO controls;
2. plan, design, and implement a regional LTCP; and
3. participate in regional and interstate watershed-based planning and analyses.

Phase One—Nine Minimum Combined Sewer Overflow Controls

The EPA's nine minimum controls[5] for CSOs do not require significant engineering studies or major construction and can be implemented in a relatively short time frame (EPA, 1995b). These include (1) proper operation and maintenance of the sewer system, (2) maximum use of the collection system for storage, (3) modification of the pretreatment program, (4) maximization of wastewater flow to the treatment plant, (5) elimination of chronic dry weather overflows, (6) control of solids and floatables, (7) pollution prevention, (8) public notification of overflow occurrences and impacts, (9) and monitoring to characterize sewer overflow impacts.

Phase Two—LTCP

According to the draft LTCP, ALCOSAN (1999, p. 1-2, 3-1) proposed to use the presumption approach guidance as outlined by the EPA to address its CSO problem. This approach permits meeting regulatory requirements by the indirect method of reducing the amount of combined sewage overflow and presuming that this action will meet water quality standards. In brief, the presumption approach as outlined in the draft LTCP includes expanding

---

[5] For further information on EPA's nine minimum controls for CSOs, see *http://cfpub.epa.gov/npdes/cso/ninecontrols.cfm?program_id=5*.

the existing ALCOSAN wastewater treatment plant over a 20-year period from the current 225 million gallons per day (mgd) to a total wet weather capacity of 875 mgd and also bringing about infrastructure changes to the sewerage system in the form of interceptors to significantly increase the proportion of wet weather flows arriving at the treatment plant instead of discharging untreated into streams. Of this increased flow, 310 mgd would receive full secondary treatment and the remaining 565 mgd of wet weather flow would receive primary treatment and disinfection only. ALCOSAN estimates that these proposed changes will permit capture of 85 percent of the wet weather combined sewage flow—the majority of which will be given primary treatment only.

A major portion of the ALCOSAN interceptor sewer system roughly parallels the three major rivers in the Pittsburgh area, the Allegheny, Monongahela, and Ohio Rivers. Flows within the sewer system in excess of those planned for conveyance to the main treatment plant would undergo high-rate flow regulation and primary treatment by five of the system's swirl (vortex) separators (see more below). The other major portion of the ALCOSAN interceptor sewer system roughly parallels four major tributary streams: Chartiers Creek, Saw Mill Run, Turtle Creek, and Thompson Run (see Figure 1-2). These interceptors are installed in relatively shallow excavations, and under the draft LTCP (ALCOSAN, 1999, pp. 1-2, 3-5) the excess wet weather flow in these interceptors would be handled differently from that of the main river interceptors. For these interceptors, up to 85 percent of the wet weather flow would be handled by a combination of interceptor upgrades, peak flow storage, and two of the system's vortex separators.

A portion of ALCOSAN's existing interceptor sewer system, built for the most part at the upper extensions of the main interceptors, is designed for collection of sanitary sewage only. Unfortunately, at the lower ends of the interceptors this sanitary sewage becomes mixed with combined sewage such that much of the benefit of the separate sewage infrastructure is lost. In addition, the wet weather flow in the separate sewers averages 1,000 gallons per capita per day (gpcd), while the dry weather flow averages only 190 gpcd (ALCOSAN, 1999). This indicates that an excessive amount of runoff from precipitation enters the separate sewage system, leading to separate system overflows, which are illegal under the federal Clean Water Act. According to the draft LTCP, ALCOSAN expects that member municipalities will commit to a long-term (approximately 50-year) effort to reduce this inflow and infiltration.

The estimated costs of the interceptor system and associated regulator/grit treatment upgrades under the draft LTCP are provided in Table 5-3. The total construction cost in 1998 dollars is approximately $922 million, with an annual operating cost of $3.51 million. However, this does not include the cost of implementing the previous (1996) ALCOSAN plans for upgrades to expand ALCOSAN's plant to 875 mgd wet weather flow in accordance with the requirements of Pennsylvania Act 537. Total 2002 costs for the LTCP, expansion of its treatment plant ($210 million), and upgrades to the non-ALCOSAN-owned collection system ($1.9 billion) are expected to exceed $3 billion (TPRC, 2002).

In order for the LTCP to be successful, an extensive rehabilitation and/or reconstruction of the overall ALCOSAN sewerage system must ultimately be accomplished to reduce infiltration and inflow. However, ALCOSAN controls only a portion of the sewers and a portion of the total watershed area of the rivers and streams flowing into its service area; thus, the success of much of this effort will depend on action by other entities in the region. The long-

TABLE 5-3 Estimated Capital and Operations and Maintenance Costs for Interceptor and Nonplant Treatment Modifications

Planned Modifications	Capital Costs ($ million)	Operation and Maintenance Costs ($ million/year)
Three Rivers Interceptors		
Allegheny	146.7	0.512
Monongahela	105.4	0.333
Ohio	38.1	0.170
Subtotal	334.1	1.965
Tributary Interceptors		
Saw Mill Run	37.7	0.283
Thomson Creek/Turtle Creek	121.5	0.71
Chartiers Creek	366.0	0.145
Subtotal	525.2	1.136
Other	62.4	0.410
Totals (1998 dollars)	921.7	3.513
Totals (2002 dollars)	1,030.5	3.928

SOURCE: ALCOSAN, 1999.

range reconstruction and rehabilitation of the collection systems by member municipalities is estimated to cost between $1.2 billion and $4.2 billion (in 1999 dollars) over a 50-year period (ALCOSAN, 1999).

Phase Three—Regional and Interstate Participation

Although the third phase of ALCOSAN's LTCP was not discussed in detail in the March 1999 draft plan, that report does state (p. 1-2) that "As a stakeholder in the watershed planning process, ALCOSAN is willing to participate in regional and interstate watershed management activities." The ALCOSAN third party review (TPRC, 2002) briefly describes the role of regional and interstate participation in the LTCP in that a "regional approach" is referred to in summaries of activities to be undertaken in Phases I, II, and III. The "regional approach," however, appears to be limited to ALCOSAN, the 83 member communities served by ALCOSAN, the Three Rivers Wet Weather Demonstration Program (3RWW), and the regulatory community—those entities identified as the principal stakeholders. That wording suggests that this regional approach encompasses only Allegheny County, rather than southwestern Pennsylvania as a whole. Chapter 6 of this report addresses the roles of local, regional, state, and federal organizations in addressing water quality improvement in southwestern Pennsylvania and, consistent with this chapter, specifically recommends that a much larger watershed area be considered. Furthermore, Chapter 7 includes a discussion of several related issues that have broader implications beyond southwestern Pennsylvania.

Summary of Third Party Review of ALCOSAN's Draft LTCP

Following the establishment of a Review Committee in 1998 and an Engineering Peer Review Committee in 1999, ALCOSAN formed a third party review committee (TPRC) in 2001 to address several questions raised by affected member communities concerning ALCOSAN's draft LTCP (see also Appendix B). Toward this end, the TPRC, consisting of eight prominent residents of southwestern Pennsylvania, directed an independent assessment of the LTCP by a team of environmental engineering consultants (Greeley and Hansen LLC, HydroQual, Inc., and McGuire Woods, LLP). The collective result of their effort was a June 2002 report (TPRC, 2002; see also Appendix B) entitled *Third Party Review of the ALCOSAN Regional Long Term Wet Weather Control Concept Plan.*[6] That report addressed questions raised about the LTCP and suggested a way forward. Although this committee did not conduct a comprehensive review of the TPR report, it acknowledges the overall value of that report and its findings in helping the committee to prepare the present report.

As noted previously, the draft LTCP includes modification and cxpansion of the wet weather treatment capacity of the ALCOSAN wastewater treatment plant to 310 mgd peak secondary treatment capacity and 565 mgd wet weather treatment for a total of 875 mgd (ALCOSAN, 1999). The TPRC concluded that the portion of the LTCP to expand the treatment plant, an activity that has already advanced to the facilities-planning stage, is a cost-effective component of the plan and should move forward. However, many concerns with other aspects of the draft LTCP were raised in that report.

The draft LTCP calls for reducing infiltration and inflow into separate sewer systems connected to the ALCOSAN interceptors from peak flows of 1,000 gpcd to a Pennsylvania state design standard of 250 gpcd at an estimated cost of $2 billion (ALCOSAN, 1999). The TPR report questions, however, whether such an investment would be adequate to reduce flows to this level (TPRC, 2002). Furthermore, the TPRC raised the question as to why such a reduction would be needed since the LTCP facilities were planned as if no reduction in flow were achieved.

The draft LTCP does not address costs of specific facilities required for systems other than the ALCOSAN-owned system, except for the Municipal Collection System Rehabilitation and/or Reconstruction Program (ALCOSAN, 1999). In order to convey large amounts of sewage to the ALCOSAN wastewater treatment plant during wet weather events, additional trunk sewers in partner communities may be required, but costs for this are not included in the LTCP. If paid for separately by the partner communities, there are likely to be inequities in payment schedules for the different communities and the TPRC (2002) recommended that this be addressed.

As noted previously, several recent developments in law, policy, and guidance have changed the regulatory requirements for addressing CSOs and SSOs. Because of these changes, the TPR report suggests that detailed monitoring and modeling are necessary to assess water quality impacts of wet weather events. They are also necessary to determine what actions and controls will be most cost-effective and timely, essentially favoring activities similar to those required in the demonstration approach. The TPRC report notes that other cities have invested large funds in overflow controls but have nevertheless faced subsequent and stringent enforcement actions because of public dissatisfaction that priority problems were not addressed. Public participation in the selection of facilities is recommended in the TPR report, together with a wider range of alternatives to better help determine how pollution prevention is related to facilities costs. A full financial capability analysis, including consideration of financial impacts on disadvantaged communities, is needed.

The TPR report raises concerns about heavy reliance on the use of vortex separators, because their effectiveness for water quality improvement has been inconsistent across the

[6] Available on-line at *http://www.alcosan.org/directory/third_party.htm.*

country (TPRC, 2002). In the draft LTCP, vortex separators are proposed for use without disinfection, so the separators may be ineffective for reducing the concentration of microorganisms in combined sewage. Also, questions about the acceptability of the many proposed upstream treatment and storage facilities have been raised. Additionally, the draft LTCP does not address many other water quality problems in the region, such as urban stormwater, acid mine drainage, upstream sewer overflows from other communities, wildcat sewers, failing septic systems, and agricultural runoff (see Chapters 3 and 4).

Based on these criticisms, the TPRC (2002) questions the ability of the presumption approach proposed by the draft LTCP to meet water quality standards. In fact, a conceptual screening model approach provided in that report indicated that bacterial water quality standards were unlikely to be met by the draft LTCP. The TPRC recommended that a watershed approach be adopted in order to determine what watershed needs are most important to stakeholders and how they can best be addressed and funded. Such an approach may affect the relative significance of CSOs and SSOs. The TPR report also recommends that a phased approach be used. The first phase would involve implementing the portions of the LTCP that are clearly cost-effective and would be part of any final plan for the region. This includes inspecting priority areas of the collection system and correcting structural deficiencies, gathering information needed to complete comprehensive facilities plans, conducting a comprehensive financial capability analysis, and beginning a process to determine the ultimate wet weather water quality requirements. In the second phase, additional abatement projects that are identified as a result of system inspections would be implemented. In the third and final phase, the extended LTCP actions identified and prioritized through use of a watershed approach would be undertaken. Lastly, the TPRC recommended that to develop a workable approach, several social, institutional, and financial realities must be addressed.

Cost Comparisons of ALCOSAN's Draft LTCP

Other regions of the United States have experience with addressing wet weather-related water quality problems, and the cost of these efforts has been documented. The ALCOSAN draft LTCP includes a cost estimate of $3 billion, which includes activities by ALCOSAN within its own systems at $1 billion and costs for rehabilitation in the municipal collection systems and of private homeowners' laterals at $2 billion (in 1999 dollars). These costs lead to an estimate of a total cost for the LTCP of $9,000 per ALCOSAN customer. They do not include costs for the more than 500,000 sewered customers in the region who are served by organizations other than ALCOSAN or costs associated with rehabilitation of OSTDSs in southwestern Pennsylvania. However, these costs represent an important point for comparison between the region and other areas of the country with wet weather water quality issues.

The third party review of the ALCOSAN draft LTCP (TPRC, 2002) provides EPA data on CSO program costs for representative U.S. municipalities (see Table 5-4). Although not directly comparable since these programs did not include costs for lateral repair, the cost per household indicated is lower than anticipated in the ALCOSAN draft LTCP. Based on the proposed multibillion-dollar investment in reducing untreated sewage overflows, regional investments to improve water quality must be prioritized.

TABLE 5-4 EPA Data on CSO Program Costs[a] for Representative U.S. Municipalities

Community	Cost to Date	Total Cost Expected	Grants	Population	Total Cost per Person	Local Cost per Person (cost to date minus grants)
Bremerton, WA	$23.00	$44.00	$7.20	36,000	$1,222	$439
Burlington, LA	$14.80	$35.10	$7.00	27,500	$1,276	$284
Rouge River, MI	$350.00	$1,300.00	$193.00	1,600,000	$813	$98
San Francisco, CA	$1,472.00	$1,472.00	$692.00	800,000	$1,840	$975

[a] Costs are in million dollars.
SOURCE: Adapted from TPRC, 2002.

As noted previously, the estimated cost of the ALCOSAN draft LTCP is in excess of $3 billion over approximately 50 years. This translates to an investment of about $9,000+ per household in the ALCOSAN service area (see Figures 1-1 and 1-2) or approximately $3,400 per person (TPRC, 2002). Because of the different infiltration and inflow problems in different wastewater collection systems and the uneven distribution of needed lateral repairs in the region, these costs will not be divided evenly over all ALCOSAN customers. However, an average cost over the customer base (amortizing the $3 billion bill over 50 years at a 5 percent interest rate) would be $40 per month. Some customers would pay much less and some much more; however, since this value is already at the affordability level suggested by the EPA for the median income in the United States (EPA, 2001a), the estimated LTCP costs are considered high. It is likely that this additional cost, however it is distributed among the population, would result in wastewater and water bills for some that would be substantially higher than in most communities around the country. Furthermore, comparisons of quarterly water and wastewater bills for Pittsburgh, detailed in Chapter 6, indicate that even before customers pay for the challenge of solving CSO problems, Pittsburgh's water and sewer bills are higher than those of many U.S. cities with comparable population sizes.

Past federal grants have been a substantial part of the solution for many communities nationwide, thus reducing the local burden, but such federal grants are no longer likely to be available. Grant funding, which existed at a national annual rate of $8 billion to $10 billion at today's values, has been replaced with annual appropriations of $2 billion to $3 billion directed primarily to loans to small communities with severe water quality problems and limited affordability capability. Extensive 20-year efforts by various water utilities and communities to obtain significant funding have not been successful, but federal funds remain available for research, security, and development of specific projects that received congressional approval based on the national significance of the project. Although such opportunities exist in southwestern Pennsylvania, in the committee's judgment the likelihood of obtaining major federal grants for wet weather projects is remote.

Current Status of ALCOSAN Wet Weather Plans

The *Narrative Summary of Active Projects: 2004 ALCOSAN Capital Budget* (ALCOSAN, 2004) summarizes many ongoing and planned activities and their relationship to the ongoing consent decree negotiations, the draft LTCP (ALCOSAN, 1999), and the TPR report (TPRC, 2002) that are of direct relevance to this report, including CSO and SSO abatement

Programs; interceptor system improvements; planned (Phase II) wet weather plant expansion; distributed control system upgrade; the Satellite Treatment Facilities Demonstration Program; and the Flow Monitoring/Modeling/Water Quality Sampling Program. These important projects will be of central importance in improving water quality in the region's urban core areas and critical to the success of a broader water quality management plan for southwestern Pennsylvania.

Regulatory Compliance Strategy

Regardless of the regulatory approach used in ALCOSAN's LTCP for controlling CSOs, the committee concludes that it is necessary to address watershed-wide problems and sources of contaminants other than CSOs and SSOs. CSO abatement elements must operate as part of a system of controls, meaning that the individual abatement systems outside ALCOSAN's authority and the ALCOSAN system must be reasonably optimized. The institutional and management needs and options to administer and finance such a system are presented in Chapter 6. Any solution must be cost-effective, and previously estimated costs (e.g., TPRC, 2002) have resulted in a high cost per person. Even so, the risk of failure to achieve water quality standards in the receiving streams is high. The system-wide solution must also include considerations of affordability and be cost-effective. Finally, ALCOSAN's draft LTCP does not adequately consider the potential for innovative technologies and approaches for controlling wet weather discharges such as those described later in this chapter.

Economic Benefits of Reducing Water Quality Impairment

As discussed in Chapter 2, abundant water for human populations, industrial production, and transportation has been a cornerstone of the historic development of southwestern Pennsylvania. Since the early 1970s, there have been significant improvements in water quality, but problems remain. Addressing these problems would benefit the region's current and future residents, as well as people who visit for recreational or other purposes; however, there will also be significant costs. Accordingly, a key issue is striking a balance between benefits and costs of water quality improvement. Cost and financing issues are addressed in more detail in Chapters 6. In this section, the general economic benefits of water quality improvements are examined.

Economic Services of Water

The contributions of water resources to the economy and the quality of life of the region derive from the services that water resources provide to people directly (e.g., for drinking, recreation, as an amenity) and indirectly from the use of water in the production of goods and services that people consume or value and that provide income and employment. The services that water resources provide to society are incredibly diverse. For a comprehensive review of this rapidly growing field, see *Valuing Ecosystem Services: Toward Better Environmental Decision-Making* (NRC, 2004b). Historically, water resources have been economically valued and managed for conventional uses such as public drinking water supplies, power generation, irrigation, transportation, recreation, fisheries, and waste disposal and assimilation. Until recently, decision makers have tended to take for granted (i.e., place no value on) many other services provided by water resources that can be difficult to value economically such as providing habitat for aquatic ecosystems and aesthetics. However, these types of services are increasingly being recognized, leading to new demands on water management.

Mitchell and Carson (1986) provide a useful typology for considering the economic value of services from freshwater, and the sources of benefits or costs from changes in water quality (see Table 5-5). The values of water are broadly categorized as use values and nonuse (existence) values. The use class consists of current direct and indirect ways in which individuals or other entities make physical use of water. Examples include domestic uses (drinking, cleaning, preparing food, watering trees and lawns), recreational uses (swimming, fishing, boating, near-water recreation), and use for the production of goods and services (e.g., water used in industrial production, power generation, irrigation of crops, and hotels and restaurants). Use values can be subdivided into in-stream uses ("nonconsumptive" uses such as swimming, boating, fishing) and uses that require withdrawal ("consumptive" uses such as drinking, irrigation, cooling). The indirect category is subdivided into aesthetic and ecosystem values. Nonuse values (also called existence values or passive use values) are values that people hold for a resource without using or visiting that resource and do not require direct experience of use of a resource (Freeman, 2003; NRC, 2004b). Thus, people who make no direct or indirect use of particular water resources may still value their existence and condition.

Water quality degradation as found on various rivers and stream segments in the southwestern Pennsylvania region impairs water for various uses and possibly for existence values. Economic losses are associated with these impairments, and economic benefits are to be gained from their abatement and removal. Estimates of the benefits and costs of water quality protection projects could be very useful for helping to set water quality priorities and selection of projects that make economic sense (see Box 5-1). There are, however, no comprehensive estimates of the economic benefits of addressing the remaining water quality problems for southwestern Pennsylvania or from proposed projects to address the region's water quality

TABLE 5-5 A Typology of Possible Benefits Resulting from Improvements in Freshwater Quality

Benefit Class	Benefit Category	Benefit Subcategory (Examples)
Use	In-stream	Recreational (fishing, swimming, boating)
		Commercial (fishing, navigation)
	Withdrawal	Municipal (drinking water)
		Agriculture (irrigation)
		Industrial/commercial (process treatment)
	Aesthetic	Enhanced near-water recreation (hiking, picnicking)
		Enhanced routine viewing (office and home views)
	Ecosystem	Enhanced recreation support (duck hunting)
		Enhanced general ecosystem support (food chain)
Nonuse (Existence)	Vicarious consumption	Significant others (relatives, close friends)
		Diffuse others (general public)
	Stewardship	Inherent (preserving remote wetlands)
		Bequest (family, future generations)

SOURCE: Adapted from Mitchell and Carson, 1986.

BOX 5-1
Economic Benefits of Water Quality Improvements: Basic Concepts

Concepts and methods for estimating the benefits and costs of environmental improvements are well-developed (see for example, EPA, 2000a; Freeman, 2003; NRC, 2004b). The fundamental theoretical concept for valuing the benefits of water quality improvement is willingness to pay (WTP). An individual's WTP for a water quality improvement is the maximum amount of money the individual would voluntarily exchange for an improvement in water quality rather than have water remain in the existing condition. The underlying principal is that the economic value of goods and services, whether sold in a market or made available through other means, is defined by the economic trade-offs individuals are willing and able to make in order to consume the good or service. When the amount that an individual is willing to pay exceeds the cost of the good or service to the individual, the individual realizes an "economic surplus." Benefit estimation essentially entails measurement of changes in economic surpluses. The total benefit of a water quality improvement is the sum of the individual benefits.

For example, suppose a particular individual considers a good day of fishing to be worth $50. Presently, the individual must travel an hour to find a spot that will provide a good day of fishing, the cost of which, including time, is $40.00. Thus, the economic surplus per day of fishing is $10.00. Now suppose that a stream restoration project nearer to home reduces the travel time and cost for a good day of fishing in half. The economic surplus for a fishing day at the new site is $30.00, resulting in an economic surplus gain of $20.00 per fishing day. Suppose this hypothetical individual fished 10 days a year at the original site but would fish 15 days a year at the closer, less expensive site. The total benefit to the individual is then $350.00 per year.

While the concept of WTP may not have immediate meaning or appeal to the noneconomist, some of the measurements to infer WTP may have utility. These measurements would include the following: increased value of residential properties adjacent to streams, rivers, or lakes that would benefit from water quality improvements that increase the amenities of living on or near such water resources; reduced expenditures on medications and medical care that would result from reduced incidence of waterborne disease transmitted in drinking water or by exposure through in-stream recreation; reduced expenditures on water treatment or procurement by households, municipalities, and industries to achieve mandated or desired water quality levels in drinking water or water used in industrial processes; and increased profits to businesses that realize productivity gains from water quality improvements.

It is important to emphasize that environmental valuation for project and policy analysis generally is not concerned with the total economic value of an existing or prospective environmental condition, but rather in the benefit or cost that results from a change in condition, because this is the relevant information in assessing whether a change is economically beneficial.

An assessment of the benefits of a water quality project or policy should in principle begin with a set of questions about current conditions without the policy and the conditions projected to exist with the policy. In the typical "effect-by-effect" approach to benefit assessment (EPA, 2000a; Ribaudo and Shortle, 2001), key data would include the following:

1. Existing water quality conditions, as described by various physical, chemical, and biological parameters; it is apparent from the discussion above that information on current conditions is limited.
2. Projections of changes in these conditions as a result of the project or policy. As indicated in this chapter, an essential step toward assessing the relationship between water quality stressors and water quality conditions—and therefore, reliable valuation of the benefits of water quality improvement projects—is modeling the relationships between water quality conditions for various uses and pollution loads (see, for example, the approach to benefit and cost assessment described in Ribaudo and Shortle, 2001; see also NRC, 2001 on the TMDL approach to water quality management).
3. Identification of use or existence categories that would be affected by the projected change in conditions.
4. Projections of the changes beneficial impacts within the affected benefits categories. For example, consider a project that reduces waterborne disease risk in a particular stream segment used for swimming. One set of beneficial impacts would be the reduced likelihood of waterborne disease among existing users. Another set of beneficial impacts may be increased use of the stream for swimming by people who are currently unwilling to take the risk. This example highlights a key aspect of estimating beneficial impacts—namely, that it generally requires projecting behavioral changes that may accompany water quality changes.

problems. Nevertheless, potentially large benefits from addressing remaining water quality problems in the region can reasonably be anticipated. To obtain a greater sense of the potential economic benefits of water quality protection and improvements, the benefits of drinking water protection and recreational benefits are discussed in the following sections.

Benefits of Reducing Drinking Water Contamination

Given the frequent discharges of untreated sewage upstream of drinking water intakes and frequent findings of pathogens and their microbial indicators in source waters discussed in Chapters 3 and 4, it can be expected that there would be considerable economic benefits from reducing microbial contamination of source waters, contingent on a positive probability of treatment system failures. Use of contaminated source waters for drinking imposes at least three types of economic costs. The first is the cost of water treatment to eliminate or reduce the presence of disease-causing agents in drinking water to acceptable levels. The second is cost of the risks of disease from potential exposure to contaminated drinking water in case of treatment system failure. Economic benefits would accrue from reduced discharges of untreated sewage if the reduction permits treatment costs to be diminished or waterborne disease risks to be reduced. Costs are also incurred in monitoring water quality and in treating water to meet federal and state drinking water standards.

Generally, the risks of endemic waterborne disease in a well-managed system are very low (see NRC, 2004a). However, in some instances, treatment system failures can result in episodes where finished water is grossly contaminated and poses a significant threat to the health of customers. In such instances, water system authorities must notify customers that the water is unfit for consumption and issue advisories to boil water or take other steps to protect their health (see also Box 3-4). Costs associated with measures to prevent disease are referred to as avoidance costs and can be very high depending on the number of people affected and the length of advisory (e.g., Abdalla, 1994; Harrington et al., 1991; Laughland et al., 1996). The costs associated with morbidity from waterborne diseases include those of medical treatment, days lost at work, and related pain and suffering. The costs of morbidity and mortality from waterborne disease outbreaks can be substantial depending on the number of people affected and the severity of illness.

The costs of a water contamination event can be illustrated by a giardiasis outbreak in Pittston, a town in northeastern Pennsylvania. The PADEP issued a boil water advisory for the community in December 1983. The advisory was lifted for some residents in March 1984 but remained in effect for nearly nine months. A study by Resources for the Future estimated the costs of the waterborne disease outbreak at between $23 million and $55 million (in 1984 dollars) (Harrington et al., 1991). Medical costs, lost work time, lost work productivity, and lost leisure time of those who became ill were estimated at $5.6 million. The costs of actions taken by community residents to avoid drinking contaminated water (e.g., boiling water, hauling or purchasing water) were estimated at $12.9 million. Costs to area businesses (e.g., restaurants) from measures to contend with contaminated water or from business losses were estimated at $3.6 million. It is apparent from this example that the economic costs of a widespread contamination event resulting in a boil water advisory and illness affecting one of the many community water supply systems in southwestern Pennsylvania could be quite substantial.

Recreational Benefits

In addition to the potential economic benefits resulting from drinking water protection, economic benefits may also be derived from improvements in water quality for other uses. As discussed in Chapter 3, approximately 26 miles of streams and rivers have been listed as impaired for contact recreational use in southwestern Pennsylvania. Research on water quality benefits has demonstrated that there can be substantial economic value in restoring water quality to levels that permit recreational uses. These benefits can accrue from reductions in the risks of waterborne disease and their related costs; from increases in the opportunities and levels of swimming, boating, and fishing; and from improvements that enhance the quality of recreational experiences (see, for example, EPA, 2000a; Mitchell and Carson, 1986).

The benefits of water quality improvements at recreational sites on the Monongahela River in the early 1980s were estimated by Smith et al. (1986). Depending on the specific valuation methodology, the study found the value per trip per person for an improvement in water quality from boatable to swimmable to be between $12.95 and $56.39 (updated to 1998 dollars as reported in Koteen et al., 2002). These findings in conjunction with the significant improvements in water quality in southwestern Pennsylvania over the past 30 years would suggest that the region has already realized possibly large economic benefits from water quality protection and improvement.

The essential issue now is the additional gains that may be obtained from further reductions in recreational impairments. Clearly lacking for such an analysis at this time are estimates of the increased recreational use of waters that would occur with possible water quality improvement projects. There is, however, ample evidence that current residents are willing to pay for recreational water quality improvements. For example, Heberling et al. (2004) estimate the economic benefits of remediating AMD in Clearfield Creek in Clearfield County, Pennsylvania, to the east of the study region. The study estimated that, on average, Clearfield County households would be willing to pay at the margin $54 per year to increase the quality of the stream from its currently highly degraded baseline level to fishable (holding all else constant). Farber and Griner (2000) estimate the benefits of improving the water quality of Loyalhanna Creek and the Conemaugh River, both subwatersheds of the Allegheny, that are degraded primarily by AMD. For one scenario, improving stream condition from severely degraded to unpolluted, they find valuations ranging from $75.63 to $112.44 per household per year. Both studies were conducted in the mid-1990s.

Less place specific, but still indicative of the economic values people place on recreational use of waters, are estimates of the economic value of various water-based recreational activities adapted from a literature review by Rosenberger and Loomis (2001). These are presented in Table 5-6. The values are per person per activity-day and can be used to provide a rough guide to possible gains resulting from increased recreational use along with reasonable projections of the increased recreational use of water resources associated with water quality improvements.

Summary

There are no comprehensive estimates of the economic benefits of addressing the remaining water quality problems of southwestern Pennsylvania or from proposed projects to

TABLE 5-6 Selected Average Consumer Surplus Values per Activity-Day per Person from Recreation Demand Studies (1967-1998)[a]

Activity	Mean of Estimates	Range of Estimates
Swimming	$21.08	$1.83-49.08
Motorized boating	$34.75	$4.40-169.68
Nonmotorized boating	$61.57	$15.04-263.68
Waterfowl hunting	$31.61	$2.06-142.82
Fishing	$35.89	$1.73-210.94

[a] All amounts are in 1996 dollars.
SOURCE: Adapted from Rosenberger and Loomis, 2001.

address the region's water quality problems. Nevertheless, the region would be expected to benefit economically from measures that significantly reduce drinking water risks and enhance recreational opportunities. Whether benefits would exceed the costs of water quality projects cannot be known without analysis of specific options. Estimates of the benefits, in addition to the costs, of specific water quality projects could be very useful for helping to set water quality priorities and selecting projects that make economic sense considering both benefits and costs. Further, to the extent that it can be shown that benefits exceed costs, benefit estimates could be used as a tool for building public support for regional water quality initiatives.

COMPREHENSIVE WATERSHED ASSESSMENT AND RESPONSE PLAN

The committee believes that the evaluation of causes and sources of water quality impairment in southwestern Pennsylvania must be refined and integrated with the decision-making structure for water quality improvement actions. Further, the appropriate scale for this evaluation and for water quality improvement decisions is the watershed level. In this regard, the flow of water from the Allegheny and Monongahela Rivers to the Ohio River is a single hydrologic regime. Prudent management of water quality in this system should consider all factors that affect water quality and anticipate future uses of the water resources of the region. To achieve this, it is necessary to develop both a technical and an institutional-financial approach. The institutional and financial approach is discussed in Chapter 6, and the technical approach is embodied in what the committee calls the "Comprehensive Watershed Assessment and Response Plan," or CWARP is discussed in this chapter. It is important to note that the "Three Rivers CWARP" described in this report is not a single document or program. It is an intentionally flexible umbrella concept identifying activities that can be carried out by the organizations that are most technically and institutionally capable of achieving the desired results. Therefore, activities will be conducted at various levels and by various organizations depending on existing and potential capabilities.

The CWARP relies on the acquisition of sufficient knowledge about the Three Rivers watershed's response to various contaminant loads at different temporal and spatial scales to prioritize problem agents, develop specific action plans, predict water quality benefits, and estimate related costs and affordability. It is especially important that this plan be followed in the ALCOSAN service area and neighboring communities where very large capital outlays have been proposed to reduce or eliminate CSOs.

Planning and Implementation Framework

The framework recommended for planning and implementation of CWARP consists of five basic steps as follows:

I. Problem identification
II. Assessment of existing conditions, including quantification of loads and modeling their relationships to water quality
III. Projection of future loads, their timing, location, and impacts on streams
IV. Formulation and evaluation of alternative management strategies, including assessment of the effects of alternatives on future conditions and ordering and scheduling of various elements of the preferred strategy
V. Adaptive implementation of elements of the strategy, relying on feedback from implementation of each element to inform reformulation and evaluation of subsequent elements.

Water quality problems in the region occur at several different temporal and spatial scales, arise from several different classes of sources, and will require very different planning and management strategies and different kinds of organizational arrangements. At one end of the spectrum are problems related to long-distance transport of contaminants and their impacts within river basins at the scale of the Allegheny, Monongahela, and Ohio Rivers over which state and interstate agencies have jurisdiction. At the other end of the spectrum are threats to health of individual families served and/or affected by nearby failing on-lot wastewater disposal systems, the regulatory responsibility for which has been delegated to local governments. In between are the problems of contamination from CSOs, SSOs, and stormwater runoff during and after rainstorms in high-density urban areas. In this report, planning and management for water resources is viewed as consisting of needs at four interrelated scales:

- river basin;
- multicounty/metropolitan scale;
- high-density urban areas; and
- rural areas.

The five-step process must be adapted to address each of these scales.

The committee recognizes that the region is not starting with a blank slate, but nonetheless it is useful to provide this framework. Step I has been largely completed for all four scales. The assessment of water quality and sources of contamination as discussed in Chapters 3 and 4 provides a reasonably complete picture of the problems. Substantial progress has been made on estimation of loads and development of predictive models as called for in Step II, but as noted earlier, significant gaps remain. Because the problems are largely associated with existing conditions and there is only modest growth in the region as a whole, Step III may be less important, but changes in land use that are occurring in suburban areas (see Chapter 2) cannot be ignored. Some of the alternatives called for in Step IV of CWARP have been formulated in ALCOSAN's draft LTCP (ALCOSAN, 1999) and focus on the urban core areas of the region, but that set of options is limited in scope. Many steps toward water quality improvement in the region have already been identified and implemented, but it is clear that many more will be required under Step V for all scales of CWARP.

River Basin Planning and Management

Southwestern Pennsylvania lies primarily within the Allegheny, Monogahela, and Ohio River basins that transport contaminants into the region and export remnants of its waste and stormwater to downstream reaches of the Ohio. Several water quality problems in the area appear to transcend regional boundaries and are logically the jurisdiction of state and interstate agencies. The two most prominent large-scale water quality problems appear to be: (1) metals related to AMD, and (2) mercury and persistent organics, both of which may be legacy contaminants, resulting from past agricultural, industrial (especially mining), stormwater, and sewage practices. The microbiological contaminant linkages are uncertain, despite the current regulatory environment. Given the lack of established linkages at the river basin scale, CWARP should be focused on Steps I and II, which are discussed together below.

Steps I and II of the planning process at the basin scale should seek to enhance data collection and analysis by various agencies, ongoing or planned, to better define long-distance transport and fate of contaminants into and out of the region. One program that shows promise is EPA's rejuvenated Environmental Monitoring and Assessment Program (EMAP).[7] The EPA has recognized that comprehensive, large-scale monitoring information on large river ecosystems is vital but lacking. One component of that program, called Great Rivers,[8] has as its objective to improve the scientific basis for water quality assessment at the large scale. It uses a probability sampling approach to generate statistically valid inferences about the status and trends of water quality in those rivers.

Preliminary designs for monitoring and analysis have been completed for 3 of 11 river systems nationwide, including the Ohio River from the City of Pittsburgh to its confluence with the Mississippi. The Three Rivers CWARP should take advantage of the Great Rivers component of EMAP to help design and fund the needed additional water quality monitoring. The Great Rivers effort presents an opportunity for EPA and Pittsburgh region water quality authorities to collaborate on water quality data collection, because there is a shared interest in generating water quality status and trends information for improved decision making, especially as it relates to provisions of the Clean Water Act. While the program is still in development, it would seem wise to extend it spatially to capture the lower reaches of the Allegheny and Monongahela Rivers. This could assist in developing a validated probabilistic approach to sampling as well as in eliciting valuable direction and critique from EPA and the Ohio River Valley Water Sanitation Commission (ORSANCO; see Chapter 6 for further information).

Several programs of the U.S. Geological Survey (USGS) should also provide valuable sources of water quality information for CWARP. The USGS is widely recognized for its national stream gauge and water quality programs, but it is a major contributor to regional water quality programs, as well. The USGS has provided valuable water quality information for more than a century in the Ohio River basin in Pennsylvania and in the greater Pittsburgh region. More specifically, USGS administers both historic and current data for more than 400 water quality sampling sites in southwestern Pennsylvania, where it tests for various and multiple parameters of water quality across the region. Although most of the historic USGS sampling efforts targeted chemical (e.g., nutrients, pH), physical (e.g., temperature), biological (e.g., dissolved oxygen), inorganic, radiologic, or sediment attributes of water quality, beginning in

---

[7] Further information about EMAP is available on-line at *http://www.epa.gov/emap/index.html*.
[8] For more information on EPA's Great Rivers program, see *http://www.epa.gov/nheerl/arm/greatrivers.htm*.

2001, sampling in some areas now includes testing for fecal coliforms, *Escherichia coli*, enterococci bacteria, and turbidity (see Chapter 3 for further information).

Some of these multiple-test sites represent collaborative efforts in the region to address water quality issues from the USGS National Water Quality Assessment Program (NAWQA), the National Stream Quality Accounting Network, the U.S. Army Corps of Engineers and other federal agencies; ORSANCO; and Carnegie Mellon University. Southwestern Pennsylvania provides a good example of how a national-level program can generate water quality monitoring data at regional and local scales in a timely, responsive, and reliable manner that can be used in support of decision making. The NAWQA program, in particular, for several years sought to gain understanding of water quality conditions, trends, and stressors in the Allegheny and Monongahela basins (see Anderson et al., 2000 and Appendix B) but unfortunately, that focused effort has been terminated for budgetary reasons (see NRC, 2002). Nonetheless, data collected on water quality are maintained by the USGS, are available to the public, and provide a long-term, rich source to track changes and trends in water quality in the Allegheny, Monongahela, and Ohio Rivers.

Biological Water Quality Monitoring

Because regional information on the biological quality of receiving waters is scant (see Chapter 3), its collection during and in support of CWARP at the river basin scale is critical. To be comprehensive, it is necessary to assess and evaluate not only physical and chemical, but also biological water quality information to determine ecosystem health (see Box 5-2). Biological water quality indices, and their change over time, can yield important information about ecosystem change and help quantify the environmental benefits affected by pollution abatement. Thus, information collection for CWARP should include biological data to assist in ecosystem health assessment benchmarking and to help document changes to the ecosystem that occur as a result of changing stressors. At a minimum, CWARP should be designed to establish an Index of Biotic Integrity (IBI) benchmark so that changes in environmental health can be linked to changes in the IBI for the regions' three main stem rivers.

Step III should focus on models to track the fate and transport of contaminants within the basins. It should be noted that EMAP's Great Rivers program is limited to an assessment of status and trends of aquatic ecosystems. The preliminary design for the Ohio River component is focused only on effects; it does not include any tasks to track causes of degradation or to relate cause and effect through water quality or ecosystem models.

Sources and Linkages

Given existing evidence and that which may be forthcoming from the EMAP Great Rivers program and other monitoring, there is a need to identify probable sources of contaminants that places stress on the aquatic ecosystems in southwestern Pennsylvania. It would be desirable to use an aquatic ecosystem model such as EPA's AQUATOX in such an effort. This tracks concentrations of nutrients and organic chemicals, temperature, and their

BOX 5-2
Overview of Biological Water Quality Monitoring

The use of biological indices as surrogates for in-stream environmental quality has become increasingly common across the United States. The modern water quality monitoring criteria were developed in order to provide a system for measuring the "biological integrity" sought by the Clean Water Act (Karr et al., 1986). Davis and Simon (1995) describe the development and application of modern indices of biological integrity.

The most widely applied methods for assessing biological integrity involve collecting, counting, and enumerating species within a taxonomic group (typically fish or benthic macroinvertebrates such as aquatic insect larvae) and comparing the measured characteristics of the sampled community to a reference community using a scoring method. Standardized collection methods are codified for any local or regional area. Watershed areas are generally characterized within some defined biogeographic area (typically ecoregions). Habitat characterization is performed and usually compared to some concept of an expected habitat to support the assessed fish or macroinvertebrate community.

Scores calculated by these methods are unit-less numbers that represent the relative health and performance of the chosen community in that location. The most common versions of these scores are generally termed Indices of Biotic Integrity for stream fishes in North America (Karr et al., 1986). The IBI is a "multimetric" index based on adding subscores for different characteristics of the distribution and abundance of species within a sample. The indices incorporate measures of pollution tolerance, pollution intolerance, distribution of different feeding groups, and other characteristics of resident fish that provide measures of community composition and performance. The original IBI has been modified and incorporated across a range of watersheds and has been developed for other groups such as macroinvertebrates and even some terrestrial communities such as birds (see O'Connell et al., 1998).

The multimetric approach has been applied successfully in a regional and local context, such as the long term control plan for the City of Akron, Ohio.[1] The methods are most successful where a good knowledge base exists about expected community composition within a regional or watershed area. Although some of the methods used to develop these indices can be standardized nationally, it is not possible to develop an index that can universally be applied across widely varying regions because of the different biota normally present in different habitats and climates. The National Research Council included multimetric indices in its review of biological and ecological indicators as a local or regional indicator (NRC, 2000). The Heinz Center has also included multi-metric indicators as a possible gauge of freshwater conditions nationally (by aggregation of regional scores) in its recent effort to provide a measure of ecosystem resources for the nation (Heinz Center, 2002).

The Ohio River Valley Water Sanitation Commission Compact (1948)[2] states that the Ohio River should support and maintain a balanced, diverse, and healthy ecosystem. To determine whether pollution control efforts have achieved this objective, ORSANCO routinely conducts biological assessments that include the following components: (1) macroinvertebrate population reviews, (2) fish population studies, and (3) a fish tissue contaminants program. Macroinvertebrate populations provide perspective on aquatic life conditions in the river because many species are highly sensitive to pollution and are relatively immobile. The ORSANCO conducts macroinvertebrate sampling each year including site-specific studies, pool-wide studies, and river-wide surveys. Fish population studies have been a major component of ORSANCO monitoring activities throughout the history of the organization. Studies in 11 of 20 pools on the Ohio River below Pittsburgh have resulted in the collection of data for the development of biological criteria, or biocriteria, for the Ohio River. Methods employed included sampling fish in the chambers of locks at dams on the Ohio River and electrofishing. The fish tissue contaminants program of ORSANCO characterizes the levels of certain contaminants, such as pesticides or other organic chemicals, in Ohio River fish and provides a basis for determining the need for human health or fish consumption advisories.

[1] CSO Sytemwide Study Final Report, 1995, prepared by a consulting team to support the Akron Facilities Plan, which was finalized in 1998.

[2] The Ohio River Valley Water Sanitation Compact is available in its entirety on-line at http://www.orsanco.org/orsa/compact.pdf.

effects on selected species in the ecosystem, including fish, invertebrates, and aquatic plants.[9] Outputs are the concentration or masses of species distributed over time and space. This approach has its limits. While it can track effects on individual species, mechanistic ecosystem models have not advanced to the point of being able to predict community structure or biotic integrity.

Even in the absence of formal models that relate cause and effect at the river basin scale, it is important that the most likely source of ecological stresses be identified through best available evidence. Processes are in place for Step IV of basin-scale planning. They include the establishment of assimilative capacities of streams consistent with water quality standards through the development of total maximum daily loads (TMDLs). Federal and state agencies have this authority under the Clean Water Act. They also have the authority to allocate these loads to sources. As discussed in Chapter 6, federal and state financial assistance can also be targeted to priority areas. Thus, planning at the river basin scale establishes priorities and performance standards that must be addressed by sources within the region, whether they be urban or rural in character.

Multicounty/Metropolitan Planning and Management

A second scale included in CWARP is the multicounty or metropolitan level. Regional planning for transportation and economic development at the multicounty scale is being conducted by the Southwestern Pennsylvania Commission (SPC; see Chapter 6 for further information). This organization is the officially designated metropolitan planning organization for transportation planning. It has developed an extensive database on the regional landscape, economic activities, land use, and transportation systems. However, water resources planning at that scale is not well-developed. Either SPC or an alternative organization should formulate regional water resource plans and integrate them with transportation and land use plans.

The CWARP planning process should focus on at least two aspects of water management at the multicounty level. First, large-scale transportation planning is occurring at that scale. Decisions about transportation infrastructure have significant effects on land use and watersheds. At the very least, water quality planning at this scale should be sufficient to inform regional interests of the potential effects (including constraints if any) of water quality conditions on future development; consequences of development on water quality where it occurs; and how these effects and consequences can and should be modified. Second, planning at this scale should also result in the identification of opportunities for economies of scale in the delivery of water and wastewater services through cooperative arrangements among local governments. Those opportunities are likely to exist primarily among clusters of local governments located in a common drainage area, but substantial benefits of cooperative arrangements also may accrue to sets of small communities and rural areas that do not and cannot afford to have the technical or administrative expertise to manage their water resources properly.

Step III should be an assessment of impacts of growth on the water resource, existing infrastructure for water supply and water quality, and opportunities to achieve economies of scale in the provision of water supply and wastewater management. A useful technique to support this assessment is development of alternative growth scenarios and their relationships to

---

[9] For further information on AQUATOX and other ecosystem models, see *http://www.epa.gov/waterscience/models/*.

regional transportation systems. In addition, water demands, wastewater generation, stormwater, nonpoint source pollution, and conversion of land uses should be estimated for each scenario.

Priority areas should be identified in Step IV based on results of Step III, and strategies should be formulated to address those priorities. Strategies should include both positive incentives to guide development away from critical areas and toward cost-effective infrastructure using a variety of techniques, including but not limited to public education, zoning, investments in infrastructure, transfer of development rights, conservation easements, and outright purchase of especially sensitive areas.

Finally, the multicounty process should include as Step V a continuing monitoring, assessment, and reporting process that keeps the public informed about the state of water resources and needed corrective actions as development occurs.

High-Density Urban Area Planning and Management

As noted earlier, the most pressing water quality problem in southwestern Pennsylvania from a regulatory perspective, especially in the region's urban core areas, is degradation of the microbiological quality of streams due to CSOs, SSOs, and discharge from separate stormwater systems in wet weather conditions. Clearly the most significant of the high-density areas is the contiguous urban area centered on the City of Pittsburgh. An approximate boundary for that area is shown in Figure 5-2. It covers much of Allegheny County and portions of Beaver, Butler, Washington, and Westmoreland Counties (see also Box 1-2). More than 1.25 million people live in that area, and it is affected by both SSOs and CSOs. There are other high-density urban areas outside the contiguous urban core. Several smaller cities and boroughs not directly linked to the urban core have densities in excess of 3,300 persons per square mile—far above the criterion of 1,000 used by the U.S. Census Bureau for defining urban areas. Among them are Indiana, Greensburg, Uniontown, Washington, Butler, and New Castle. These urban areas may be experiencing SSOs, excessive infiltration and inflow, and other water quality problems, but they are not plagued by the problem of widespread CSOs. In the discussion that follows, all urban areas are treated alike, but steps in the process related to CSOs would not be applicable to the outlying urban areas. Some of the outlying urban areas also do not offer the same opportunities for cost-effective collaboration as in the contiguous urban core.

As discussed in Chapter 3, elevated levels of fecal indicator bacteria (e.g., coliforms) far in excess of water quality standards have been observed during and immediately after storm events that caused CSOs. What appears to be lacking is an adequate quantification of sources of fecal indicator bacteria and an effective strategy to reduce their levels to acceptable concentrations. Although Step I of the planning process is reasonably complete at the urban scale, Steps II-V are not and are discussed in the following sections.

Step II—Assessment of Existing Conditions

Step II of the CWARP process for high-density urban areas should include simultaneous monitoring of (1) wet weather discharges into the region's streams and rivers and (2) impacts on these receiving streams. It should also include development of models that are sufficient to disentangle effects of multiple sources on water quality in receiving streams. For example, models that track flows of stormwater and sanitary sewage from contributing watersheds into and through the array of collection systems to points of discharge are also included under Step II. The models and data should be available for public review, and data from these technical studies should be reduced and translated into needed corrective actions in a manner that is understandable to decision makers and the public in general.

Pollutants of primary concern in this context include the following:

- pathogenic microorganisms such as enteric viruses, *Cryptosporidium, Giardia*, and their surrogates (indicators) including total coliforms, fecal coliforms, and *Escherichia coli*;
- oxygen-demanding substances including carbonaceous and nitrogenous organics, suspended solids and sediments; and
- "conservative" toxic substances such as metals and toxic organics, including pesticides and herbicides.

Monitoring of Discharges of CSOs and Separate Stormwater Systems. The primary sources of wet weather contamination in the high-density urban areas are CSOs, SSOs, and surface water runoff, including direct runoff to streams and runoff that is transported through separate stormwater sewer systems (i.e., stormwater drainage).

An initial task is to delineate the areas that contribute to these sources. This includes not only the high-density urban areas, but lesser-developed areas of watersheds from which stormwater flows are mingled with urban runoff. Delineation of these areas will require a detailed examination of watershed boundaries and urban development in the area. A preliminary delineation is shown in Figure 5-2 for illustrative purposes with an overlay of contiguous urban areas and watershed boundaries in and adjacent to Allegheny County.

Within these watersheds it is important that monitoring stations be established and operated so as to provide estimated temporal and spatial patterns of pollutant loads from CSOs and from separate stormwater sewer system discharges. As discussed in Chapter 4, urban stormwater discharges are associated with numerous pollutants (including fecal indicator bacteria) and effects on receiving waters, though the impact of stormwater on public health has largely been ignored.

Montoring and Modeling of Receiving Water. A basic task in Step II of implementing CWARP in high-density urban areas is to construct a credible relationship between the sources of stressors (discharges from CSOs and separate stormwater sewer systems) and quality of water in receiving streams. This requires simultaneous monitoring of discharges and instream responses during storm events. It also requires that concentrations of contaminants in streams flowing into the system from upstream sources be accounted for by monitoring at system boundaries. A detailed monitoring system should be designed using specific information about locations of streams and stormwater management within contributing watersheds.

However, a substantive discussion of such information is considerably beyond the scope of this report. Thus, only a tentative delineation of the area to be monitored and modeled is provided (see Figure 5-3). To isolate the greater Pittsburgh area from upstream sources of pollution to the Monongahela, Youghiogheny, and Allegheny Rivers, boundary monitors would

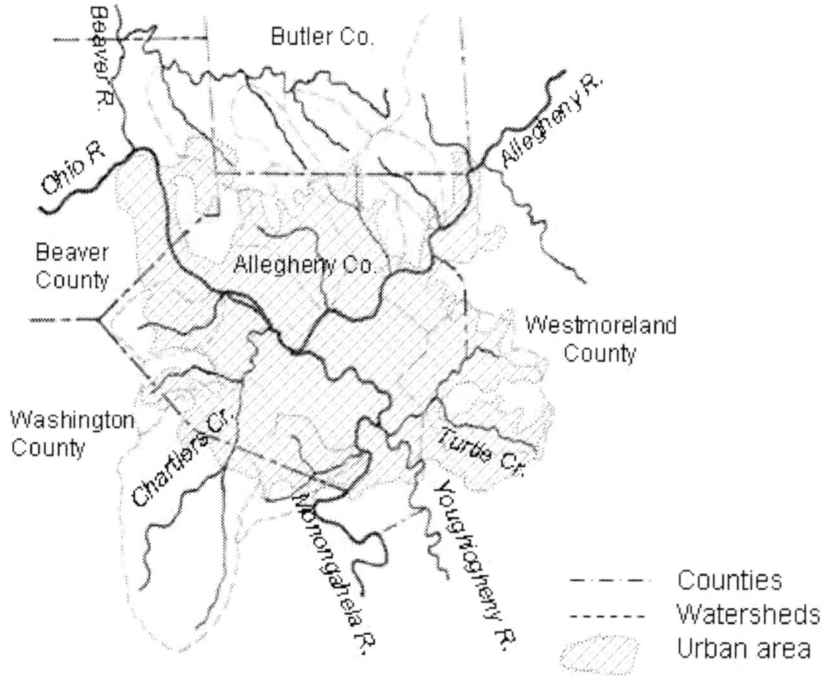

FIGURE 5-2 Watersheds and contiguous urban areas in and adjacent to Allegheny County. SOURCES: U.S. Census Bureau, *http://ftp2.census.gov/geo/maps/urbanarea/uaoutline/ UA2000/ua69697/* and *http://tiger.census.gov/cgi-bin/mapbrowse-tbl.*

have to be established somewhere in the vicinity of those shown in Figure 5-3. Downstream of those locations, there are approximately a dozen major watersheds that also contribute urban runoff and CSOs that affect water quality in the Monongahela, Allegheny, and Ohio Rivers. Each of these watersheds should be monitored to determine stresses placed on the major receiving streams. Instream monitoring should extend as far below the CSO and WWTP outfalls as necessary to determine the fate of contaminants entering the three major rivers. Wet weather monitoring should be focused on a short list of pollutants that are either known to cause or are suspected to cause significant water quality degradation during storm events. Fecal coliform bacteria or other indicators of waterborne pathogenic microorganisms would be high on that list.

Receiving water models should be able to simulate the fate of conservative (e.g., salt, metals), nonconservative (e.g., oxygen demand, nutrients), and microbiological (e.g., coliforms, *Cryptosporidium*) pollutants. The CWARP data collection program must be designed to quantify pollutants of concern discharged to the rivers as well as to calibrate the models with selected data on river concentrations of these same pollutants within the study area.

Wet weather events occur over relatively short periods of time (days), and there is considerable variation in both temporal and spatial patterns of loads and impacts within each event. Therefore, models covered by this discussion must be capable of representing both temporal and spatial variations, and they must include hydrologic and hydraulic behavior as well as water quality. Receiving water models exist for almost every type of waterbody such as wetlands, streams and rivers, lakes, reservoirs, and estuaries. Like collection system models, receiving water models have been available for some time and for almost every type of conceivable application. Some of the first, and still most frequently used, receiving water

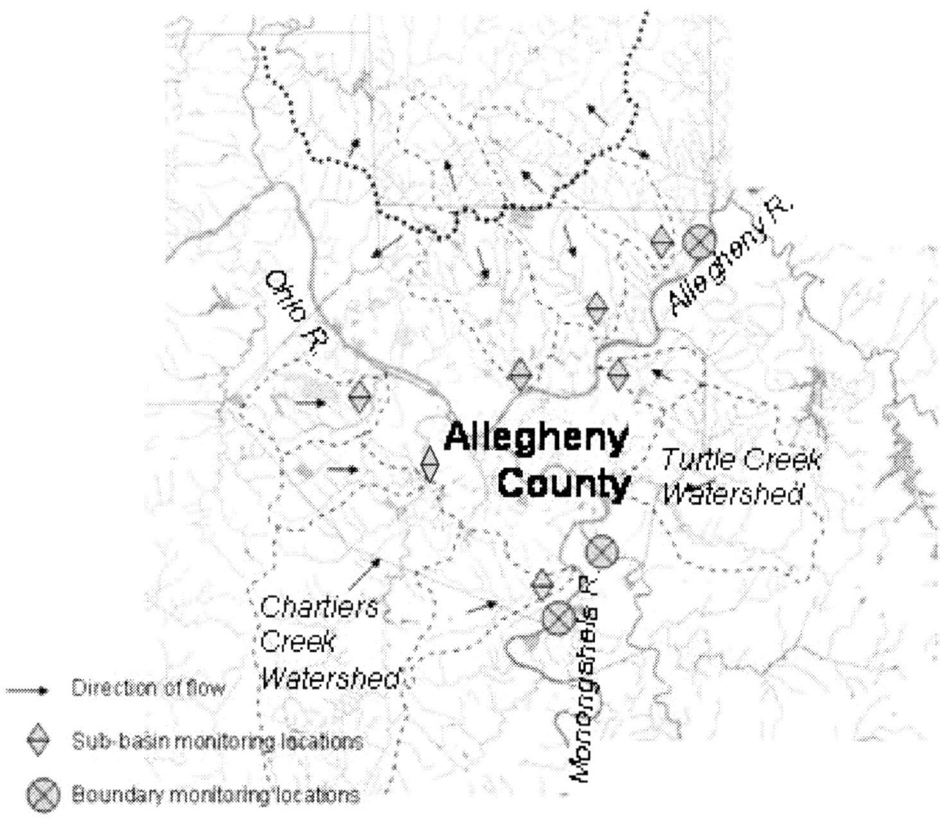

FIGURE 5-3 Preliminary representative sites for CWARP monitoring activities in the region's urban core areas.

models were developed by the U.S. Army Corps of Engineers, EPA, and the U.S. Department of Agriculture Natural Resource Conservation Service. Water quality models of this type, including the EPA's WASP6 and AQUATOX models, tend to be more complex and require much more input data than the steady-state models frequently used by regulatory agencies for waste load allocations. Despite their complexity, these models are continually being supported by federal agencies, and there is a body of experience with their use.

As with collection system models, the most recent improvement in the application of receiving water models is coordination with a geographic information system (GIS). This allows users to quickly generate input files, run modeling scenarios, and graphically and spatially view the modeling results through a GIS interface.

The type of water quality model necessary to simulate the Three Rivers watershed system for short-term CSO flow and load impacts is categorized as a multidimensional, time-variable water quality model of the finite difference or finite element mathematical process. Assuming the Three Rivers hydrodynamic regime is vertically well mixed (nonstratified), then a two-dimensional, advective (forward flow) hydrodynamic model coupled with a fully capable water quality model, including dissolved oxygen deficit and nutrients and sediment oxygen demand, would be necessary. The time increment used to advance the model must be sufficiently short to address the impact of short-term CSO wet weather flows and load.

Although receiving water quality modeling appears to be extremely limited in the region's three main stem rivers, the committee recommends that it be used to estimate impacts of pollution loadings on the receiving streams and to help prioritize alternatives for pollution control. Programs that have used similar models include ORSANCO's work in the Ohio River basin (see, for example, ORSANCO, 2002) and EPA's Great Lakes Initiative.[10]

It is important to recognize that in one of the first activities in CWARP planning and implementation (i.e., data/information collection), expert consultation is necessary to more fully define the particular model, the exact hierarchy of pollutant investigation, and the monitoring program necessary to calibrate and validate the mathematical model simulations prior to using the predictive components of the model.

Sewer System Routing Models. The use of an additional set of models from that used in Step II will be necessary to support Steps III and IV of CWARP in the high-density areas of southwestern Pennsylvania. Driving forces on the most severe wet weather events in the Pittsburgh region that have adverse effects on water quality are rainstorms. Although it is likely that melting snow packs also cause problems, they do not occur during seasons of intense recreational use of the region's streams and rivers. Both types of events could adversely affect any downstream water quality and supply, and aquatic ecosystems.

An understanding of the temporal and spatial patterns of the flow of stormwater and sanitary sewage into and through the extensive network of separate and combined sewer systems in the urban core of the Pittsburgh region is essential to the formulation and evaluation of management strategies. It is necessary to understand these flow patterns so as to identify key locations for which in-line or off-line storage and treatment options may be used to achieve cost-effective strategies. Flow patterns must be understood for a range of typical storm events to support design of collection systems. There is also a need to have monitoring and models of real time flows in the networks to support management of these flows during storm events that minimize adverse effects on receiving streams.

The suite of models for this task must include the following:

- rainfall-runoff models for all tributary watersheds, including both urbanized and non-urban areas that generate both hydraulic and contaminant inputs to combined sewer systems and separate stormwater sewer systems;
- dynamic stormwater routing models that transport stormwater inflows and their pollutant loads from points of entry to points of discharge; and
- dynamic combined sewer models that perform similar functions.

Dynamic Sewer System Modeling. Dynamic sewer system modeling can be utilized to simulate hydraulic and pollutant loading characteristics of extensive sewage collection systems, including CSO regulators. Mathematical models such as XP-SWMM (a stormwater and wastewater modeling package) have been used extensively for such purposes. ALCOSAN and its consultants utilized dynamic modeling in the development of the draft LTCP (ALCOSAN, 1999). The committee recommends that this application be expanded to simulate watershed-

[10] For more information on the EPA's Great Lakes Initiative, see *http://www.epa.gov/waterscience/GLI/*.

wide collection system behavior under CWARP. Hydraulic and pollutant load data developed from ALCOSAN's partner communities should be evaluated for inclusion in such an expanded model, if necessary. It is important that sewer system hydraulic modeling include analysis of express sewer contributions to the collection system.

Dynamic Stormwater Modeling. Dynamic (time-variable) stormwater models can combine hydrological (storm runoff hydrographs) and flood routing mechanics to calculate flows associated with precipitation from storms. These simulations are important for estimating the rate of stormwater flow at various locations and times over land and in the collection system, especially if sanitary sewer separation is recommended as one option to reduce or eventually eliminate CSOs. Understanding the integrated relationships between CSOs and separate stormwater flow is necessary for dealing with the safe and effective transport and discharge of each.

Real-Time Sewer Flow Control Modeling for Analysis and Operation. A key feature of a well-structured, long term CSO control plan is effective operation of the wastewater system in order to maximize in-system storage as well as flow to and through the treatment plant. Not only are these concepts embodied in EPA's CSO regulations (the nine minimum control provisions; see EPA, 1995b), but by effectively achieving these objectives, a municipality can substantially reduce the cost of compliance.

At the core of a successful operational control technology for both collection and treatment is the ability to monitor the system and then to respond to changes in the system in a manner that maximizes system capacity and minimizes adverse impacts on the environment resulting from the discharge of untreated sewage. Historically, most combined sewer systems have relied on static controls—such as fixed overflow weirs and off-line storage tanks—to regulate flows within the system. In most cases however, such methods will not meet EPA's nine minimum control criteria (EPA, 1995b).

To satisfy control requirements of the federal CSO program, while maximizing treatment and minimizing costs, a municipality must rely on the benefits inherent in real-time control (RTC) of the collection and treatment system. RTC modeling is a recent addition to the mathematical modeling toolbox for managing wastewater treatment facilities and collection systems that carry stormwater, whether in a combined system or in a separate stormwater collection system. Real-time control expands the utility of software—for example, through the use of a GIS interface—to coordinate predictive software models with artificial intelligence to "learn" from previous events, predict future events, and control the wastewater system accordingly. Applications include coordination of collection system models with collection system controls, and integration of treatment facility models with instrumentation and control equipment at the plant. Three basic approaches may be employed:

1. *Manual control.* Conceptually, this is the simplest approach. It requires that an operator remotely control a regulating device (e.g., an influent gate at a WWTP), in response to changing conditions. However, due to the inability of operational personnel to apply this response mechanism in a consistent and timely manner, this approach generally is not feasible for all but the most limited applications.

2. *Supervisory control.* This approach applies reactive control logic, in which system conditions are monitored and controlled in response to observed conditions, generally without direct operator intervention. If the control logic responds to monitored conditions near the flow control device (directly upstream or downstream of the gate), it is called local reactive control (LRC). If the control logic responds to remote measurement, it is termed extended reactive control (ERC). These supervisory control approaches, LRC and ERC, are the most common types of RTC for sewer systems and offer far greater control than manual approaches.
3. *Automatic control.* This approach involves the most complex type of RTC and uses rule-based systems (e.g., heuristic, expert system, fuzzy set theory, learning production theory, neural networks) or model-based systems to control system operation. However, the entire approach is collectively termed "optimal global predictive" (OGP) RTC. Although more complex than supervisory control, automatic control via OGP can maximize the capture, storage, and conveyance of combined flow by the sewer system. It simultaneously minimizes the cost of infrastructure improvements necessary to achieve these objectives. For these reasons, OGP provides the most efficient and effective method to meet the nine minimum control requirements of EPA's CSO program.

Using RTC, the network system has the means to be controlled more efficiently according to system feedback and responding to variations monitored during wet weather events. In this regard, RTC improves the operation of flow-regulating devices by way of automation and makes better use of available capacity, maximizing both the flow to the treatment plant and the use of available storage (in-line and off-line) in order to reduce the volume of CSOs.

It is beyond the scope of this report to provide more specific information for a given wastewater application without a control strategy study and modeling evaluation of network behavior. CSO control goals can also be a significant factor in the ultimate choice in technology. It is strongly recommended that RTC be thoroughly evaluated concurrently with the potential of storage/treatment of CSO in abandoned mines in the region described later in this chapter. Although the future of RTC modeling for managing wastewater appears very promising, the success of such modeling will rely heavily on the accuracy and quantity of input data and the ability to accurately link cause-and-effect relationships. The same type of pipe physical metrics (diameter, slope, roughness, etc.) and flow data are also needed to run these simulations.

Step III—Projection of Future Loadings

Southwestern Pennsylvania is not a rapidly growing or shifting urban area, but neither is it static; some areas of the region are growing while others have declining populations (see Chapter 2). Projections of changes in the regional landscape are important in the planning and implementation of Step III of CWARP. Planning studies at the multicounty/metropolitan scale should be sufficient for this purpose and should include the following:

- projected transportation infrastructure at appropriate time intervals, perhaps every 5 years, over the 25-year planning horizon for the regional transportation plan;
- spatial patterns of land use related to the regional transportation plan; and
- probable patterns of water supply and wastewater disposal for future developments and the effects of alternative urban growth scenarios.

Step IV—Formulation and Evaluation of Alternatives

Steps II and III of the Three Rivers CWARP for high-density urban areas in the Pittsburgh region are preparatory to the formulation, analysis, and evaluation of alternative management strategies for addressing the problems of degraded water quality in receiving streams associated with wet weather discharges. Hydraulic, organic, and biological stresses imposed on receiving waters during wet weather events originate from CSOs, separate stormwater sewer system discharges, and SSOs transported by stormwater runoff. Management strategies should recognize the potential of each of those sources for significant impacts.

Elements of the Strategy. At least six components of a strategy to implement Step IV of CWARP for high-density urban areas should be considered, some of which are mandated under provisions of the Clean Water Act. These include the following:

1. Rehabilitation of separate and combined sanitary sewers to reduce inflow and infiltration
2. On-site measures to reduce rates of stormwater runoff before it enters either a combined sewer, a separate storm sewer, or direct discharge to a receiving stream
3. Segregation of flows in combined sewers into separate sewers in areas where it is cost-effective to do so
4. Real-time control of combined and separate sewers to control rates at which discharges occur and control of inflow rates to treatment systems
5. Storage and treatment of combined sewage and separate stormwater
6. Price incentives and enforcement actions to encourage implementation of the strategy.

1. Rehabilitation of Wastewater Collection Systems: Capacity, Management, Operations, and Maintenance (CMOM) Programs. Monitoring and modeling activities in Step II should be used to support an initial phase of the corrective action plan for urban areas served by sanitary and combined sewer systems. All municipalities and special districts that operate separate sanitary and combined sewers should review and, as necessary, upgrade their operation and maintenance programs to current standards. Much of the existing infrastructure described in this report was built early in the last century (see Chapter 2 for further information). Concern about deterioration of these facilities has led to a renewed focus on their maintenance and orderly rehabilitation or replacement where necessary. CMOM has come to represent a careful plan to maximize the use of existing infrastructure and maintain its utility for as long as possible. CMOM infrastructure assessments are generally undertaken as part of utility plans for control of wet weather overflows. Currently, the federal CMOM program is not codified in federal regulations for controlling SSOs, but a memorandum[11] dated April 27, 2002, from the EPA assistant administrator, Office of Enforcement and Compliance Assurance, encouraged the agency to incorporate elements of a CMOM program in actions to address CSOs and SSOs. Elements of a CMOM program include the following:

- *General standards*. Properly manage, operate, and maintain all parts of the wastewater collection system; provide adequate capacity to convey base and peak flows; stop and mitigate

[11] See *www.epa.gov/Compliance/resources/policies/civil/cwa.strat312.pdf* for further information.

the impact of SSOs; provide notification to parties with a potential exposure to pollutants; and develop a written summary of the CMOM program.

- *Management program.* Develop a CMOM program to comply with general standards.
- *Overflow response plan.* Develop and implement an overflow response plan.
- *System evaluation and capacity assurance plan.* Prepare and implement a plan for system evaluation and capacity assurance if peak flow conditions are contributing to an SSO discharge.
- *CMOM program audits.* Conduct an audit, and submit a report of such an audit, evaluating the facility's CMOM program and its compliance.
- *Communications.* Communicate regularly with various interested parties on the implementation and performance of the facility's CMOM program.

All future CMOM programs will be linked with the enforcement provisions of the Clean Water Act and EPA's SSO policy,[12] which remains under development. This differs significantly from the post-1972 cost-effectiveness analysis procedure established as part of the federal construction grants program because they were linked only tangentially with the enforcement provisions of the Clean Water Act.

As a result of recently signed consent orders (see more below), communities that are served by ALCOSAN will be implementing programs (see Table 5-7) that closely resemble a CMOM program for improving the operation and maintenance of their collection systems (see below). The committee recommends that the first priority for all wastewater collection systems located in the watershed, particularly in the urban core areas of southwestern Pennsylvania, is to be fully compliant with EPA's CMOM policy or an equivalent program.

As noted in Chapter 1, on October 14, 2003, PADEP sent consent orders to 26 municipalities in Allegheny County to address problems related to CSOs. Furthermore, both the City of Pittsburgh and its Water and Sewer Authority signed a consent order and agreement to address CSOs on February 24, 2004.[13] These consent orders set forth two phases for implementing corrective measures. In the first, affected municipalities will have to inventory and evaluate the condition of all sewers in their ownership and repair major defects. In the second phase, the municipalities must install flow monitors and implement operation and maintenance plans. They must also continue to implement the nine minimum controls required under EPA's CSO policy. The committee recommends that similar actions be undertaken by all other municipalities that contribute sanitary sewer flows to the ALCOSAN system. These actions should be integrated with and supported by the dynamic sewer system modeling discussed earlier.

2. On-site Measures to Reduce Rates of Stormwater Runoff. Implementation of measures required under Phase I and Phase II of the federal stormwater regulations and more complete coverage of authorities under Act 167 should be utilized to reduce rates and volumes of stormwater runoff. Consideration should also be given to identification of high-priority sites where existing runoff could be reduced in a cost-effective manner.

---

[12] For further information about EPA's developing SSO policy, see *http://cfpub.epa.gov/npdes/home.cfm ?program_id=4.*

[13] See *http://www.dep.state.pa.us/NewsReleases/?ID=2778* for further information.

TABLE 5-7 Comparison of Consent Order Versus CMOM Requirements for 83 Communities Serviced by ALCOSAN

Basic CMOM Requirements	Consent Order Requirements
1. More prescriptive characterization of collection system requiring detailed maps and extensive record keeping	1. Hire an engineer to prepare a sewer investigation plan
2. Televising of collection system includes 180 specific questions	2. Inspect the collection system through physical surveys and TV cameras; conduct dye testing of roof leaders, driveway drains, springs, and catch basins
3. Asset management program ties value of aging infrastructures and long-term repair costs	3. Write ordinances and develop enforcement program to eliminate illegal laterals; develop a plan to determine capacity and remediate the problems in the collection system
4. Requirements of program tied to NPDES permits	4. Implement the plan and coordinate with ALCOSAN and the 83 communities
5. Completion dates more long-term and comprehensive	5. Completion dates not less than 5 years

As noted in the previous section, the second part of consent orders directed local governments to continue to implement the nine minimum controls required under EPA's CSO policy. Several of these measures are directed toward reducing stormwater flows entering collection systems. As for CMOM, the committee recommends that similar actions be undertaken by all other municipalities that contribute sanitary sewer flows to the ALCOSAN system.

These measures might also include a wide range of alternative technologies, including urban retrofitting, low-impact developments, and recently considered ecological techniques that may offer potential stormwater flow reductions and are being tested in other communities.

3. Segregation of Combined Sewer Flows into Separate Sewers. In some portions of the combined sewer systems, it may be cost-effective to separate stormwater from sanitary sewage by constructing parallel collection systems. In some instances, construction of separate sanitary sewers may be appropriate. Segregated stormwater could then be treated and discharged through decentralized treatment systems, including wet and dry detention basins and other technologies discussed in later sections of this chapter. In some cases, the quality of stormwater runoff may be sufficient to allow rerouting of segregated stormwater through open channels with direct discharge to streams.

4. Real-Time Control of Combined and Separate Sewers. The use of RTC of flows in separate sanitary and combined sewers and stormwater conveyance systems should be investigated and evaluated as a means to reduce peak loading on centralized and decentralized treatment systems. The October 2003 consent orders required affected municipalities to install flow monitors. Those monitors should be integrated into a system-wide, real-time monitoring and control system.

5. *Decentralized and Centralized Storage and Treatment.* After initiatives have been taken to reduce peak rates and volumes of flow and after investigating the potential for real-time controls to maximize storage in existing collection systems, consideration should be given to decentralized and centralized options for storing and treating flows in combined sewer and separate stormwater sewer systems. Decentralized storage could be used in conjunction with real-time flow control to reduce peak flow, resulting in fewer CSO discharges and reducing the loading on ALCOSAN's treatment plant. Reduced loads on that plant could result in reduction in the size of planned expansions needed at the central WWTP. In some instances it may be possible to treat and discharge stormwater to nearby streams before it enters combined systems.

During the course of the committee's deliberations, various technologies and approaches (including some discussed in ALCOSAN's draft LTCP and the TPRC report) were discussed to determine what, if any, role they might play as remedial actions for high-density urban areas of the watershed. All of these technologies could be characterized as innovative because they are still in the developmental stage at the national scale; others may have proven to be effective in other locations but cannot prudently be designed for effective operation without site-specific performance data. In this regard, the committee recommends at least the following technologies and approaches for consideration in the planning and implementation of Step IV of CWARP for the region's high-density urban core:

- CSO storage and treatment in local underground, abandoned mines;
- vortex separators;
- ballasted flocculation; and
- in-river CSO storage using the flow balancing method.

There is a need to develop further information on the feasibility, performance, and costs of these technologies via prototype or full-scale studies. For example, uncertainties in the hydraulic performance of vortex separators could be investigated through the construction of large-scale physical hydraulic models or full-scale field (demonstration) units located in critical CSO control locations. Each of these innovative technologies or approaches is discussed later in this chapter, including the potential adoption of wet weather water quality standards, to help improve water quality in southwestern Pennsylvania.

6. *Pricing and Enforcement Incentives.* In addition to the technical strategies suggested, it is strongly recommended that pricing and enforcement incentives be adopted. Some of the municipalities contributing flow to the ALCOSAN system have contracts that prescribe pricing of services in proportion to flow; others have contracts that allow them to discharge to ALCOSAN for a fixed fee regardless of flow. The latter practice should be phased out, and consideration should be given to peak load pricing as well as volumetric pricing as recommended and discussed in Chapter 6. Any new pricing policies should be developed using concepts of marginal costs of providing the services. In addition to pricing incentives, enforcement of stormwater Phase I, Phase II, CMOM, and CSO regulations should be an active part of the

strategy. Without effective periodic inspections and enforcement, the regulations can be easily ignored. Alternative organizational arrangements for the establishment and enforcement of these regulations are also discussed in Chapter 6.

Step V—Adaptive Implementation of Strategy

Successive steps for implementing CWARP in high-density urban areas should be implemented in an adaptive manner. Before each step is implemented, the best available monitoring, modeling, predicted effectiveness, and cost information should be used to guide its design and implementation. After implementation of each step, monitoring and modeling should be updated to incorporate information obtained from implementing that action, and revised predictions of performance and cost should be used to modify designs for the next step. A graphical representation of such a process is provided in Figure 5-4.

Rural Area Planning and Management

Step I—Problem Identification.

Testimony before the committee from several persons throughout the course of the study cited numerous problems of inadequate water supply and waste disposal in rural areas. The most frequently cited issue was failing OSTDSs and straight-pipe discharges from individual homes, and even entire communities, to nearby streams. Although no systematic survey data were available, these anecdotal accounts were compelling. Given the large number of OSTDSs in southwestern Pennsylvania and the nature of soils in the region (see Chapter 2), a conclusive finding that widespread failures were occurring would not be surprising.

Nonpoint sources of pollution that contribute fecal bacteria to surface waters can be difficult to identify. In evaluating potential pathogen and indicator microorganism sources in watersheds with multiple nonpoint sources, investigators and watershed managers must be careful to avoid biases toward implicating any single source without sufficient data to support such conclusions. To target the most critical public health and environmental needs, to most effectively utilize the funding available for water quality improvements, and to avoid unwarranted financial burdens on private individuals, farms, and businesses, it is crucial to determine the major sources of microbiological contamination in a watershed and to prioritize efforts, funding, and/or user costs accordingly. Many rural watersheds may have several potential nonpoint sources of pathogens and indicator microorganisms, including single-home domestic wastewater, livestock, pets, wildlife, and biosolids (manure and treated wastewater sludge). All of these sources can contribute to microbial loading, although in some watersheds a particular source or sources may dominate.

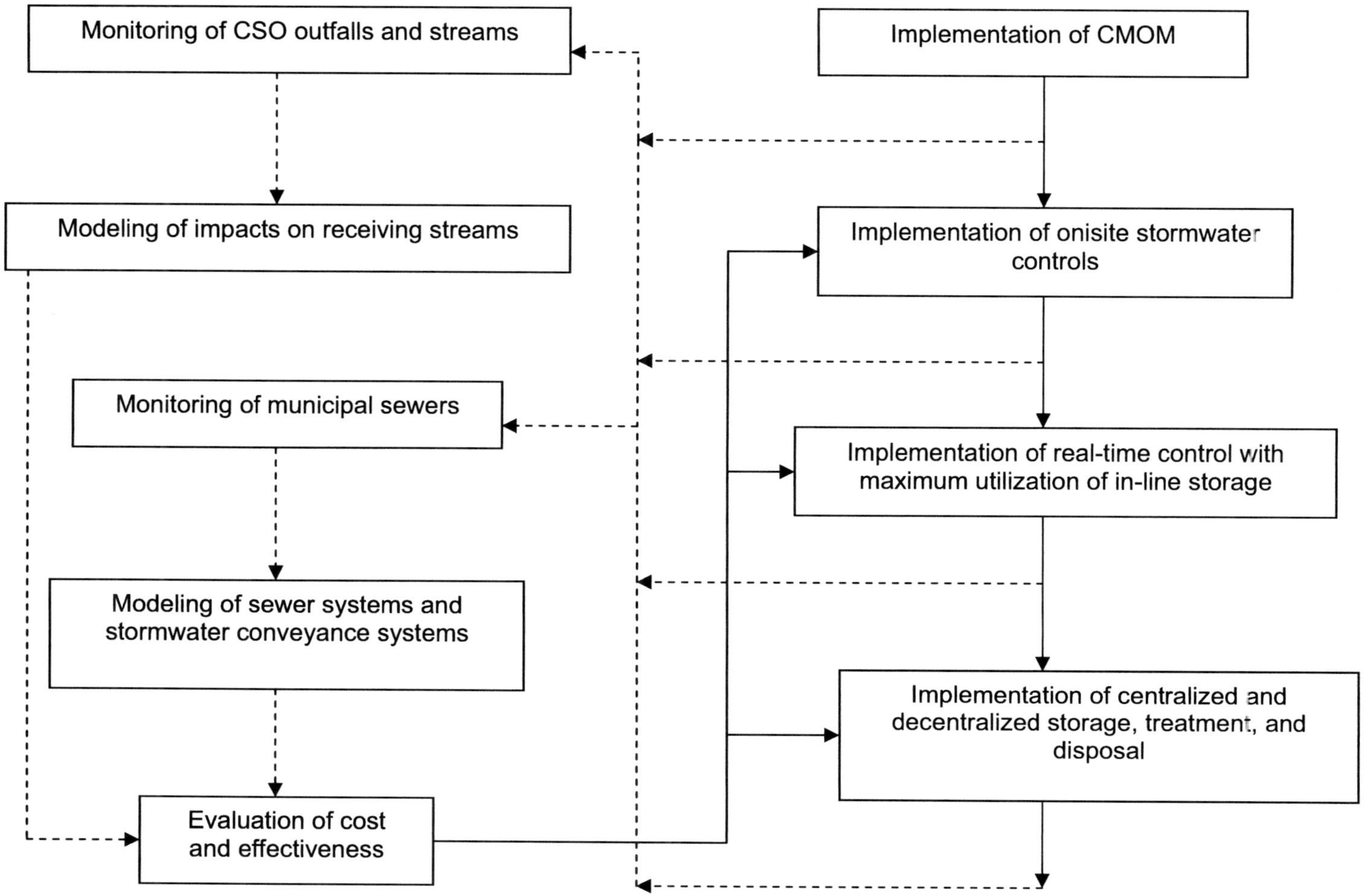

FIGURE 5-4 Adaptive implementation (Step V) of CWARP for high-density urban areas of southwestern Pennsylvania.

The following sections describe how the CWARP process can be used to improve water quality in the predominantly rural areas of southwestern Pennsylvania.

Step II—Assessment of Existing Conditions

It is advantageous to use the approaches practiced in wellhead and source water protection, identifying the nature and spatial locations of microbial contaminant sources. As indicated in the Mud River case study (Box 5-3), it is not sufficient to sample only the main stem or downstream reaches of a stream; water quality samples and land use analysis in tributaries and headwaters are also needed for thorough watershed analysis. In some instances, more detailed, targeted monitoring and analysis of specific stream reaches may be needed.

In recent years, watershed scientists also have used a variety of techniques to attempt to track specific sources of bacterial contamination; some of these methods are based on microorganisms themselves, others rely on distinct chemical "identifiers" of specific sources. Box 5-4 discusses two cases of the successful application of microbial source tracking techniques in Virginia. Some of the techniques are costly, some require refinement and additional validation, and some are still considered experimental (see NRC, 2004a). It is important to note that none of them should be considered stand-alone methods; rather, they should be regarded as tools to be used in conjunction with other careful watershed analyses of use impairment. These two examples of the use of microbial source tracking in Virginia watersheds illustrate the importance of determining the dominant source of fecal indicator organisms.

Although there is no reason to doubt anecdotal data of the extent of OSTDS failure in parts of southwestern Pennsylvania, the extent of the problem and the location of concentrations of failures that pose the most serious problems cannot be determined without systematic assessments. Pennsylvania Act 537 requires registration of all septic systems since that law was passed, but it does not require registration of previously installed systems.

The committee recommends that best management practices for OSTDSs should be implemented throughout southwestern Pennsylvania within the CWARP framework, along with the following related actions:

- within each county, register all individual on-site and cluster OSTDSs with the appropriate SEO; registration should include residences, houses of worship, schools, business establishments, and other locations;
- institute a program of periodic mandatory inspection and certification (or decertification), either by a public entity or a qualified/licensed private contractor;
- conduct statistically valid surveys of septic tank and absorption field conditions, residence by residence, to identify communities that should be given high priority for funding by PENNVEST (Pennsylvania Infrastructure Investment Authority) or the Rural Utilities Service (RUS) of the U.S. Department of Agriculture for remediation of failed and failing OSTDS throughout the region.

Step III—Projection of Future Loadings

Data gathered from Step III of CWARP for rural areas should be used to identify areas of high concentrations of OSTDS failures and their potential impacts on public health and the water quality (especially microbiological) of nearby receiving streams. Failure rates alone may provide an adequate indicator of potential public health effects. Modeling and monitoring may be necessary to determine impacts on receiving streams into which OSTDS drainage may be discharged.

Step IV—Formulation and Evaluation of Alternative Management Strategies

Although individual OSTDSs are permitted locally and current technical standards are available to ensure proper performance, they may be ignored. Furthermore, prevention of the discharge of untreated sewage into local waterways or ditches is difficult to enforce. The region needs a coordinated, well-funded program for oversight and routine maintenance of OSTDSs. Such a program can be self-sustaining through user charges, provided they are applied on a cooperative regional or county basis (see also Chapter 6). Such a registration and inspection program should also be used to identify and order elimination of illegal direct discharges of human waste into streams and to identify areas where cluster OSTDSs may be feasible. Several examples of approaches to these kinds of rural water quality problems could be reviewed and cited; one successful program is summarized in Box 5-5.

These recommendations are consistent with a detailed guidance published by the EPA in 2003 that provides voluntary national guidelines and a draft of a handbook for managing on-site or clustered wastewater treatment systems in rural areas (EPA, 2003a). Basic elements of a management program are listed in Box 5-6. EPA guidance suggests alternative models of the "who and how" of wastewater management, depending on the complexity of appropriate wastewater treatment systems and the nature of the risk they pose to human health and the environment (EPA, 2003b). Those models are summarized briefly in Box 5-7.

Step V—Adaptive Implementation

Wastewater treatment systems (e.g., cluster OSTDSs) for sparsely populated rural areas of southwestern Pennsylvania should be designed and implemented over a time horizon that allows for evaluation of outcomes of initial remediation projects before subsequent projects are undertaken. Lessons learned from initial projects should be used to modify later projects in the schedule.

Economic Evaluation of Strategies and Projects

The CWARP process can yield a list of alternative management strategies and projects that are technically feasible and capable of addressing the region's water quality problems at a variety of scales, but the question remains: Which is the better option? Comprehensive evaluation of options under CWARP (especially Step IV for high-density urban areas) requires

BOX 5-3
Left Fork of Mud River

In Lincoln County, West Virginia, the Mud River's Left Fork—a small headwaters watershed in the Guyandotte River watershed—illustrates the critical information gaps that may exist when bacterial water quality sampling is confined to locations on a main stem river. In such cases, the nature, distribution, and sources of microbial contamination throughout the watershed may be poorly understood, leading programs for water quality improvement to target the wrong areas or sources or to achieve only partial reduction of contamination. Because the Left Fork's land uses and terrain resemble those of many isolated rural watersheds throughout southwestern Pennsylvania, its watershed may serve as an example of the severity and complexity of localized bacterial contamination that can affect smaller, upstream tributaries in that region.

The watershed of the Left Fork of the Mud River includes headwaters streams that are part of the much larger watershed (1,680 square miles) of the Guyandotte River, a tributary to the Ohio River at Huntington, West Virginia. The Mud River joins the Guyandotte in its lower reaches (approximately 20 miles east of Huntington), and the lowermost Mud River lies within this urban-suburban-industrial corridor.

Similar to southwestern Pennsylvania, the geology of the Left Fork watershed comprises flat-lying or gently tilted layers of Pennsylvanian age sandstone, shale, clay, coal, and limestone. Terrain consists of steeply sloping hills and narrow valleys. The area is sparsely settled and dominantly forested, with scattered homes, small farms, and several small, unincorporated communities. In the late 1990s the U.S. Army Corps of Engineers constructed a dam at the confluence of the Left Fork and the main stem of the Mud River; the resulting Upper Mud River Lake is managed by the West Virginia Department of Natural Resources (WVDNR) as a fishing and recreational lake. Lincoln County maintains a youth camp and multipurpose recreational area on the lake's shoreline near the mouth of the Left Fork watershed.

West Virginia's 1996, 1998, and 2002 Section 303(d) lists (see Chapter 3 for further information) included 123 impaired waterbodies within the Guyandotte watershed. Impairments included pH, iron, manganese, selenium, fecal coliforms, and biological impairment. TMDLs were completed in March 2004 for the Upper and Lower Guyandotte hydrologic unit codes (HUCs). Of the 123 waterbodies in the Guyandotte watershed, only the main stem Guyandotte is listed as impaired due to excessive fecal coliforms and sources are listed as unknown. Although individual grab samples for coliform were collected throughout the watershed during May 1998, only the main stem had a sufficient number and seasonal range of samples to support the 303(d) listing. Examination of WVDEP's grab sample values throughout the watershed indicates that most of the extremely high fecal coliform counts (in the thousands per 100 mL) were obtained in more densely developed urban-suburban areas with WWTPs, CSOs, and urban runoff, but several areas of extremely high fecal coliform values are located in very rural subwatersheds tributary to the Mud River. One of these subwatersheds is the Left Fork headwaters, where three grab samples collected in May 1998 had fecal coliform counts of 5,000, 5,200, and 2,900 per 100 ml (EPA, 2004).

During 1998, water quality in the Left Fork and Upper Mud River Lake raised concerns locally. High fecal coliform counts caused numerous closings of WVDNR's public beach on the lake, and

additional considerations, including costs, benefits, and fairness. It continues throughout the process of selecting a preferred long-term management strategy for a particular water quality problem at given scale. It is essential in the formulation of alternatives to provide feedback as to how initial designs have to be modified or discarded in the search for a cost-effective strategy. It is also essential to the process of establishing priorities among the several elements that may comprise the management strategy. Finally, it must be a continuing process during implementation to evaluate how well each element has performed. A variety of evaluation frameworks are available; some of the more prominent are discussed below.

water in the lake near the youth camp on the Left Fork suddenly acquired a reddish color for a short period of time. As a result of these concerns, a water quality investigation was undertaken with funding from the Service Learning Program at West Virginia University (WVU). In collaboration with the Lincoln County WVU Extension Office and science students from two local high schools, professors and students from WVU sampled 53 locations in the Left Fork watershed during April 1999 to April 2000. At 42 of the locations (79 percent), samples exceeded the West Virginia State Board of Health's one time sample total coliform limit of 2,400 and at 16 locations (30 percent) samples exceeded the one time sample fecal coliform limit of 400 for primary contact recreation. During April 2000, the fecal coliform standard was exceeded in most of the watershed. Counts were exceptionally high (many in the tens of thousands) during a high-flow sampling in that month. Very high values were also obtained at clusters of houses. In one small tributary with no houses, several samples exceeded standards, but no reason for the high counts was determined.

In a subsequent investigation of possible links between watershed bacterial contamination and septic systems (presence, condition, and maintenance), the investigating team examined county health department and National Small Flows Clearinghouse survey information. Notably, of 8,000 homes in the county, 7,000 have on-site septic systems. Failure rate is estimated to be 50 percent, and repairs are common due to system age, inadequate size, impermeable or saturated soils, damage to the system, inadequate removal of surface water, and improper maintenance. The team conducted a survey of 77 of the 250 households in the watershed. While 90 percent of residents reported that a septic system handled all (76 percent) or part (14 percent) of their wastewater, about 50 percent did not know the size or condition of their system, and only 30 percent had performed any maintenance on their systems. However, 85 percent of the respondents perceived that few problems exist with their systems, and 80 percent perceived no effects on their water wells, neighbors' water wells, groundwater, or surface water. Local health department records indicate that in the 12 years prior to the study, only 2 permits for septic systems and 6 permits for home aerobic units had been issued, and the team surmised that many households either had systems constructed without permits or had little actual knowledge of the wastewater system (if any) serving their homes (Collins et al., 2000).

Using that study as a basis for a proposal, in 2003, Lincoln County and the WVU Extension Service were awarded a grant from EPA for a collaborative demonstration project involving a detailed, comprehensive study of bacterial contamination in the Left Fork and installation of alternative on-site wastewater treatment technologies. Although the earlier study of the watershed suggested strong links between inadequate household wastewater treatment and bacterial contamination in the watershed, the investigators did not specifically examine the possible role of other bacterial sources. As part of the EPA demonstration project, preliminary examination of land uses in the watershed, using aerial photographs and field surveys, indicates several areas where cattle are confined immediately adjacent to a tributary stream or where a small tributary flows through a farm's feedlot. In addition, noticeable concentrations of Canada geese waste were observed at several locations on the shore of the lake. These observations, as well as the previous study's "anomalous" tributary, suggest the need for thorough investigation of sources and careful documentations of household wastewater-related contributions, to effectively prioritize and correct watershed contamination.

Cost-Effectiveness

Cost-effectiveness analysis ranks projects based on their relative costs in achieving a specified outcome. Essential to this analysis is consideration of long-term costs of alternatives that may have varying useful lives and the actual improvements expected to be achieved. The term of the analysis should be 20 or more years, and appropriate criteria should be used to consider inflation and the time value of money. Economic efficiency (cost containment) is a

BOX 5-4
Microbial Source Tracking Techniques

Water quality monitoring in the Page Brook watershed in Clarke County, Virginia, exhibited use impairment due to excessive concentrations of fecal coliforms. Land uses in the watershed suggested two likely nonpoint sources of fecal contamination: (1) relatively widespread livestock (cattle) farming and (2) widespread use of single home septic systems. Routine water quality surveys in this watershed were unable to determine the relative contributions of these two sources. To field test a microbial source tracking method using patterns of antibiotic resistance in fecal streptococci, Hagedorn et al. (1999) initiated a watershed improvement project in 1996. Application of the antibiotic resistance method at three highly contaminated sites within the watershed identified the dominant source (79 percent of the isolates) as cattle, with small proportions of waterfowl, deer, and unidentified sources. Based on these results, cattle access to the stream was restricted through BMPs (installation of fencing and in-pasture watering stations). Fecal coliforms were subsequently reduced at the three sites by an average of 94 percent, from pre-BMP average populations of 15,900 per 100 mL to post-BMP average populations of 960 per 100 mL. After fencing, less than 45 percent of fecal streptococcus isolates were classified as being from cattle, indicating that the dominant source of contamination had been successfully remediated

Septic systems were suspected as a major source of fecal coliform impairment contamination of shellfish beds at a tidal inlet on Virginia's Eastern Shore. Using shoreline survey techniques and discrete sampling over small areas, Simmons et al. (1995) tracked nonpoint sources of fecal coliform at tidal inlets on Virginia's Eastern Shore and at an uninhabited island. Based on DNA fingerprinting methods, they characterized *Escherichia coli* from raccoon, goose, otter, and muskrat. These results, along with examination of land use and wildlife management patterns in the area, led them to conclude that fecal contamination of tidal inlets, bays and estuaries on Virginia's Eastern Shore could be largely attributed to fur-bearing animals. The populations of these animals have increased over the past several decades due to lack of predation and land development patterns that concentrate their populations in undeveloped shoreline areas. Simmons and colleagues also determined that septic systems, when sufficiently elevated above the water table and not mechanically damaged or overloaded, are effective at removing *E. coli*, thus avoiding the need for costly replacement systems that would likely have had minimal impacts in reducing fecal contamination.

serious issue in the region, given that resources for water quality protection and other public objectives are scarce. In colloquial terms, cost-effectiveness analysis is about getting the "biggest bang for the buck." Simply stated, if two projects yield the same outcome, the preferred project is the less expensive one. Cost-effectiveness is particularly useful for optimizing theachievement of a well-defined policy target. In this context it is used to identify a single project or set of projects that minimize the costs of achieving the target. The use of cost-effectiveness analysis for ranking projects becomes limited when the outcomes are not comparable.

Benefit-Cost Analysis

Benefit-cost analysis ranks projects based on the relationship between benefits and costs. The economic benefits of measures to reduce or eliminate discharges of untreated sewage into the region's source waters would include the reduction in the likelihood of such events and the associated costs of averting activities and disease. Other things being equal, desirable projects yield benefits in excess of costs, with projects being ranked according to their relative net benefits. Benefit-cost analysis is more powerful than cost effectiveness analysis for project evaluation because it allows for comparison of projects with otherwise noncomparable outcomes.

BOX 5-5
Guest River Watershed Restoration Program

Water quality improvements in southwestern Virginia's Guest River watershed provide an excellent example of a successful rural watershed partnership between grassroots community groups and government agencies. Draining approximately 100 square miles of Appalachian Plateau in Wise County, Virginia, the Guest River watershed is a tributary of the Clinch River, which is in turn part of the larger Tennessee River basin. Environmental, economic, and social issues in the watershed are typical of those affecting many rural, coal-mining-impacted communities in the central Appalachians. Although the watershed's rugged terrain, coalfield history, and rural Appalachian folk life offer potential for tourism and recreation development, contamination of the watershed's streams, as well as inadequate water and sewage service, creates barriers to attracting new economic development.

In 1995 several groups of watershed residents and more than 15 local, state, and federal government agencies formed the Guest River Group, an informal coalition dedicated to the protection and restoration of the Guest River watershed. Efforts of this group have led to the development of the Guest River Restoration Project, an integrated program of multiple projects to address a variety of pollution sources for the entire watershed. Projects have included septic system pumpouts and inspections, repairs, and replacements; illegal solid waste dump cleanups; stream bank restoration and slope revegetation; porous paving projects; abandoned mine cleanups; and outdoor classrooms for the school system. Financial support for the Guest River Restoration Project has come from Clean Water Act Section 319 grants; Virginia Water Quality Improvement Fund grants; EPA's National Onsite Demonstration Project (NODP); and other local, state, and federal sources.

Elimination of sewage problems is a major part of the Guest River Restoration Project. Typical of many small communities in the central Appalachian coalfields, the former coal camp of Imboden had community sewers leading to two "community straight pipes" on a small tributary stream and limited land available for conventional septic systems. Because of its rugged topography and relative remoteness from other communities, Imboden had no realistic possibility of connecting to any existing municipal sewage systems. With a combination of NODP funds, local funds, and in-kind contributions, the community has completed a "cluster system" consisting of new septic tanks (two households per 1,500 gallon tank), small-diameter sewers, and a central treatment system consisting of a recirculating textile filter and a community drainfield. Approximate cost (including in-kind worth) per household was $7,000. The municipal sewer utility in the town of Appalachia will provide system management for operation and maintenance. For further information, see Clean Water Action Plan Partners (2000).

BOX 5-6
Elements of a Decentralized Wastewater Management Program

- **Public education and participation**—communicate risks and develop responses
- **Planning**—based on impacts on human health and water resources
- **Performance requirements**—system design and technology selection
- **Site evaluation**—system sizing and design
- **Designs**—based on site conditions, loadings, and performance requirements
- **Construction**—oversight for compliance with siting and design
- **Operation and maintenance**
- **Residuals management**
- **Training and certification/licensing** of regulators and all service providers
- **Inspections and monitoring**—document performance and initiate remediation
- **Corrective actions and enforcement**
- **Record keeping, inventory, and reporting**
- **Financial assistance**

SOURCE: EPA, 2003a.

BOX 5-7
Management Models

Management Model 1 "Homeowner Awareness"

- Treatment systems are owned and operated by individual property owners in areas of low environmental sensitivity.
- Treatment technologies are limited to conventional systems that require little owner attention.
- Regulatory authority mails maintenance reminders to owners at appropriate intervals.

Management Model 2 "Maintenance Contracts"

- More complex designs are employed to enhance the capacity of conventional systems to accept and treat wastewater.
- Contracts with qualified technicians are needed to ensure proper and timely maintenance.

Management Model 3 "Operating Permits"

- Applicable sustained performance of treatment systems is critical to protect public health and water quality.
- Limited-term operating permits are issued to the owner and are renewable for another term if the owner demonstrates that the system is in compliance with the terms and conditions of the permit.

Management Model 4 "Responsible Management Entity (RME) Operation and Maintenance"

- Frequent and highly reliable operation and maintenance of decentralized systems is required to ensure water resource protection in sensitive environments.
- An operating permit is issued to an RME instead of the property owner to provide the needed assurance that appropriate maintenance is performed.

Management Model 5 "RME Ownership"

- This provides the greatest assurance of system performance in the most sensitive of environments; program elements and activities for treatment systems are owned, operated, and maintained by the RME.

SOURCE: EPA, 2003b.

In the context of the Pittsburgh region, projects with comparable outcomes would be those that produce essentially the same impacts on water quality for all water quality criteria. However, many feasible projects may fare better by some water quality criteria than by others. If multiple water quality criteria are important, projects with differing water quality outcomes may not be comparable for the purposes of cost-effectiveness analysis. Benefit-cost analysis would allow comparisons between such projects. Further, benefit cost-analysis would allow inclusion of ancillary benefits if any. Importantly, to be useful in setting priorities for projects and policies, estimates of benefits and costs must be tied to the water quality gains from specific initiatives.

Although different techniques of varying data intensiveness exist to estimate the benefits of reducing the risks of waterborne disease, essential information is lacking for estimating the relationships between CSO control and the likelihood and severity of contamination events. For example, a key question that cannot be answered with current data on sources of pathogens and the effectiveness of prospective control measures (see also Chapter 4) is the extent to which the likelihood of contamination events in various systems will be diminished. Recent work by Casman et al. (2000) illustrates an approach to modeling drinking water risks that would offer a good starting point for systematic analysis of risk management in the region. Further, for the purposes of drinking water protection, consideration of the relative cost-effectiveness of different strategies merits attention. For example, to what extent should limited financial resources be allocated between water system upgrades and source water protection in the region?

The use of cost-effectiveness as the primary method of evaluating options for achieving water quality objectives in the region is recommended. It should include an analysis of incremental costs to achieve elimination of low-probability contamination events. The committee also recommends the use of benefit-cost analysis in the evaluation of water quality improvement projects in the region, and in helping to set priorities. Although "state-of-the art" studies can be very expensive, it may be possible to gain useful information using "lower cost" methods. For example, "benefits transfers" is a method for estimating benefits of environmental quality improvements using results from other regions. The EPA's (2000b) *Guidelines for Preparing Economic Analyses* describes methods and procedures for cost-effectiveness and benefit-cost studies.

Multicriteria Methods

Decision makers in the region may want to apply criteria beyond long-term cost-effectiveness and benefit-cost analysis to project evaluation. They may also have interest in metrics of performance beyond costs, benefits, or water quality indicators, such as equity, social justice, and other social objectives. To the extent that this is the case, multicriteria methods offer tools for systematic evaluation of water quality improvement projects and related decisions when there are multiple competing objectives. A brief introduction to the array of methods is presented in Janssen and Munda (1999).

TMDL PROCESS FOR SIGNIFICANT CONTAMINANTS

As the CWARP process is being implemented, it is essential that it be integrated with the ongoing process of establishing total maximum daily loads (TMDLs) for impaired streams being conducted by PADEP under requirements of the Clean Water Act. There are many parallels between CWARP and the process for establishing TMDLs. A TMDL defines the pollutant load that a waterbody can assimilate without causing violations of its water quality standards. The term TMDL also refers to a plan for those waters in violation of their water quality standards, in which the excess pollutant loading is allocated between contributing point sources and nonpoint sources and subsequent actions are taken to control and eliminate these excesses.

Although the TMDL program originated from the Clean Water Act (Section 303d), it was largely overlooked during the 1970s and 1980s as states focused on controlling point sources of pollution through compliance with NPDES permits (NRC, 2001). Beginning in the 1980s, citizen lawsuits forced EPA to develop guidance for the TMDL program, which is now considered pivotal in achieving the nation's water quality goals.

The NRC (2001) review of the scientific basis of the TMDL program recommended that the program meet certain objectives, some of which are also an integral part of the CWARP approach recommended in this report. These objectives include the following:

- improve the condition of waterbodies as measured by attainment of designated uses;

- encompass all stressors that determine the water quality conditions of the waterbodies;
- develop appropriate use designations for waterbodies prior to TMDL development;
- apply adaptive implementation consisting of monitoring, modeling, design, and construction of facilities followed by further monitoring to assess whether expected improvements were attained;
- utilize biological criteria, in conjunction with physical and chemical criteria, to determine whether a waterbody is meeting its designated use, and define all chemical and some biological criteria in terms of magnitude, frequency, and duration of exceedences of criteria; and
- utilize reasonably obtainable monitoring data to assess attainment of water quality standards.

As shown in Figure 5-5, "adaptive implementation" is a cyclical process in which TMDL plans are assessed periodically for their achievement of water quality standards, including designated uses. If the implementation of a TMDL plan is not achieving attainment of the designated use, scientific data and information should be used to revise the plan. Adaptive implementation is needed to ensure that the TMDL program is not halted because of a lack of data and information, but rather progresses while more and better data are collected and analyzed with the intent of improving upon initial TMDL plans. Adaptive implementation is an important concept in CWARP at all scales of its planning and implementation.

The TMDL process recommended in the 2001 NRC report—supplemented by additional analyses of constituents that may not be readily subject to the rigorous TMDL approach, including biological, environmental, and other measurable factors—should be combined with watershed, regional, and subregional analysis of beneficial uses to provide the basis for selection of remedial actions in the study area. Table 5-8 summarizes the current schedule of TMDLs to be completed in Pennsylvania, divided into AMD and non-AMD categories.

IMPLEMENTING CWARP

The CWARP effort should be completed quickly to provide timely support for those water quality improvements that are required and others deemed in the public interest. It is difficult to estimate the cost of implementing CWARP, but in the committee's judgment it should be low compared to the cost of improvements and more than offset by potential savings. It is important to reemphasize that CWARP is an adaptive implementation process. Large tasks and responsibilities to be started first in the multiyear program include the following:

- advance plan and design approach for the construction of pilot and/or demonstration vortex separators in the ALCOSAN system;
- implement CMOM requirements for all wastewater collection systems in the watershed, especially in the high-density urban areas of southwestern Pennsylvania;
- plan and implement the monitoring and modeling components of CWARP; and
- begin feasibility studies for innovative technologies and approaches.

In addition, the major effort to better define and implement the institutional regionalization needed to manage CWARP at all scales of implementation as described in Chapter 6 should proceed immediately.

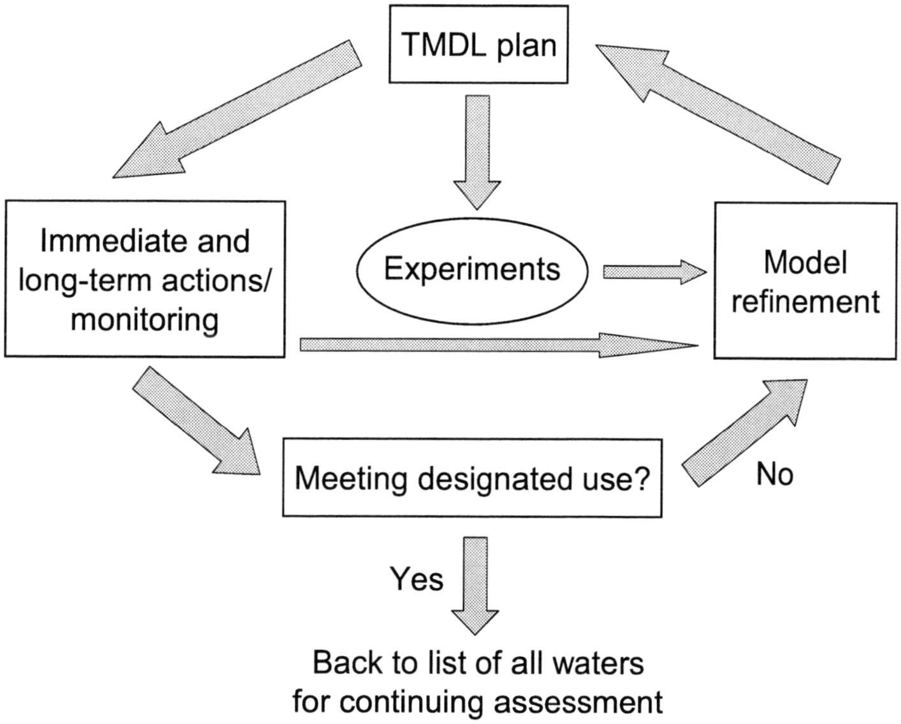

FIGURE 5-5 Adaptive implementation flow chart for TMDL plans in southwestern Pennsylvania under CWARP. SOURCE: Adapted from NRC, 2001.

TABLE 5-8 Schedule of TMDLS to Be Completed in Pennsylvania Through 2009

Year	Non-AMD	AMD
2005	30	85
2007	30	85
2009	—	169

SOURCE: Adapted from PADEP, 2004.

Lastly, ALCOSAN's draft LTCP for controlling CSOs should be reevaluated in the context of the overall CWARP approach to reflect ongoing consent order negotiations, CMOM, and information from CWARP as it is developed in the future. The CWARP approach is recommended as a framework for development of the LTCP and similar documents because of the circumstances (especially data limitations discussed throughout this report) that exist in southwestern Pennsylvania and, in principle, would apply to other regions of the United States with similar water quality problems and circumstances. The implementation of CWARP, the utilization of real-time control of CSOs, the possibility of CSO storage and treatment in nearby abandoned mines, and other innovative technologies and approaches discussed in the following section, could potentially result in reducing the need to expand existing wastewater treatment plants. As a result, the costs and time factors associated with these activities should be evaluated to ensure that potentially unneeded investments are not made.

INNOVATIVE TECHNOLOGIES AND APPROACHES

As noted previously, there are at least five innovative technologies and approaches to water quality improvement in southwestern Pennsylvania, especially for addressing wet weather water quality problems in the region's urban core areas. Although not specifically recommended, the committee believes these should be explored, and each is discussed below.

CSO Storage and Treatment in Underground, Abandoned Mines

One possibility for reducing the cost of CSO collection and treatment is to take advantage of existing, abandoned caverns from past coal mining activity to store very large volumes of CSO for subsequent treatment at a remote location where inexpensive land is plentiful. This innovative approach to CSO abatement could potentially allow for minimum disruption of surface infrastructures, high reliability and availability, reduced maintenance and operational costs, and large and easily expanded capacity. Potential advantages of this concept include minimizing concerns with existing inflow and infiltration, no need for conveyance to the ALCOSAN plant, and expansion of decentralized treatment facilities to handle short-term high flow rates.

It is important to note that a feasibility study for short-term CSO storage during wet weather in abandoned coal mines in the Pittsburgh region is currently being conducted under the sponsorship of the Township of Upper St. Clair (Gateway Engineers, Inc., and GAI Consultants, Inc., 2003). The evaluation is focusing on the Lower Chartiers Creek watershed, with specific emphasis on the possibility of temporary mine storage of overflows from the McLaughlin Run interceptor. The stored combined sewage, together with the acid mine drainage with which it would become mixed, would be pumped to the surface and sent by interceptor to the ALCOSAN treatment plant for subsequent processing and discharge to the Ohio River. The project goals include determination of the feasibility of the proposed technology with respect to performance criteria and regulatory requirements, specifically the following:

1. mining and geotechnical issues, including mine subsidence, mine sealing, mine storage capacity, and mine ownership;
2. gas generation and health and safety issues, including an evaluation of the potential for gas generation, migration, and odor control;
3. water quality impacts, including the potential effects of sewer overflow storage on surface and groundwater quality and interaction with mine water;
4. treatment system requirements and impacts, including long-term generation of solids within the mines and removal requirements, and impacts and requirements for the existing collection and treatment system for handling the wastewater after the end of the wet weather event;
5. hydrologic and hydraulic analysis, including evaluating inflow to the mine during wet weather and overflows and groundwater seepage;
6. economic analysis including life-cycle costs; and
7. legal and institutional analysis addressing requirements for construction, operation, and administration.

An even closer look into the potential for acid mine storage of CSOs is warranted. As discussed in Chapter 2, the Pittsburgh Coal Seam has been extensively mined in southwestern Pennsylvania for decades and offers a potential set of appropriate "reservoirs." It is a vast, 1,600 square mile elliptical basin with the northern end of its long axis ending in Pittsburgh and the southern end in Charleston, West Virginia. The lowest point of the basin is near sea level and is reached in northern West Virginia. The Pittsburgh Seam outcrops along the hills immediately south of the Monongahela and Ohio Rivers. The seam dips below the elevation of surface streams around Cannonsburg, Pennsylvania.

For illustrative purposes, acid mine storage and treatment of CSOs could be attempted in the Montour #4 coal mine located about 10 miles south of Pittsburgh, where mining ceased in the 1980s. Since that time, the mine has been allowed to flood to within 30 feet of the base of Chartiers Creek near Cannonsburg, Pennsylvania. Its water level is maintained by Consol Energy's pump and treatment plant along Chartiers Creek, which is pumping the mine at the rate of 5.5 million gallons per day. Peak CSO discharges in the region are estimated to be about 400 mgd. The flooded area of Montour #4 is about 15 square miles, providing about 8 billion gallons of void space; it is surrounded by other, similarly situated mines (see Figure 5-6). Montour #4 produces net alkaline water with about 70 mg/L of ferrous iron that is treated with aeration and iron flocculate handling.

To store CSO flows in abandoned mine voids, CSOs would discharge into a series of room-and-pillar main headings developed in the underlying Freeport Coal Seam along the main rivers, with laterals collecting additional CSOs as needed. The headings could be constructed similarly to main haul ways in coal mines with heavy unmined coal pillars designed to prevent subsidence. There is an existing Freeport Coal Seam tunnel under the Allegheny River at Springdale, Pennsylvania, indicating good roof conditions and precedence for tunneling under the main rivers in the Freeport. The mains could be constructed using conventional mining equipment and would be three to four entries wide to provide access, ventilation, and conveyance of the mined coal. The extracted coal could be marketed, to help offset development costs.

The preceding strategy might meet design criteria at costs likely to be significantly lower than surface storage and treatment because the excavation is already done and the need for multiple pump stations is greatly reduced due to the availability of gravity flow. It is recommended that an expert panel be established to carefully evaluate and further assess the feasibility of this type of approach to CSO abatement. The scenario outlined above would have to be subject to intensive review and, if feasible, designed by mining and civil engineers as well as tested in a pilot or demonstration project before application. Further details of coal seam CSO storage are provided in Box 5-8.

Vortex Separators

One remedial option for treating mixed rainwater and sewage is to use vortex separators, which target the settleable portions of the fecal matter. Vortex separators are designed to screen and remove all settleable solids from any flows that exceed WWTP capacity during precipitation events, including bacterial solids. They are cylindrical tanks that allow water to enter tangentially and move in a centrifugal fashion before exiting. The centrifugal motion forces suspended particles to the center of the device, where they settle to the bottom. In addition to concentrating settleable solids, vortex separators also provide flow regulation.

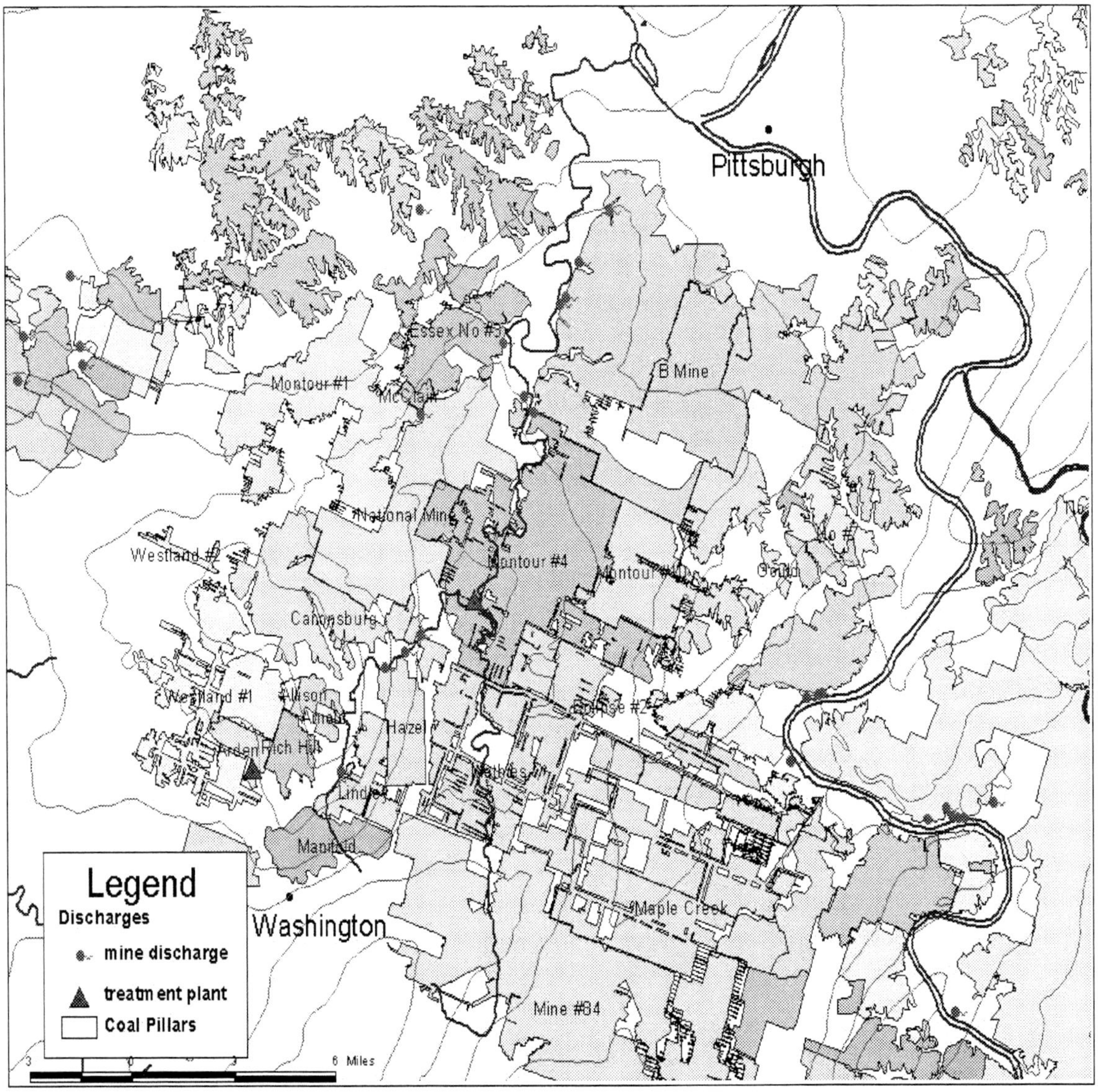

FIGURE 5-6 Pittsburgh seam coal mines south of the City of Pittsburgh.
NOTE: Montour #4 is in the center of the map; only Maple Creek and Mine #84 are still active.

There is currently no consensus opinion regarding the effectiveness of vortex separators. One advantage is that they do not typically require a lot of regular operator attention. In addition, although preliminary expenses are high, operation and maintenance costs are low. Vortex separators can be coupled with chemical disinfection (e.g., chlorine, sodium hypochlorite), which increases the separators' overall effectiveness at meeting water quality standards for microbial indicators. Finally, they have no moving parts except for pumps to remove settled solids.

Unfortunately, removal efficiencies of the settleable solids are not as high with vortex separators as with other available technologies. Vortex separators are good at removing grits and heavier particles, but not as effective at removing lighter particles. For example, they provide 5-15 percent net BOD (biological oxygen demand) and TSS (total suspended solids) removal, as opposed to enhanced high-rate clarifiers, which remove 80-90 percent of BOD and TSS.

BOX 5-8
Coal Seam Storage as a Remedial Option for CSOs

A typical, large underground coal mine in the Pittsburgh Basin is between 10 and 30 square miles in area. On average, the Pittsburgh Coal Seam is 6 feet thick. In southwestern Pennsylvania, between the Monongahela and Ohio Rivers, about 50 percent of the surface area is underlain by worked-out underground coal mines. After mining, much of that area will consist of collapsed roof rock that will form a "caved zone" up to 20 feet above the mine floor. Although exhibiting significant porosity, the caved zone will consist of large, broken fragments of shale from several inches to many feet long. Parts of the mine, such as main transportation corridors and ventilation entries, will be protected by large, intact coal pillars. In these areas, there may be little or no roof collapse decades after mining ceases. The estimated void volume of this mined area is 8 billion gallons.

Groundwater entering the mine during operations is currently pumped to the surface and treated for release to a stream. Mines that are below the deepest point of the surface landscape will completely fill with groundwater, a process that is generally complete 5 to 10 years after pumping stops. Initially, the mines fill with strongly acidic water (pH ~2.7, acidity up to 2,500 mg/L), with high concentrations of Fe^{+2} (up to 1,000 mg/L) and SO_4^{2-} (up to 3,000 mg/L). After flooding, this water is gradually replaced with net alkaline water (pH~6.3, alkalinity up to 600 mg/L). Fe^{+2} declines to between 20 and 100 mg/L, while SO_4^{2-} will remain greater than 1,000 mg/L. The shift in acid-base balance is accompanied by a transition from oxidizing to reducing conditions in the mine water. Flooded mines will have a very large though finite capacity to store the water and leakage of groundwater into the mine will force an equal amount out at a rate that correlates with depth of the mine. Any additional water will accelerate the rate of discharge.

The current excess stormwater flow for the Pittsburgh region is estimated to be a high of 400 mgd. The estimated total overflow volume is 11.5 million gallons per year, which is equivalent to an average flow rate of 32 mgd. If all of the excess stormwater was diverted into this abandoned mine, the storage time would be about 0.7 years. Using abandoned, underground mine voids for storage and treatment of sewage will require resolution of several complex issues—regulatory, financial, and technical. The chemistry itself will be complex and largely biologically controlled. In its simplest form, the addition of small community sewage to a flooded, anoxic mine pool will involve the following chemical reactions: sulfate reduction to iron and hydrogen sulfides, ferrous precipitation as iron sulfide, reduction of nitrates, and formation of ammonium compounds. All of these reactions will generate alkalinity, and all are probably operating in existing flooded coal mines. Addition of sewage will likely stimulate the existing microbial populations and accelerate most of these reactions. However, the fate of waterborne pathogens in these environments is poorly understood—especially for protracted periods of storage time. Thus, with this approach, many difficult issues would have to be addressed.

However, vortex separators may not effectively handle widely varying flows, such as the transient flows of CSOs. They have a large footprint: the land area required for this process equipment is large, compared to other available treatment technologies. Finally, because vortex separators attain solids separation by physical means, they do not remove colloidal-sized particles. Therefore, indicator bacteria and pathogens contained in very small fecal particles or unattached to fecal particles are not removed and are passed on to receiving waters unless effective disinfection is applied. As might be expected from the preceding discussion, the effectiveness of vortex separator technology has been inconsistent across the United States. Although this technology has been an operational success in some places, demonstration projects in other locations have concluded that the technology does not meet local needs.

The ALCOSAN draft LTCP relies heavily on the use of vortex separators for treatment of CSOs, with the proposed use of 10 of these separators (TPRC, 2002). The total treatment capacity of the 10 vortex separators recommended in the ALCOSAN draft LTCP is 481 mgd

(ALCOSAN, 1999). While the LTCP specifies vortex separators as "treatment facilities" that treat and discharge all flow to adjacent waterways during periods of peak wet weather flow, they actually only split flow into two effluent streams—a "clarified discharge" stream (approximately 85 to 95 percent of the inflow at design flow conditions) and a "foul underflow" stream (approximately 5 to 15 percent of the inflow at design flow conditions). Although the clarified stream is in fact discharged to the water body, the foul underflow must still be conveyed to the WWTP for subsequent treatment.

The topographical and physical features of Allegheny County seem to be compatible with vortex separators because the steep slopes provide adequate energy to drive wet weather flows through the vortex devices. Moreover, the location of the deep interceptors in certain sections will also allow construction of vortex facilities that do not require effluent or foul flows to be pumped. However, the disadvantages discussed above justify a review of the ability of vortex separators to effectively improve water quality before a significant investment is made. If a desktop review is favorable, evaluation of the technology on a pilot or demonstration scale is an appropriate follow-up action.

Reservations about vortex separators were also expressed in the TPR report (TPRC, 2002). It noted that the number of remote treatment plant facilities should be reduced. Where possible, outfalls to centralized facilities should be consolidated. Siting of these facilities is likely to be a problem in Allegheny County as it has been in other densely populated areas. That is, a "not-in-my-backyard" attitude may develop in the community once ALCOSAN begins to attempt to acquire sites for the 10 vortex facilities currently in the draft LTCP. In addition, the operation of multiple facilities at remote locations will not be easy. The TPRC also noted that where vortex units are to be used, they should be designed so that typical hydraulic overflow rates remain below 10 gallons per minute per square foot (gpm/sf) and average 5 gpm/sf so that they act as primary treatment devices. The foul underflow will have to be conveyed from the vortex facilities in transmission conduits, some of which appear to have no additional capacity. This capacity limitation may be another reason to consider retrofitting some (if not many) of the regulators with automated gates and computer controls (RTC) so that room can be made in the flow-limited sections for vortex foul flows. Finally, the TPRC advised that during facility planning, chemical storage should be included in the facility layout, so some form of disinfection may be added in the future, if needed.

The committee shares the concerns about the performance of large-scale vortex separators described in the TPR report (TPRC, 2002). As this report was nearing completion, ALCOSAN advertised a request for proposals to plan and design the first two vortex separators. It is recommended that at a minimum, these projects be implemented prior to widespread application in the ALCOSAN system.

The draft LTCP also calls for the construction of 5 previously planned and 10 new storage basins away from the main ALCOSAN treatment plant, which (with existing basins) represent a total capacity of 110 million gallons (not including an additional 43 million gallons of new storage facilities in the local systems) (ALCOSAN, 1999). The operation and maintenance of such remote facilities can be burdensome and sometimes impractical. Although the remote facilities approach has been successful and economical in some communities (e.g., Columbus, Georgia), the committee also recommends that this aspect of the ALCOSAN draft LTCP be considered carefully before proceeding.

Ballasted Flocculation

Ballasted flocculation is a process in which small particles in water or wastewater are mixed with bridging agents and tiny, high-density sand particles to increase their original size and weight. These heavier, "bulked-up" particles settle much faster and, therefore, can be removed faster and with much smaller surface area of process vessels. This process train—consisting of two stages of mixing for flocculating particles with coagulant and then attaching them to sand grains with polymer, an inclined-plate settling tank, and centrifugal recovery of the sand ballasting agent—is marketed by vendors for treatment of drinking water, SSOs, and to a lesser extent CSOs (see more below). Numerous facilities of this type have been installed in Europe, where the technology was developed, and they are now being used throughout the United States. Ballasted flocculation is an option for treating excess flow from CSOs because the process is very compact, operates at high rates, and can be started on short notice when wet weather occurs and flow increases.

Use of the ballasted flocculation concept for at-source treatment of CSO is relatively new and is not practiced extensively in the United States. However, the small unit footprint and the "instant-on" nature of the process seem to be well matched with the intermittent, high-volume nature of CSOs. This suggests that some hydraulic stability and dependable, consistent solids removal relative to vortex separators might be achieved in this type of application.

In 2003, a ballasted flocculation treatment facility was installed in Lawrence, Kansas, to handle up to 40 mgd of excess sanitary sewer flow associated with heavy rainstorms (Wagner et al., 2003). Studies projected that by 2020, the peak flow to the city's 25 mgd WWTP could reach 65 mgd during 10-year storm events. Localized storms can cause high sanitary sewer flows even though the receiving water, the Kansas River, may be in a low-flow condition due to low rainfall in the upstream watershed.

In the facility at Lawrence, excess flow is diverted at an excess flow splitter at the WWTP, where ferric chloride coagulant is added prior to the wastewater going to ballasted flocculation for treatment. Coagulated wastewater enters a mixing tank where a powder-fine sand (microsand) and polymer are added. Polymer addition results in attachment of coagulated wastewater to the microsand during a short period of flocculation that is employed before the settling process. The microsand causes the floc that forms to be much denser than ordinary floc, so the settling tank can be operated at overflow rates up to 60 gpm/sf—40 to 60 times higher than conventional overflow rates. This very high settling rate permits the use of a much smaller settling basin, enabling the size of the treatment facility to be a small fraction of the size of a conventional facility. The clarified ballasted flocculation effluent is chlorinated and then dechlorinated before discharge to the receiving water, while the settled sludge is separated from the microsand in a cyclone separator and then treated in the existing wastewater treatment plant.

A ballasted flocculation facility can be started up and treatment optimized in about 15 minutes. To aid operators in anticipating the need to operate the facility, a monitoring system was installed in collection sewers and at pumping stations, so the Lawrence treatment plant personnel will know when high flows are coming and when to turn on the auxiliary facility.

In June 2003, performance testing was conducted in each of the two 20 mgd ballasted flocculation treatment trains, using primary wastewater plant effluent. Total suspended solids removal rates ranged between 65 and 80 percent (Wagner et al., 2003). Shortly after the SSO treatment facility was tested in the summer of 2003, it was operated during a major storm event in which Kansas City and the surrounding area received about 10 to 13 inches of rain in a two-day period. The ballasted flocculation facility was operated for 32 hours, with flows ranging from 5 to 28 mgd (Cindy Wallis-Lage, Black & Veatch Corporation, personal communication, 2003). During this period of operation, total suspended solids removal ranged from 80 to 90 percent, and after the first three hours of operation, effluent total suspended solids ranged from 28 mg/L to 11 mg/L. For the entire project, construction costs for all of the facilities to monitor and treat excess flow were about $9 million, or $0.22 per gallon of installed capacity (Wagner et al., 2003).

The applicability and feasibility of ballasted flocculation should be evaluated as an alternative to vortex separators for CSO source treatment in the ALCOSAN system, and laboratory or pilot-scale studies should be conducted.

In-River CSO Storage Using Flow Balancing

In-river CSO storage using the flow balancing method (FBM) may be feasible for certain point sources. The basic concept is that a volume of CSO can be contained in a tank, consisting of flexible plastic curtains placed in a receiving stream for the temporary storage of CSOs. Combined sewer overflow that results during and immediately after wet weather enters the tank and displaces river water contained in the tank. After the storm event, the stored CSO is pumped back into the sewer to be transported to the treatment plant, and river water flows back into the tank. The plastic curtains forming the tank are suspended by pontoons and anchored to the riverbed (thus forming the base of the tank) by concrete weights.

A pilot-scale FBM, following a concept developed in the late 1970s in Sweden, was constructed and evaluated by the EPA in conjunction with the New York City Department of Environmental Protection in the 1990s at Fresh Creek in New York City (Field et al., 1994, 1995; Fordran et al., 1991). The Fresh Creek FBM was somewhat different from the original Swedish concept, which was originally designed for installation in a lake. For that purpose, a series of tanks or bays was used, with the first one receiving the CSO and the last one discharging to the lake in a "plug-flow" manner. The series of tanks helped reduce mixing between the CSO and the lake water. In the Fresh Creek study, the receiving water consisted of seawater, which had a higher density than the low-salinity CSO and only a single tank was used. The CSO that entered the tank floated on top, displacing the seawater, which then passed out into Fresh Creek through openings in the tank bottom.

The initial capacity of the tank used in the Fresh Creek study was 0.41 million gallons, and this was later expanded to a final capacity of 2 million gallons (Field et al., 1994, 1995). Because the CSO volumes at this location were generally much larger, in the 5 million to 10 million gallon range, the volume of the pilot system was insufficient to contain much of the CSO. Nevertheless, the pilot study was sufficient to demonstrate the principles of operation and the ability of the system to withstand marine environmental conditions. No damage resulted due to stresses caused by saltwater, tidal exchanges, CSO events, or coastal storms. A phase-one study was conducted using the smaller-capacity system to determine the efficiency with which the CSO was captured by the FBM. Notably, 77 percent of CSO was captured for one wet weather event in which the CSO volume was less than the volume of the tank. However, an operational difficulty with the Fresh Creek FBM was that a portion of the suspended solids within the CSO tank settled to the creek bed. For this reason a system of sediment pumps was needed to capture the settled solids.

An FBM design to help control periodic CSOs in southwestern Pennsylvania—and, more specifically, to help equalize flows to ALCOSAN's collection system and wastewater treatment plant—would need to be different because of the differences between flow in a creek in New York City and the main stem rivers in the Pittsburgh region. A design more similar to that of the original Swedish design would likely be required, though a number of questions would have to be addressed. These include how the flexible curtains would be anchored in the river bottom; how to provide for differences in river stage between low flow and flood stage, if pontoons are used to suspend the flexible curtains; the extent to which the cross-sectional area of the river would remain available for unimpeded river flow with the curtain walls in place; and the effects of river current on the flexible curtain during flood stage. Although widespread use of an FBM system would not be adequate for the majority of CSO discharges in southwestern Pennsylvania because of the typical volumes involved, there may be locations in which CSO volumes are sufficiently low, and the cost of a conveyance system to a WWTP or more conventional CSO control approaches (e.g., basin or tunnel construction) so high, that an FBM system could be a good alternative to explore. Therefore, the committee recommends that in-river CSO storage using FBM technology be explored and, if feasible, piloted at a particularly suitable location for such a system.

Wet Weather Water Quality Standards

As part of Step IV of the CWARP planning and implantation process for high-density urban areas, the Pittsburgh region, in cooperation with PADEP and EPA, may find it necessary to revisit Pennsylvania's water quality standards. Existing policy recognizes that absolute limits on water quality parameters may not be economically achievable under all hydrological and climatological events. Water quality criteria for a variety of parameters, including chemical contaminants, microbiological indicators of fecal contamination, and physical characteristics such as color and temperature are usually set and enforced to protect ambient waters during very low flows in streams. For example, in Pennsylvania, effluent limits for temperature and pH are to be established using a design flow of the lowest 30-day average that is expected to occur every year with a probability of 10 percent (25 PA Code §93, Water Quality Standards). If flow drops below that level, exceedances of numerical standards for temperature and pH are not considered to be violations of water quality standards.

In southwestern Pennsylvania, wet weather, high-flow events are some of the leading contributors to water pollution. Thus, under very high flow conditions, numerical standards for some contaminants may be exceeded but designated uses of a stream may not be impaired. For example, during flood events, numerical turbidity standards in steams are frequently exceeded, but recreational uses of the stream may be foreclosed for safety reasons, not because of water quality conditions. The stream may then return to normal uses when flood flows recede. Achievement of numerical standards during all high-flow events could be prohibitively expensive. Determining an acceptable frequency and duration for such high flow events when exceedances of numerical standards are allowed is a difficult and controversial decision.

The EPA recognized this possibility when it promulgated the national CSO policy in 1994. That policy permits modification of state water quality standards and related uses when the standards cannot be achieved because of CSOs. For example, EPA's CSO policy allows for the possibility of revisions to wet weather water quality standards as evidenced by the following excerpt from its 1999 memorandum:

> Data developed during LTCP development can inform decisions about the attainability of designated uses and the appropriateness of any WQS [water quality standard] revisions. State and federal WQS authorities need to be involved throughout the planning process to ensure that, if the LTCP is based in part on anticipated changes to WQS, those changes are appropriate and satisfy federal regulatory requirements.

Leo (1999) reviewed the history of EPA's CSO policy through 1999 and stated that the intent of the policy was to control CSOs up to the point at which maximum benefits could be achieved. Water quality standards would then be modified to allow exceedances for wet weather events that were more extreme. Leo reported that EPA had approved water quality standards in Ohio, Maine, and Massachusetts that did account for wet weather flows. In some cases, use attainability analysis would be required before the standards could be applied to particular streams.

Adoption of wet weather water quality standards is likely to be a highly controversial process. Woodworth (2000) recounts the process in Washington, D.C. All streams in Washington, D.C., were classified for primary contact recreation, and the related standard prior to preparation of a regional LTCP had both a narrative and a numerical standard. The narrative standard stated that the waters shall be free of untreated sewage, a condition that could not be attained under all high-flow events. Efforts to change the standard to account for wet weather flows were criticized as rolling back environmental standards. Woodworth was very critical of the process by which the standard was modified.

This committee recommends that changing water quality standards be considered as a last resort and concurs with Woodworth's (2000) admonition that "Water quality standards should be reevaluated only after a comprehensive long-term control plan has been designed, approved, and implemented. Provisions should be made to monitor and upgrade the plan as necessary." In addition, the committee also recommends that (1) a detailed estimate of incremental costs and an assessment of the impact on existing designated uses be included in any reevaluation of water quality standards, and (2) any reevaluation be conducted in close cooperation with PADEP and with broad public participation.

SUMMARY: CONCLUSIONS AND RECOMMENDATIONS

A fundamental prerequisite to the formulation of cost-effective plans for reducing water quality impairments in southwestern Pennsylvania is a systematic and extensive set of water quality data covering both sources of impairments and instream responses. As discussed in Chapters 3 and 4, serious water quality problems exist in southwestern Pennsylvania, but there are not sufficient data to determine the relative seriousness of the environmental and human health problems, the relative importance of potential sources of contamination, and the improvements that are likely to result from alternative pollution control measures.

The most important water quality problem in the region from a regulatory perspective and the potential for adverse human health effects is controlling microbial contamination of streams that derives from the effect of wet weather conditions on sewer systems (CSOs, SSOs, and stormwater), failing OSTDSs, and agricultural and urban runoff. Remedial actions are planned and anticipated by ALCOSAN and many of its partner communities in response to a series of consent orders that will alter the relative contribution of different sources to the water quality problems in the region. The evaluation of water quality improvements related to such activities will be critical. However, the implementation of solutions for identified impairment sources does not preclude the need for additional information related to other sources and their contributions to water quality impairment in the region. To develop better understanding of sources of contamination in southwestern Pennsylvania, water quality monitoring and modeling efforts should take place concurrently with mandated remedial activities.

It is clear that the causes and nature of water quality impairments, the parties responsible, and the individuals and waterways affected differ for each of the problem contaminants in the region. A comprehensive watershed-based approach is needed to address the spectrum of water quality problems, including wet weather problems; such a systematic approach should recognize interrelationships among problems and the need for parties responsible for each water quality problem to share in its solution. To achieve this, it is necessary to develop both a technical and an institutional-financial approach. The institutional and financial approach is discussed in Chapter 6, and the technical approach is embodied in what the committee calls the Comprehensive Watershed Assessment and Response Plan or CWARP. The Three Rivers CWARP described in this report is not a single document or program; it is a flexible umbrella concept identifying the activities that can be carried out by the organizations that are most technically and institutionally capable of achieving the desired results depending on existing and potential capabilities.

The framework recommended for planning and implementation of CWARP consists of the following five basic steps: (I) problem identification; (II) assessment of existing conditions; (III) projection of future loads; (IV) formulation and evaluation of alternative management strategies; and (V) adaptive implementation of elements of the strategy. This five-step CWARP process must be adapted to address each of the following interrelated scales: river basin, multicounty/metropolitan scale, high-density urban areas, and rural areas. The committee recognizes that the region is not starting with a blank slate, and Step I has been largely completed for each of these scales. Substantial progress has been made on Step II, but as noted in this chapter, significant gaps remain. Because the problems are largely associated with existing conditions and there is only modest growth in the region as a whole, Step III may be less important, but changes in land use that are occurring in suburban (formerly rural) areas cannot be ignored. Lastly, Steps IV and V do not appear to have been well developed at any of the scales, and these steps deserve much greater attention.

Because regional information on the biological quality of receiving waters is scant, its collection during and in support of CWARP at the river basin scale is critical. Biological water quality indices and their change over time can yield important information about ecosystem change and help quantify the environmental benefits affected by pollution abatement. Thus, information collection for CWARP should include biological data to both assist in ecosystem health assessment benchmarking and to help document changes to the ecosystem that occur as a result of changing stressors. At a minimum, the CWARP should be designed to establish an Index of Biotic Integrity for the main stem rivers. To this end, an effort should be made to expand the Ohio River component of EPA's rejuvenated Great Rivers EMAP program, with an emphasis on the biological water quality of the main stem rivers.

At least two aspects of water management are of concern at the multicounty/metropolitan scale of CWARP. First, and at the very least, water quality planning at this scale should be sufficient to inform regional interests of the potential effects (including constraints if any) of water quality conditions on future transportation and land development, the consequences of development on water quality where it occurs, and how these effects and consequences can and should be modified. Second, planning at this scale should also result in the identification of opportunities for economies of scale in the delivery of water and wastewater services through cooperative arrangements among local governments. Either SPC or an alternative organization should formulate regional water resource plans and integrate them with transportation and land use plans.

Several entities have recently estimated that solving wet weather problems in the urban core of the region by conventional means, using a combination of storage, conveyance, and treatment improvements, could cost several billion dollars. Investing large sums of capital based only on currently available data may not ultimately solve the most important problems or provide appropriate solutions. Although it is true that no amount of additional data and analyses would remove all uncertainty about water quality investments, it is clear that currently available information is lacking in several critical areas (e.g., how much surface water runoff from separate stormwater sewers affects water quality in receiving streams during wet weather events). Until these facts are known better, planning and implementation of cost-effective remedial measures will be impeded. Regardless of the regulatory approach (i.e., presumption or demonstration) used in ALCOSAN's LTCP for controlling wet weather problems, the committee concludes that it is necessary to address watershed-wide problems and sources of contaminants other than CSOs and SSOs.

Step II of CWARP at the urban scale should include simultaneous monitoring of (1) wet weather discharges into the region's streams and rivers and (2) the impacts on these receiving streams. Pollutants of primary concern in this context include pathogenic microorganisms such as *Cryptosporidium* and their surrogates (indicator microorganisms); oxygen-demanding substances including suspended solids and sediments; and "conservative" toxic substances such as metals and toxic organic chemicals. Step II should also include the development of a variety of models that are sufficient to disentangle the effects of multiple sources on water quality in receiving streams. The models and data should be available for public review, and data from these technical studies should be reduced and translated to needed corrective actions in a manner that is understandable to decision makers and the public in general.

Although receiving water quality modeling activities appear to be extremely limited currently in the region's three main stem rivers, the committee recommends that it be used to estimate impacts of pollution loadings on the receiving streams and to help prioritize alternatives for pollution control. Other modeling activities needed to implement Step II of CWARP in the region's urban core include sewer system routing models, dynamic sewer system modeling, dynamic stormwater modeling, and real-time sewer flow control modeling for analysis and operation. Projections of changes in the regional landscape are important in the planning and implementation of Step III of CWARP in the region's urban core. Planning studies conducted at the multicounty/metropolitan scale should be sufficient for this purpose and include projections for several land use, transportation, water supply, and wastewater parameters discussed in this chapter.

At least six components of a strategy to implement Step IV of CWARP for high-density urban areas should be considered and are discussed in this chapter, some of which are mandated under provisions of the Clean Water Act. The first route to successful improvement of water quality in the region is to optimize utilization of existing infrastructure. To this end, the committee strongly recommends that all wastewater collection systems located in the watershed, particularly in the urban core areas of southwestern Pennsylvania, be fully compliant with EPA's CMOM policy or an equivalent program. Thereafter, related information, approaches, and technologies recommended in this chapter and report would be available to help guide major long-term investments in improving the region's water quality.

Furthermore, ALCOSAN's draft LTCP should be reevaluated in the context of the overall CWARP approach to reflect ongoing consent order negotiations, CMOM, and information from CWARP as it is developed in the future. The CWARP approach is recommended as a framework for development of the LTCP and similar documents because of the circumstances (especially data limitations) that exist in southwestern Pennsylvania and, in principle, would apply to other regions of the United States with similar water quality problems and circumstances. In addition, in the development of a final LTCP, ALCOSAN and other wastewater treatment providers in southwestern Pennsylvania should evaluate the utilization of real-time control of CSOs. Storage and treatment of CSO in abandoned mine voids, which is currently being evaluated for the Township of Upper St. Clair, Pennsylvania, should also be evaluated. The committee also recommends consideration of the following innovative technologies and approaches for improving water quality in southwestern Pennsylvania, especially in the region's urban core: (1) at a minimum, implementation of pilot or demonstration projects prior to widespread application of vortex separators for CSO source treatment in the ALCOSAN system; (2) the feasibility of ballasted flocculation facilities and in-river CSO storage using FBM technology for controlling CSOs; and (3) the adoption of wet weather quality standards—although this is likely to be a highly controversial process and should be considered as a last resort.

Best management practices for OSTDSs should be implemented throughout the region using the CWARP framework. Although individual OSTDSs are permitted locally and current technical standards are available to ensure proper performance, they may be ignored. Furthermore, prevention of the discharge of untreated sewage into local waterways or ditches is difficult to enforce. The region needs a coordinated, well-funded program for oversight and routine maintenance of OSTDSs. Such a program can be self-sustaining through user charges provided they are applied on a cooperative regional or county basis. The committee recommends the following actions to help improve water quality in the predominantly rural areas of the region: (1) within each county, register all individual on-site and cluster disposal systems with the appropriate SEO; (2) institute a program of periodic mandatory inspection and certification (or decertification), either by a public entity or by a qualified/licensed private contractor; (3) conduct statistically valid surveys of septic tank and absorption field conditions, residence by residence, to identify communities that should be given high priority for funding by PENNVEST or the federal RUS for remediation of failed and failing OSTDSs throughout the region; and (4) use the registration and inspection program to identify and order elimination of illegal direct discharges of human waste to streams and identify where cluster OSTDSs may be feasible.

There are no comprehensive estimates of the economic benefits of addressing the remaining water quality problems for southwestern Pennsylvania or of projects proposed to address the region's water quality problems. Nevertheless, the region would be expected to benefit economically from measures that significantly reduce drinking water risks and enhance recreational opportunities. The CWARP process can identify a list of alternative management strategies and projects that are technically feasible and capable of addressing the region's water quality problems at a variety of scales, but the question remains: Which is the better option? Comprehensive evaluation of options under CWARP (especially Step IV for high-density urban areas) requires additional considerations, including costs, benefits, and fairness. It continues throughout the process of selecting a preferred long-term management strategy for a particular water quality problem at given scale. It is essential in the formulation of alternatives to provide feedback as to how initial designs should be modified or discarded in the search for a cost-effective strategy. It is also essential to the process of establishing priorities among the several elements that may comprise the management strategy. Finally, it must be a continuing process during implementation to evaluate how well each element has performed. A variety of evaluation frameworks are available; some of the more prominent are discussed in this chapter, including cost-effectiveness analysis, benefit-cost analysis, and multicriteria methods. The use of cost-effectiveness as the primary method for evaluating options for achieving water quality objectives in the region is recommended and should include an analysis of incremental costs to achieve elimination of low-probability contamination events. The committee recommends the use of benefit-cost analysis in evaluating water quality improvement projects in the region and for helping to set priorities.

As the CWARP process is being planned and implemented, it is essential that it be integrated with the ongoing process of establishing TMDLs for impaired streams being conducted by PADEP under requirements of the Clean Water Act. There are many parallels between CWARP and the process for establishing TMDLs—especially in the application of adaptive implementation. The TMDL process—supplemented by additional analyses of constituents that may not be readily subject to the rigorous TMDL approach, including biological, environmental, and other measurable factors—should be combined with watershed, regional, and subregional analysis of beneficial uses to provide the basis for selection of remedial actions in the study area.

The CWARP effort should be completed quickly to provide timely support for those water quality improvements that are required and others that are in the public interest. It is difficult to estimate the cost of implementing CWARP, but in the committee's judgment it should be low compared to the cost of improvements and more than offset by potential savings.

REFERENCES

Abdalla, C. 1994. Groundwater values from avoidance cost studies: Implications for policy and future research. American Journal of Agricultural Economics 76(5):1062-1067.

ALCOSAN (Allegheny County Sanitary Authority). 1999. Draft Combined Sewer Overflow Program Phase I Activity Report: Regional Long Term Wet Weather Control Concept Plan. Pittsburgh, PA: ALCOSAN.

Anderson, R., K. Beer, T. Buckwalter, M. Clark, S. McAuley, J. Sams, and D. Williams. 2000. Water Quality in the Allegheny and Monongahela River Basins: Pennsylvania, West Virginia, New York, and Maryland (1996-98). Denver, CO: United States Geologic Survey.

Casman, E., B. Fischhoff, C. Palmgren, M. Small, and F. Wu. 2000. An integrated risk model of a drinking water borne cryptosporidiosis outbreak. Risk Analysis 20:495-511.

Clean Water Action Plan Partners. 2000. The Guest River Watershed: River restoration in an Appalachian watershed. In Watershed Success Stories: Applying the Principles and Spirit of the Clean Water Action Plan. Washington, DC: Clean Water Action Plan.

Collins, A., R. MacDowell, M. Brooks, A. Kennedy, and J. Murphy. 2000. Water quality on the Left Fork of the Mud River: A watershed survey. Unpublished final report to West Virginia University Lincoln County Extension Office and Kellogg Community Partnership, College of Agriculture, Forestry, and Consumer Sciences. Morgantown, WV.

Davis and Simon, eds. 1995. Biological Assessment and Criteria_Tools for Water Resource Planning and Decision Making. Boca Raton, FL: Lewis Publishers, Inc.

EPA (U.S. Environmental Protection Agency). 1994. Combined Sewer Overflow (CSO) Control Policy. FRL-4732-7. Federal Register (59)75. Available on-line at *http://www.epa.gov/npdes/pubs/owm0111.pdf.* Accessed March 29, 2004.

EPA. 1995a. Combined Sewer Overflows: Guidance for Long-Term Control Plan. EPA 832-B-95-002. Washington, DC: Office of Water.

EPA. 1995b. Combined Sewer Overflows: Guidance for Nine Minimum Controls. EPA 832-B-95-003. Available on-line at *http://www.epa.gov/npdes/pubs/owm0030.pdf.* Accessed March 29, 2004.

EPA. 1995c. Combined Sewer Overflows Guidance for Funding Options. EPA 832-B-95-007. Washington, DC: Office of Water.

EPA. 1995d. Combined sewer Overflows Screening and Ranking Guidance. EPA 832-B-95-004. Washington, DC: Office of Water.

EPA, 1996. EPA Overview of the Storm Water Program. EPA 833-R-96-008. Washington, DC: Office of Water.

EPA. 1997. Combined Sewer Overflows Guidance for Financial Capability Assessment and Schedule Development. EPA 832-B-97-004. Washington, DC: Office of Water.

EPA. 1999. Economic Analysis of the Final Phase II Storm Water Rule. EPA 833-R-99-002. Washington, DC: Office of Water.

EPA. 2000a. A Benefits Assessment of Water Pollution Programs Since 1972: Part 1: The Benefits of Point Source Controls for Conventional Pollutants in Rivers and Streams. Washington, DC: Office of Water and Office of Policy, Economics, and Innovation.

EPA. 2000b. Guidelines for Preparing Economic Analysis. EPA-240-R-00-003. Available on-line at *http://yosemite.epa.gov/ee/epa/eed.nsf/webpages/Guidelines.html/$file/ Guidelines.pdf.* Accessed August 16, 2004.

EPA, 2001a. Clean Watersheds Needs Survey 2000: Report to Congress. EPA-832-R-03-001. Washington, DC: Office of Wastewater Management.

EPA. 2001b. Guidance: Coordinating Combined Sewer Overflow (CSO) Long-Term Planning with Water Quality Standards. EPA-833-R-01-002. Washington, DC: Office of Water.

EPA. 2003a. Handbook for Management of Onsite and Clustered Wastewater Treatment Systems (Draft). EPA 832-D-03-001. Washington, DC: Office of Water.

EPA. 2003b. Voluntary National Guidelines for Management of Onsite and Clustered Wastewater Treatment Systems. EPA 832-B-03-001. Washington, DC: Office of Water.

EPA. 2004. Metals, pH, and Fecal Coliform TMDLs for the Guyandotte River Watershed, West Virginia. Available on-line at *http://www.epa.gov/reg3wapd/tmdl/pdf/Guyandotte/ Sections%201-8,%20Ref/Sec-1_Guy_WV_TMDL.pdf.* Accessed June 24, 2004.

Field, R., R. Pitt, D. Jager, and M. Brown. 1994. Combined sewer overflow control through in-river water storage: An efficiency evaluation. Water Resources Bulletin 30(5):921-928.

Field, R., R. Pitt, M. Brown, and T. O'Connor. 1995. Combined sewer overflow control using storage in seawater. Water Research 29(6):1505-1514.

Forndran, A., R. Field, K. Dunkers, and D. Moran. 1991. Balancing flow for CSO abatement. Water Environment & Technology:54-58.

Freeman, M. 2003. The Measurement of Natural Resource Values: Theory and Methods, 2nd Edition. Washington, DC: Resources for the Future.

Gateway Engineers, Inc., and GAI Consultants, Inc. 2003. Progress Report: Storage of Wet Weather Overflows in Abandoned Coal Mines. Pittsburgh, PA: Gateway Engineers, Inc.

Hagedorn, C., S. Robinson, J. Filtz, S. Grubbs, T. Angier, and R. Reneau Jr. 1999. Determining sources of fecal pollution in a rural Virginia watershed with antibiotic resistance patterns in fecal *Streptococci*. Applied and Environmental Microbiology 65(12):5522-5531.

Harrington, W., A. Krupnick, and W. Spofford, Jr. 1991. Economics and Episodic Disease: The Benefits of Preventing a Giardiasis Outbreak. Washington, DC: Resources for the Future, Inc.

Heberling, M., W. Delavan, J. Shortle, and R. Brooks. 2004. Prioritizing Stream Restoration Using the Stated Choice Method. Working paper from the Department of Agricultural Economics and Rural Sociology. University Park, PA: Pennsylvania State University.

Heinz Center (The H. John Heinz III Center for Science, Economics, and the Environment). 2002. The State of the Nation's Ecosystems: Measuring the Lands, Waters, and Living Resources of the United States. Cambridge, UK: Cambridge University Press.

Karr, R., K. Fausch, P. Angermeier, P. Yant, and I. Schlosser. 1986. Assessment of Biological Integrity in Running Waters: A Method and its Rationale. Illinois Natural History Survey Special Publication No. 5. Champaign, IL: Illinois Natural History Survey.

Kocagil, P., N. Demarteau, A. Fisher, and J. Shortle. 1998. The value of preventing *Cryptosporidium* contamination. Risk: Health, Safety and Environment 9:175-196.

Koteen, J., S. Alexander, and J. Loomis. 2002. Evaluating benefits and costs of changes in water quality. PNW-GTR-548. Available on-line at *http://www.fs.fed.us/pnw/pubs/gtr548.pdf.* Accessed August 16, 2004.

Laughland, D., W. Musser, J. Shortle, and L. Musser. 1996. Construct validity of averting cost measures of environmental benefits. Land Economics 72:100-112.

Leo, William M. 1999. Wet Weather Water Quality Standards, HydroQual, Inc. Available on-line at *www.hydroqual.com/Conf/c_wml_001.htm.*

Metcalf & Eddy, Inc. 1991. Wastewater Engineering: Treatment, Disposal, Reuse, 3rd Edition. G. Tchobanoglous and F. Burton (eds.). New York: McGraw Hill Education.

Mitchell , C and R. Carson. 1986. Using Surveys to Value Public Goods: The Contingent Valuation Method. Washington, DC: Resources for the Future.

NRC (National Research Council). 2000. Ecological Indicators for the Nation. Washington, DC: National Academy Press.

NRC. 2001. Assessing the TMDL Approach to Water Quality Management. Washington, DC: National Academy Press.

NRC. 2002. Opportunities to Improve the U.S. Geological Survey National Water Quality Assessment Program. Washington, DC: National Academy Press.

NRC. 2004a. Indicators for Waterborne Pathogens. Washington, DC: National Academies Press.

NRC. 2004b. Valuing Ecosystem Services: Toward Better Environmental Decision-Making. Washington, DC: National Academies Press.

O'Connell, T., L. Jackson, and R. Brooks. 1998. The Bird Community Index: A Tool for Assessing Biotic Integrity in the Mid Atlantic Highlands. Report No. 98-4. University Park, PA: Pennsylvania State University Cooperative Wetlands Center, Forest Resource Laboratory.

ORSANCO (Ohio River Valley Water Sanitation Commission). 2002. A Study of Impacts and Control of Wet Weather Sources of Pollution on Large Rivers. Cincinnati, OH.

PADEP (Pennsylvania Department of Environmental Protection). 2002. Model Stormwater Ordinance, Draft. Available on-line at *http://www.dep.state.pa.us/dep/subject/ Proposed_regulations/SW_MS4_Model_Ordinance.pdf.* Accessed May 29, 2004.

PADEP. 2004. Pennsylvania DEP's Six-Year Plan for TMDL Development. Available on-line at *http://www.dep.state.pa.us/dep/deputate/watermgt/wqp/wqstandards/TMDL/TMDL_ 6yearplan.pdf.*

Ribaudo, M. and J. Shortle. 2001. Estimating the benefits and costs of pollution control policies. In Environmental Policies for Agricultural Pollution Control. J. Shortle and D. Abler (eds.). Oxon, UK: CAB International Publishing.

Rosenberger, R., and J. Loomis. 2001. Benefits Transfer of Outdoor Recreational Use Values: A Technical Document Supporting the Forest Service Strategic Plan (2000 Revision). Fort Collins, CO: U.S. Department of Agriculture, Forest Service.

Simmons, G., Jr., S. Herbein, and C. James. 1995. Managing nonpoint fecal coliform sources to tidal inlets. Journal of Contemporary Water Research and Education 100:64-74.

Smith, V., W. Desvousges, and A. Fisher. 1986. A comparison of direct and indirect methods for estimating environmental benefits. American Journal of Agricultural Economics 68:280-290.

TPRC (Third Party Review Committee). 2002. Third Party Review of the ALCOSAN Regional Long Term Wet Weather Control Concept Plan. Pittsburgh, PA: ALCOSAN.

U.S. Census Bureau. 1996. Table 3: Land Area, Population, and Density for Places in Pennsylvania: 1990. Available on-line at *http://www.census.gov/population/census data/places/42pa.txt.*

Wagner, D., M Schultze, and J. Keller. 2003. Kansas plant clarifies new overflow solution. Public Works 134(11):59-62.

Woodward, L. 1961. Ground water contamination in the Minneapolis and St. Paul suburbs. In Ground Water Contamination: Proceedings of the 1961 Symposium. Technical Report W61-5. Cincinnati, OH: U. S. Department of Health, Education, and Welfare, Public Health Service, Robert A. Taft Sanitary Engineering Center.

Woodworth, J. 2000. Balancing bathers and bacteria: Managing recreation, wet-weather flows and the legacy of a combined sewer. Abstract presented to the National Symposium on Designating Attainability Uses of the Nation's Waters, sponsored by the EPA. Washington, DC, June 3-4. Available on-line at *www.epa.gov/waterscience/standards/ symposium/abstracts/.*

WSIP (Southwestern Pennsylvania Water and Sewer Infrastructure Project Steering Committee). 2002. Investing in Clean Water: A Report from the Southwestern Pennsylvania Water and Sewer Infrastructure Project Steering Committee. Pittsburgh, PA: Campaign for Clean Water.

6

Water Quality Improvement: Institutional and Financial Solutions

Water quality problems and issues in southwestern Pennsylvania are both local and regional as evidenced by a variety of reports included in Appendix B, water quality assessments by the Pennsylvania Department of Environmental Protection (PADEP), and testimony received by the committee. Some of these water quality problems are associated primarily with urbanization in the immediate Pittsburgh vicinity; some are associated with activity in the Monongahela and Allegheny River basins; still others are common to the predominantly rural counties in southwestern Pennsylvania. Large differences exist among the sources of problems, their potential effects on public health and environmental quality, and their likely solutions. Further, resolution of water quality issues in southwestern Pennsylvania is affected by other regional issues such as transportation, land use, and governance of the metropolitan area.

The existing pattern of water supply and water quality services in the region is highly fragmented, with more than 1,000 providers operating in the multicounty region. In Pittsburgh's metropolitan area, like many other metro areas in the United States, large-special purpose authorities such as the Allegheny County Sanitary Authority (ALCOSAN) can achieve substantial economies of scale through joint management agencies. Although private organizations may not have direct voting power in what mix of organizations is chosen to implement the plan, they could very well influence how the public and its elected and appointed representatives make these choices. Although no single unit of government has all the necessary power to implement the Three Rivers Comprehensive Watershed Assessment and Response Plan (CWARP) recommended and discussed in Chapter 5, it is desirable to have some mechanism to facilitate continued oversight of regional progress (or lack thereof) toward clean water and its relationships to other regional goals and activities, and to help southwestern Pennsylvania realize the benefits of cooperation.

Furthermore, the situation is not static. Although the Pittsburgh metropolitan statistical area (MSA) is among the few in the nation to actually lose population during the 1990s (−1.5 percent; see Chapter 2 for further information), it is nevertheless listed by American Rivers (2002) as among the top 20 metropolitan areas in terms of "urban sprawl." This ranking is based on the percentage increase in developed land in 1997 compared to 1982. According to American Rivers,[1] the Pittsburgh MSA experienced an increase of 42.5 percent in urbanized land, accompanied by a decrease in average density of 35.5 percent over those 15 years. Planning for water quality improvement, especially where capital investment is substantial, must therefore reflect regional planning goals concerning economic development and demographic character, such as impacts of urban sprawl and (re)development.

---

[1] American Rivers is a national nonprofit conservation organization dedicated to protecting and restoring natural rivers; see *http://www.amrivers.org* for further information.

Finding the right mix of existing and new organizations that best fulfill the necessary conditions for planning, implementation, and oversight of CWARP will be a difficult and time-consuming process. Several options that the region should consider are discussed in this chapter. The discussion begins with a review of management functions necessary to deliver water supply and water quality services and criteria for evaluating alternative organizational arrangements to perform those functions. The challenge is to find the right mix of organizations that can perform the necessary functions in an efficient and politically accountable manner. The committee's examination of specific arrangements begins with existing organizations in the region. This is followed by a brief review of what other regions with somewhat similar problems have done. Future options for water resource and quality management in southwestern Pennsylvania are then explored. These options are discussed in light of existing enabling legislation and what additional legislation may be desirable. Also, two other significant factors influencing the choice of organizational arrangements are discussed: (1) potential sources of financing and (2) financial burdens that may be imposed on citizens of the region.

CRITERIA FOR EVALUATING ORGANIZATIONAL OPTIONS

Choosing an appropriate organization or set of organizations to address regional water quality problems holistically is a complex task. Criteria for guiding the formulation and evaluation of alternative arrangements usually include consideration of the following:

- efficiencies with which each organizational arrangement could carry out the various policy-making and management functions by exploiting economies of scale;
- geographic coverage sufficient to incorporate significant hydrological, biological, and chemical processes between upstream and downstream elements of the water resource system and to incorporate significant linkages in construction and operation of infrastructure that crosses political boundaries;
- capacity to integrate water systems, wastewater systems, stormwater systems, and other aspects of water resources with land use and transportation;
- legal, technical, and financial capacities of each option to perform management functions;
- capacity of each option to involve the many faces of the public and minimize conflict in decision making processes; and
- the nature of existing contracts and other commitments.

Before these criteria can meaningfully be applied, it is appropriate to describe the management functions, scale, and authorities of alternative arrangements.

Management Functions

A list of water quality planning and management functions for water systems is provided in Box 6-1. They are listed in approximate order of statutory authority necessary to perform them, beginning with the least intrusive government power and concluding with the most intrusive. Collection of data, planning, and technical assistance require only modest statutory authority. Implementing actions including financing, construction, taking of land, and adoption and

BOX 6-1
Water Quality Planning and Management Functions

- Organization of public forums to discuss and initiate appropriate activities
- Collection of basic data on water quality, sources of pollution, land use, and other relevant data
- Technical and financial assistance
- Planning for water quality improvements and related land use and transportation
- Construction of facilities
- Operation and maintenance of facilities and delivery of services
- Taking of land for public facilities
- Financing authority, including authority to incur debt and establish and implement user charges or taxes to recover costs of service
- Establishment of water quality related regulatory standards for private and public development activities and post-construction operation and maintenance
- Allocation of assimilative capacity to new and expanding regional activities

enforcement of regulations require substantially greater authority. General-purpose local governments, including municipalities and counties, usually have the broadest array of powers delegated to them by state legislatures. Therefore, they tend to face fewer legal obstacles, exercise greater power to integrate land use and water services, and have greater flexibility to implement economically efficient management programs within their limited geographical jurisdictions.

Issues of Scale

Scale is a key factor in selecting an appropriate mix of organizations to deliver services in the region. The National Research Council (NRC) Committee on Watershed Management (NRC, 1999) addressed the issue of choosing an appropriate scale for planning that includes all relevant hydrologic linkages, commenting as follows:

> Managing water resources at the watershed scale, while difficult, offers the potential of balancing the many, sometimes competing, demands we place on water resources. The watershed approach acknowledges linkages between upland and downstream areas, and between surface and ground water, and reduces the chances that attempts to solve problems in one realm will cause problems in other...Organizations for watershed management are most likely to be effective if their structure matches the scale of the problem.

Planning at the watershed scale offers the opportunity to address externalities among several parties within the basin.

That earlier NRC committee addressed the problem of incorporating hydrologic and biological interdependencies that exist in water resource systems. Unfortunately, the geographic jurisdictions of organizations with the range of necessary legal authorities seldom match watershed boundaries. For example, Figure 5-2 shows about a dozen watersheds that contribute stormwater runoff to the contiguous urban area in and adjacent to Allegheny County and the City of Pittsburgh. There are approximately are parts of five counties and 100 municipalities within that area alone. New organizational arrangements may have to be created to effectively and efficiently manage water, but development of these arrangements may entail difficult political decisions that involve the transfer of some powers and responsibilities from existing units of government. These difficulties must be weighed against the anticipated economy-of-scale benefits that new organization(s) may offer.

As discussed in Chapter 5, planning and management are needed to address the array of water resource problems at four interrelated scales in (and beyond) the Pittsburgh region, and organizational arrangements should be responsive to each of the following scales:

1. river basin, to address issues related to imports and exports to the multicounty region, including areas and states outside southwestern Pennsylvania;
2. multicounty/metropolitan scale, where decisions are being made about large-scale infrastructure and related land use in southwestern Pennsylvania that affect water resources and where opportunities exist to achieve efficiencies and avoid conflicts in regional water management;
3. urban areas in and around Allegheny County and outlying urban centers, where combined and separate sewer overflows and stormwater runoff must be addressed (see Figure 6-1); and
4. rural areas within southwestern Pennsylvania having problems of inadequate human waste disposal and water supply.

CURRENT SITUATION IN SOUTHWESTERN PENNSYLVANIA

Water quality management in the Pittsburgh region is highly fragmented, with responsibilities and authority distributed among a very large number of general purpose local governments, special districts, regional planning organizations, and the Commonwealth of Pennsylvania. For purposes of this discussion, the region is defined by the nine-county area served by the Southwestern Pennsylvania Commission (SPC). It is important to note that alternative definitions (e.g., 11 counties; see Box 1-2) are discussed elsewhere in this report.

General Purpose Local Governments and Special Districts

The 2002 Census of Governments[2] lists 526 general purpose governments within the region, distributed by county and type of government as shown in Table 6-1. In the 1997 Census of Governments,[3] boroughs, cities, and municipalities were lumped together under the heading "cities" (the number of cities in the 1997 census is the same as the sum of boroughs plus municipalities plus cities in the 2002 census), and the numbers were unchanged from 1997 to 2002. Under Pennsylvania law, each of those local governments and the nine counties are authorized to provide water supply and sewer services.

In addition to the general purpose governments, there are 154 special districts engaged in either sewer service alone or both water supply and sewer service. The special districts are distributed by county, type, and characteristics of service boundaries as shown in Table 6-2. The only special districts included in the 1997 list of "large" districts in the Census of Government finances that were clearly identifiable as delivering sewer services were the Pittsburgh Water and Sewer Authority, with an annual expenditure of about $118 million, and ALCOSAN, with

[2] See *http://www.census.gov/govs/www/cog2002.html* for further information on the 2002 Census of Governments.
[3] See *http://www.census.gov/prod/gc97/gc971-1.pdf* for further information on the 1997 Census of Governments.

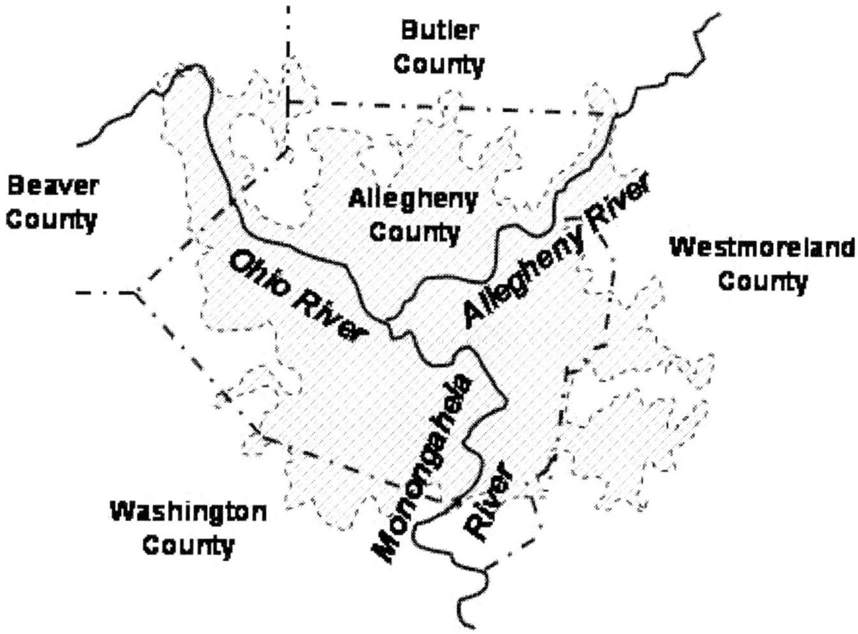

FIGURE 6-1 Approximate area of the urban core of southwestern Pennsylvania.
SOURCE: U.S. Census Bureau, *http://ftp2.census.gov/geo/maps/urbanarea/-uaoutline/UA2000/ua69697/.*

TABLE 6-1 General Purpose Local Governments in Southwestern Pennsylvania in 2002

County	Cities and Boroughs	Municipalities	Townships
Allegheny	80	6	42
Armstrong	16	1	28
Beaver	27	2	22
Butler	23	1	33
Fayette	16	2	24
Greene	6	0	20
Indiana	14	0	24
Washington	33	2	32
Westmoreland	36	8	21
Total	251	22	246

SOURCE: United States Census of Governments, 2002, *http://www.census.gov/govs/www/cog-2002.html.*

TABLE 6-2 Special Districts Providing Water and Sewer Service in Southwestern Pennsylvania

	Type of Service		Type of Boundary			
County	Sewer	Water Supply and Sewer	County	Borough, City, or Township	Within County[a]	Cross-County
Allegheny	27	5	2	7	12	3
Armstrong	6	4	1	2	4	1
Beaver	21	7	1	4	7	0
Butler	6	4	0	0	4	4
Fayette	15	2	1	4	5	1
Greene	5	3	1	3	2	0
Indiana	4	3	3	2	2	2
Washington	21	2	1	10	6	1
Westmoreland	24	2	0	5	7	4
Total	122	32	10	37	49	16

[a] Within county but not limited to borough, city, or township.
SOURCE: United States Census of Governments, 1997, *http://www.census.gov/prod/gc97/gc-971-1.pdf.*

expenditures of $284 million annually. ALCOSAN serves 83 communities, most of which are located in or immediately adjacent to Allegheny County (see Figure 1-1).

Fragmentation of sewer services in the region with its many special districts reflects the general pattern of special districts in Pennsylvania. The 1997 Census of Governments reported 2,004 single-purpose sewer districts in the United States; Pennsylvania had the highest number, 591, about 30 percent of the nation's total. Wisconsin was the next highest state with 320 single-purpose sewer districts.

Regional Planning Organizations

The Southwestern Pennsylvania Commission (SPC)[4] is the officially designated regional planning agency for the area in and around Pittsburgh. SPC's major role is "comprehensive regional planning with emphasis on transportation and economic development." It was designated in 1974 as the metropolitan planning organization for transportation (MPO; see more below). It is also the Economic Development District for southwestern Pennsylvania, as designated by the U.S. Appalachian Regional Commission and the U.S. Department of Commerce. The SPC governing board includes more than 60 members representing the 10 counties, the City of Pittsburgh, the Governor's Office, and several state and federal agencies. In addition to its primary functions, recent discussions regarding regional land use and growth decisions have pointed to the need for SPC to help address local development issues (e.g., WSIP, 2002). As a result, SPC is expected to continue to create, organize, and support public forums that bring a regional perspective to issues such as housing, sewer systems, and community development.

In 1998, SPC requested that the Western Division of the Pennsylvania Economy League[5] make a preliminary study of the region's needs. That study pointed to water supply and wastewater problems as potential impediments to future economic growth, and in 1999, the Western Division of the Pennsylvania Economy League initiated the Southwestern Pennsylvania Water and Sewer

[4] For further information about the SPC, see *http://www.spcregion.org*.
[5] For further information about the Western Division of the Pennsylvania Economy League, see *http://www.pelwest.com*.

Infrastructure Project (WSIP). The steering committee for that project included 60 public and private sector leaders from the region. As described elsewhere in this report (see also Appendix B), the WSIP report identifies several important water supply and wastewater management problems in the region, including the following:

- overflowing sewers and failing septic systems that annually discharge billions of gallons of inadequately treated or untreated sewage into the region's streams and lakes;
- lack of clean and reliable water supplies to some residents, particularly in rural areas;
- inadequate water and sewer infrastructure at otherwise desirable development sites; and
- growth limitations in many communities resulting from inadequate facilities.

The WSIP Steering Committee recommended the following:

- the SPC serve as the organization for setting regional water-related goals and priorities;
- that the Three Rivers Wet Weather Demonstration Program (3RWW) serve as the regional organization for public education and technical assistance, expanding its service area beyond the ALCOSAN area that it now serves; and
- that the Southwestern Pennsylvania Growth Alliance and the Greater Pittsburgh Chamber of Commerce serve as a regional advocacy organization.

These recommendations reflect a perspective from a knowledgeable leadership group within the region of the overall need to enhance regional water planning in southwestern Pennsylvania. The committee agrees with this need, and alternatives for meeting it are discussed later in this chapter. With its traditional focus essentially limited to economic development and transportation, SPC has not yet undertaken "comprehensive regional planning" that includes effective water planning.

Commonwealth of Pennsylvania

The Pennsylvania Department of Environmental Protection (PADEP) is the state regulatory agency charged with water quality management. In that capacity it has jurisdiction over those portions of the Ohio River basin within Pennsylvania, including the Allegheny and Monongahela River tributaries (see Box 1-2 and Figures 2-1 and 2-2). The PADEP has included the Ohio River basin among six major basins in the state (the others being Lake Erie, Genessee, Susquehanna, Potomac, and Delaware). Unlike water resource planning under Pennsylvania's Water Resources Planning Act (WRPA) of 2002 (General Assembly of Pennsylvania, 2002), PADEP does not have a planning program to guide management of water quality at the basinwide scale.

The PADEP has, however, established a watershed restoration program at a smaller scale than the Ohio River basin under its nonpoint source program—Pennsylvania's response to requirements of Section 319 of the federal Clean Water Act. The Unified Watershed Assessment was begun in 1998 to set priorities for restoration of streams where quality had been degraded by a variety of pollution sources *other* than municipal and industrial wastewater treatment plants and discharges. Included among sources are acid mine drainage, sewer system overflows, agricultural runoff, and other nonpoint sources (NPSs) of pollution. This program used PADEP's 305(b) report (PADEP, 2002a) and its 303(d) list (PADEP, 2002b) of impaired streams as a starting point. The PADEP has delineated 104 watersheds that cover the entire state, 30 of which are located in the Ohio

River basin in southwestern Pennsylvania. Each watershed was initially assigned to one of four categories (see Table 6-3) based on the percentage of stream miles assessed, the percentage of these miles judged to be impaired, and the potential for NPS pollution.

Priorities for water quality improvement were assigned to each of the 23 watersheds in Pennsylvania that fall into Category I. Watershed Restoration Action Strategies (WRASs) were then developed for priority watersheds in cooperation with federal, state and local agencies; watershed-based organizations; and the general public. Included among the 30 watersheds in the Ohio River basin for which a WRAS has been prepared are the following (see Figure 6-2): Redbank Creek, Conemaugh River/Blacklick Creek, Stony Creek/Little Conemaugh River, Lower Youghiogheny River, Upper Youghiogheny River/Indian Creek, Upper Monongahela River, Raccoon Creek, and Chartiers Creek.

Each watershed plan includes descriptions of geology and soils, natural and recreational resources, and streams classified by PADEP as being of "exceptional or high quality." Sources of water quality impairment are also discussed. Existing restoration initiatives are listed, and funding needs (to the extent they are known) are estimated. Funding from multiple sources has been provided to address some of the problems covered by these plans. Grants from Pennsylvania Growing Greener, the U.S. Environmental Protection Agency's (EPA's) Section319 and Section 104(b)3, and Pennsylvania's Watershed Restoration Assistance Program have all been received to fund restoration projects. The Pennsylvania Infrastructure Investment Authority (PENNVEST; see also footnote 8) also has made loans to local governments to address some of the problems. The PADEP Bureau of Abandoned Mines has also been an active participant in the implementation of many of these watershed plans. Table 6-4 summarizes some of the commitments already made to four of the eight watersheds listed above.

The watershed plans address important issues as identified in Pennsylvania's most recent 305(b) report (PADEP, 2002a) and 303(d) list (PADEP, 2002b) for priority watersheds (see Chapters 3 and 4 for further information), but there is no assurance that streams in these watersheds will be restored to a level that fully supports their designated uses. Section 319 requires adoption of best management practices for NPS pollution, but unlike the total maximum daily load (TMDL) process (see also Chapters 3 and 5), it does not require a demonstration using predictive models or other evidence that water quality standards will be achieved. Follow-up investigations of projects in WRAS plans will be required to assess progress toward the goal of fully restoring streams in those watersheds.

TABLE 6-3 Pennsylvania State Water Plan Watershed Categories

Category	Stream-Miles Assessed	Assessed Miles Impaired	Other Criteria
I	≥ 20%	≥ 15%	High potential for NPS pollution
II	≥ 20%	< 15%	—
III	Pristine	—	—
IV	Insufficient data	—	—

SOURCE: PADEP, *www.dep.state.pa.us*.

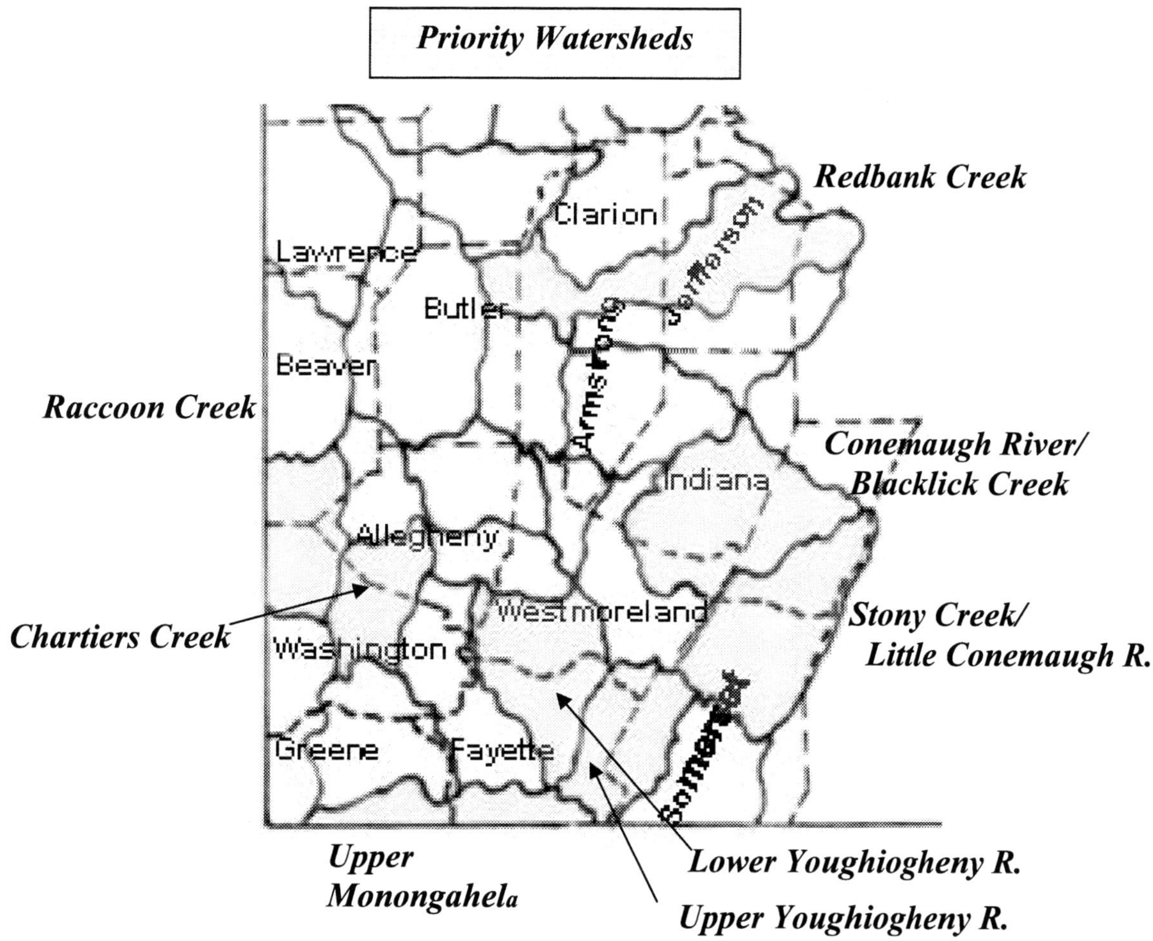

FIGURE 6-2 State-delineated watersheds in southwestern Pennsylvania. NOTE: Shows two counties (Clarion and Jefferson) not included in the study area (see also Box 1-2).
SOURCE: Data from PADEP, *www.dep.state.pa.us*.

Contaminated water supplies and improper disposal of sewage from on-site sewage treatment and disposal systems (OSTDSs) not connected to public water or sewer systems were identified in the 2002 WSIP report as being of major concern in the region, but the Unified Watershed Assessment did not include a systematic evaluation of the extent of these problems. As discussed in preceding chapters, better information is needed to make an informed assessment of the locations, magnitude, and priorities to be assigned to these water quality problems.

In contrast to PADEP's WRAS program, which focuses on priority problems within selected watersheds, water supply is being addressed on a basinwide scale that recognizes linkages among watersheds. Pursuant to the WRPA of 2002, PADEP has initiated the process to update the State Water Plan. That act establishes a Statewide Water Resources Committee (SWRC) to set guidelines and policies for the planning process and to conduct a formal review and approval of the product. Regional water resources committees are to be established for each of the state's six major basins. After conducting an open public process and consulting with the SWRC and PADEP, the Ohio Basin Committee is to recommend regional plan components to the SWRC. These areas would be

TABLE 6-4 Select Restoration Activities of the PADEP Bureau of Watershed Management's Watershed Restoration Action Strategy

Subbasin	Problem	Funding Source	Number of Projects	Project Expenditures (dollars)
Redbank Creek watershed (Allegheny River)	Abandoned mine drainage	Pennsylvania Growing Greener Grants	7	$570,000
		EPA Clean Water Act Section 319 Grants	2	$156,000
Stonycreek River and Little Conemaugh River watersheds	Abandoned mine drainage, Upgrade or expand water supply, sewers, and wastewater treatment	Pennsylvania Growing Greener Grants	17	$1,508,000
		EPA Clean Water Act Section 319 Grants	7	$1,014,000
		Pennsylvania Watershed Restoration Assistance Program	2	$54,500
		PADEP Bureau of Abandoned Mine Reclamation	4	$2,755,000
		EPA Clean Water Act 104b3	3	$518,000 (grants)
		PENNVEST	4	$5,607,000 (loans)
Upper Youghiogheny River	Abandoned mine drainage	Pennsylvania Growing Greener Grants	3	$1,371,000
		EPA Clean Water Act Section 319 Grants	4	$587,000
		Pennsylvania Watershed Restoration Assistance Program	1	$261,000
Chartiers Creek watershed	Point and nonpoint pollution, combined sewer overflow, abandoned mine drainage	Pennsylvania Growing Greener Grants	10	$497,000
		EPA Clean Water Act Section 319 Grants	7	$476,000
		Pennsylvania Watershed Restoration Assistance Program	1	$29,300
		EPA Clean Water Act 104b3	1	$49,200
		Pennsylvania Department of Conservation and Natural Resources Rivers Conservation Grants	3	$178,000
		PENNVEST	3	$3,030,000 (loans)

SOURCE: PADEP, *www.dep.state.pa.us/dep/deputate/watermgt/wc/Subjects/Nonpointsourcepollution/Initiatives/Wraslist.htm*.

designated, "critical water planning areas," and identified on a multimunicipal watershed basis. Areas in which demand is expected to exceed supplies would be so designated, and more detailed critical area resource plans, or "water budgets," would be established.

The WRPA does not have a similar mandate for water quality. Nevertheless, the planning process it establishes for water supply provides an excellent opportunity for PADEP to exert administrative leadership to better integrate water quality and water supply into a broader framework of planning for water resources at the basin scale. Basin plans should at a minimum indicate the water quality effects on public water supplies and the water quality effects of flood control activities.

Significant legislation enacted by the Pennsylvania General Assembly in 2000 could influence water planning among neighboring local governments. Among other provisions, Pennsylvania Acts 67 and 68 of the 1999-2000 legislative session, Article XI state the following:

> For the purpose of encouraging municipalities to effectively plan for their future development and to coordinate their planning with neighboring municipalities, counties and other governmental agencies, and promoting health, safety, morals and the general welfare...powers for the establishment and operation of joint municipal planning commissions are hereby granted.

Local governments were given additional powers to regulate growth. Included in those powers were authority to limit development in specially designated "growth areas" and to implement a program of transferable development rights. Municipalities were given authority to enter into intergovernmental cooperative planning and implementation agreements. Municipalities located within the county or counties were also enabled to enter into intergovernmental cooperative agreements to develop, adopt, and implement comprehensive water resource plans for entire counties or any area within counties. Such agreements also enabled participating municipalities to share tax revenues and fees.

The legislation also included incentives for municipalities to enter into such agreements. State agencies were directed (1) to consider multimunicipal plans when reviewing applications for the funding or permitting of infrastructure or facilities, and (2) to consider giving priority to applications for financial or technical assistance for projects consistent with the county or multimunicipal plan.

Former Pennsylvania Governor Ridge issued an executive order in January 1999 directing PENNVEST to take land use into consideration when evaluating water project proposals; Acts 67 and 68 of 2000 had similar implications. Among other actions, PENNVEST established as an eligibility requirement that funding of proposed projects be consistent with applicable municipal, multimunicipal, or county comprehensive land use plans and zoning ordinances (see *http://www.pennvest.state. pa.us/pennvest/cwp/* for further information). How effective this incentive will be in promoting cooperation remains to be seen.

WHAT OTHERS HAVE DONE

Southwestern Pennsylvania has many problems of water planning, delivery of services, and governance in common with other regions of the country. Knowledge and discussion of similar experiences in some of these regions may be instructive to those who will make decisions in the Pittsburgh region.

Metropolitan areas across the United States have adopted a variety of arrangements to perform water management functions that transcend boundaries of local government. These arrangements range from consolidation of city and county governments to intergovernmental contracts. In the middle of this range are special purpose service districts.

Consolidation of City and County Governments

Consolidation of city and county governments has been adopted in Jacksonville-Duval County, Florida; Nashville-Davidson County, Tennessee; Indianapolis-Marion County, Indiana; and Louisville-Jefferson County, Kentucky. Such an arrangement has several advantages. It offers opportunities to capture economies of scale in capital investments, operating expenses, and administration. General purpose local governments, such as counties and municipalities, are empowered not only to exercise the water quality functions listed in Box 6-1, but also to integrate them with comprehensive land use and other aspects of urban development. Consolidation of city and county governments has the further advantage that both entities remain politically accountable for their actions.

In *Cities Without Suburbs*, David Rusk (1995) argues that establishment of a metropolitan government is much better than alternative strategies that seek to make multiple local governments act like a metropolitan government. He contends that regions where that possibility is most viable are those in which a central city could be consolidated with suburban communities within a single county that would include at least 60 percent of the total metropolitan population. Such criteria are satisfied in Pittsburgh-Allegheny County. Consolidation would not address all water quality problems in the region, but it would offer the benefits of economies of scale, incorporation of upstream-downstream linkages, incorporation of infrastructure linkages among neighboring political jurisdictions, enhanced comprehensive planning and management within the urban core, and a strong and more flexible financial base with greater employment capability.

Louisville-Jefferson County, Kentucky, is a case of city-county merger that leaders in the Pittsburgh region may want to examine more closely. The City of Louisville and Jefferson County governments were merged, effective January 2003. Like the Pittsburgh region, Louisville and Jefferson County were served by a large special purpose sewer district. The Metropolitan Sewer District (MSD) was formed in 1946 to provide sewer services across municipal boundaries for the metropolitan area. After the merger, it remained as a separate unit of local government, but its eight-member board is now appointed by the newly formed Metropolitan Council, the elected local government for Louisville-Jefferson County. The MSD also created the Louisville and Jefferson County Regional Sewer Corporation to provide services to a portion of neighboring Oldham County and a state facility in Shelby County.

There are about 680 miles of combined sewers with 115 combined sew overflow (CSO) outfalls in a heavily urbanized area of more than 38 square miles in Louisville-Jefferson County. According to the MSD, the agency began to address CSO-related water quality problems in the early 1980s, beginning with mapping and modeling of the collection system. Monitoring was initiated in 1991, and a long-term control plan was developed as required by the EPA's 1994 CSO policy (see Chapter 5 for further information). Several infrastructure improvement projects have since been implemented, including in-line storage, separation facilities, storage basins, and pilot CSO treatment facilities (EPA, 2001). The MSD has also instituted a backup prevention program to eliminate damage to homes where stormwater creates surcharges on combined sewers. MSD also operates a stormwater utility.

At the larger, multicounty scale, the Kentuckyiana Regional Planning and Development Agency (KIPDA) is the MPO for the Louisville area, with jurisdiction over seven counties in Kentucky and two in Indiana. KIPDA's primary responsibilities are transportation, social services, and public administration. Like SPC in the Pittsburgh region, KIPDA historically has had a very limited capability in water planning.

Kentucky's Department of Environmental Protection adopted a watershed-based management approach in 1997. Five groupings of river basins and minor tributaries were identified, and assessments of these basins are made on a five-year rotating schedule. Reports generated for the Salt River and Minor Ohio River Tributaries include a 1998 status report, a 1999-2000 strategic monitoring plan, a 2001 assessment report, and a 2002 priority watershed reports. Although these reports provide substantial information about water quality in the area, they do not appear to provide very specific action plans. For those watersheds within Jefferson County, deference appears to have been given to watershed management activities initiated by the MSD.

Multiple-Purpose Metro Councils

A variant on general-purpose metropolitan government is multiple-purpose metro councils that are operating agencies as well as planning agencies. This option delegates limited authority held by general-purpose local governments to regional agencies that better match appropriate scales for water quality management. Examples of strong regional mechanisms with powers beyond planning are those in Portland, Oregon; the Twin Cities (Minnesota) Metro Council; and the Atlanta Regional Transportation Authority.

The Twin Cities Metro Council (TCMC) was created by the Minnesota state legislature in 1967 to coordinate planning and development in the seven-county metropolitan area. Through a series of additional acts, three separate agencies—the Metropolitan Transit Commission, the Regional Transit Board, and the Metropolitan Waste Control Commission—were merged into a single agency. The TCMC is governed by a 17-member council, with 16 members each representing a geographic district. All members are appointed by the Minnesota governor subject to confirmation by the state legislature.

In addition to its planning functions, the TCMC operates the region's largest bus system, collects and treats wastewater, provides affordable housing, and acquires and funds a regional park system. The TCMC's wastewater collection and treatment services are operated through its revenue-funded Environmental Services Division (ESD),[6] which operates 8 wastewater treatment plants, treating about 300 million gallons per day from more than 2 million residents in 103 communities. It acts as a wholesale supplier of wastewater collection and treatment services to those communities, which in turn provide retail services to their customers. The TCMC also provides direct services to some customers. The ESD is also active in NPS pollution management and conducts monitoring and planning for stormwater runoff.

[6] For further information about the ESD, see *http://www.metrocouncil.org/services/environmental.htm*.

Special Districts

Several metropolitan areas have addressed intergovernmental management through the formation of special districts. Yaro (2000), reflecting on the history of service delivery in New York, argues that given the very limited acceptance of the formation of metropolitan governments, attention should be focused on more modest initiatives such as regional service districts. He observed that shortly after the five boroughs were consolidated to form the city in 1898, growth and development continued at a rapid pace beyond the boundaries of the enlarged city. Within two decades, and with no prospect for further expansion of city boundaries, the leadership created several new special-purpose authorities that transcended existing city limits. Among them were the Port of New York Authority (1921), toll road and bridge authorities (beginning in 1931), the Metropolitan Transportation Authority (1968), several regional park commissions in the 1920s, and the Interstate Sanitary Commission (also in the 1920s). Similar approaches have been used with success (and some failures) in many other metropolitan areas. Notable examples in Pittsburgh are ALCOSAN (1954; see Chapter 2 for further information) and the Port Authority of Allegheny County (1956), which operates the regional mass transit system.

Special districts of this kind are important in the delivery of sewer services throughout the United States. Of $26.7 billion spent by local governments in 1997, as reported by the U.S. Census Bureau, $5.3 billion (22 percent) was spent by special districts. An analysis of data from the 1997 Census of Governments indicates that 12 other U.S. metropolitan areas were served by special sewerage districts comparable to ALCOSAN with expenditures in excess of $50 million. Those that provide both sanitary sewer and combined sewer services are listed in Table 6-5. Notably, all of them serve core urban areas within metropolitan statistical areas (although very few serve an entire MSA).

Several factors should be considered when evaluating the option of a special district for water management that serves significant portions of a metropolitan area, including its

- relationship to comprehensive regional planning;
- capacity to integrate planning and management of public water supply, wastewater collection and treatment, combined sewer overflow, stormwater management, and aspects of the water resource system; and
- relationship to general-purpose local governments within its service area.

The relationship between a special district and the regional comprehensive planning process can be important. All metropolitan areas throughout the country have some form of regional planning as mandated by various federal initiatives. For example, under the authority of the Federal-Aid Highway Act of 1962, the Bureau of Public Roads required the formation of planning agencies to carry out the mandated planning for urban areas. At about the same time, the Housing and Urban Development Act of 1965 established a grant program to encourage the formation of MPOs to be controlled by elected officials from the jurisdictions they serve. The role of MPOs was strengthened by the Intermodal Surface Transportation Efficiency Act of 1991 (ISTEA) and further reinforced by provisions of the Transportation Efficiency Act (TEA-21) passed in 1998. As noted previously, SPC is the designated MPO for the Pittsburgh MSA.

TABLE 6-5 Selected Special-Purpose Districts in Metropolitan Areas

Organization	Service Area	Services Provided	Founded	Supervisory Body
East Bay Municipal Utility District	Parts of Alameda and Contra Costa Counties, CA	WS,S,CSO	1921	7-member elected board
Louisville and Jefferson County Sewer District	Louisville, Jefferson County, and a small portion of Oldham County, KY	S,CSO,SW	1946	8-member appointed board
Metropolitan St. Louis Sewer District	St. Louis and St. Louis County, MO	S,CSO	1954	Appointed board; 3 by city; 3 by county
Milwaukee Metropolitan Sewer District	Milwaukee and 28 other cities, most of Milwaukee County and 10 cities in other counties, WI	S,CSO,SW	1982	11-member appointed commission
Water Reclamation District of Greater Chicago	Chicago and 124 suburban communities, mostly in Cook County, IL	S,CSO	1889	9-member elected board
Allegheny County Sanitary Authority	Pittsburgh and 82 cities in Allegheny County, PA	S,CSO	1946	7-member appointed board

NOTE: S = sewer; SW = stormwater; WS = water supply.

Rusk (2000) argues that with the new investments in transportation under TEA-21, MPOs will largely determine the growth and shape of urban areas, providing an impetus for regional land use planning. That is especially true in metropolitan areas such as Pittsburgh, where new transportation arteries are influencing the shape of development and this development affects both the supply and the demand for water-based services.

Milwaukee is a model of a metropolitan area with both strong regional planning and a large special sewer district. Regional planning is conducted by the Southeastern Wisconsin Regional Planning Commission (SEWRPC), established in 1960 as the official area-wide planning body for a seven-county urbanized and urbanizing area. Its scope includes planning and design of public works systems, such as highways, transit, sewerage, water supply, and park and open space facilities. SEWRPC has also been progressive in its regional approaches to flooding, air and water pollution, natural resource deterioration, and changing land use. The Milwaukee Metropolitan Sewer District (MMSD) is the special sewer service district.

The SEWRPC has been quite active in water quality planning as well as regional land use and transportation systems. Since 1990, SEWRPC has among its other activities produced a stormwater management plan for the MMSD service area. It has also produced about 40 geographically specific subarea, watershed, and lake management plans.

The capacity of a special district to integrate the multiple elements of water resource management is also important. As evidenced in part by Table 6-5, most large special sewer districts like ALCOSAN serve a central city and many outlying communities in a single county. It is especially important in areas with a CSO problem that a solution to this problem not create additional problems for separate stormwater or sanitary sewer systems. Notably, only two of the six special-purpose districts shown in Table 6-5 have jurisdiction over both.

The effectiveness of a special district may well be determined by its relationship to general-purpose local governments within its service area. An example of what can be accomplished with strong cooperation between a special district and its constituent communities is provided by the wet weather program of East Bay Municipal Utility District's (EBMUD) in northern California. The cost of that program in 2004 dollars is estimated to be about $600 million (Jerry Gilbert, J. Gilbert, Inc., personal communication, 2004); it is described as follows:[7]

> In the 1980s, deteriorated community sewer pipes and improper storm drain connections allowed rainwater to enter local sewer systems during the heaviest storms, causing overflows at more than 175 locations. In 1986, EBMUD signed a joint powers agreement with Alameda, Albany, Berkeley, Emeryville, Kensington, Oakland, Piedmont, and portions of El Cerrito and Richmond to fix the problem. The communities have spent $200 million on sewer improvements and have a long-range program to complete improvements. EBMUD expanded facilities to provide more treatment capacity for high wet-weather flows. The communities' sewer improvements will reduce the "peak" regional wastewater flows from 1.1 billion gallons per day to 775 million gallons per day (MGD). EBMUD's treatment capacity will increase from 290 MGD to 775 MGD.

An approach similar to that taken in the EBMUD service area could be attempted in the Pittsburgh area with ALCOSAN as the central planning and management agency. The ultimate success of such an effort would depend in large part on relationships between ALCOSAN and its constituent communities.

ORGANIZATIONAL OPTIONS FOR IMPROVING WATER QUALITY MANAGEMENT IN SOUTHWESTERN PENNSYLVANIA

The cases discussed in the preceding section provide examples of organizational arrangements that other metropolitan areas have adopted to address their water resource and quality management problems. In many ways they are quite similar to the Pittsburgh region where there is a well-established special district providing sewer service to 83 communities, located primarily in a single urbanized county, and a designated MPO or regional planning agency exists. There are, however, several respects in which the Pittsburgh area lags some of the more established arrangements found elsewhere. First, comprehensive planning for stormwater management is relatively new in the region's urban core, and there appears to be limited expertise to manage beyond capturing sewage overflows and transmitting them to a central wastewater treatment facility. Second, there is no comprehensive basinwide planning to address issues that transcend regional boundaries and legal jurisdictions. Finally, water quality and water resource planning at the metropolitan multicounty scale is poorly developed.

---

[7] Available on-line at *http://www.ebmud.com/wastewater/wet_weather/default.htm.*

Although comments by selected leaders of the Pittsburgh region who have been consulted by the committee do not constitute a scientifically representative poll of interests in the region, they reveal several important clues about the direction that should be taken to address the organization of water resource and water quality planning and management in the region. First, there does not now appear to be a consensus on what that direction ought to be. If the region is to take an initiative toward new organizational arrangements, serious further discussion around several specific alternatives will be necessary. A consensus must emerge from that discussion. Second, comments suggest that the Pittsburgh region shares many of the views held by other metropolitan areas that have addressed similar issues. They reflect the view that problems in the urban core are different from those in surrounding counties and multiple organizations will be needed to address such problems. The commentators seemed to suggest that water resource planning at the multicounty metropolitan level could be helpful, but it should be limited to an advisory role. They also seemed to agree that ALCOSAN is the appropriate agency for transmission and treatment of sanitary sewage to the extent that they believe current and planned expansions (see Chapter 5) are appropriate. Several commentators pointed to the positive role played by 3RWW in Allegheny County.

Options to address water resource and quality management needs at the basin scale, metropolitan/multicounty level, urban core and outlying urban centers, and rural areas are discussed in the following sections.

Basin Scale

Acid mine drainage, polluted agricultural runoff, mercury, and microbiological contamination, are among water quality problems identified in the Pennsylvania 305(b) report (PADEP, 2002a), discussed in Chapters 3 and 4, that may be exported out of individual watersheds into the region's main stem rivers and across state boundaries. Monitoring, modeling, and the formulation of remedial policy have to be done at an appropriate scale that incorporates all significant sources impacting southwestern Pennsylvania and those downstream segments that may be affected by the Pittsburgh region. The most appropriate scale is likely to be at the river basin level and portions of tributary basins that cross state boundaries.

The most likely organizational options to investigate and resolve basinwide linkages are the Ohio River Valley Water Sanitation Commission (ORSANCO) and the Commonwealth of Pennsylvania. Created in 1948, ORSANCO is an interstate commission representing eight states and the federal government to address water quality problems in the Ohio River and its tributaries. Pennsylvania is a member, along with Illinois, Indiana, Kentucky, New York, Ohio, Virginia, and West Virginia. ORSANCO performs several water quality planning and monitoring functions, including establishing effluent standards for wastewater dischargers, conducting biological assessments, monitoring chemical and physical properties of streams, and executing special studies. Its staff has expertise in general administration, data management, water quality monitoring and modeling, pollutant reduction and NPS pollution programs, public information programs, wet weather projects and CSO abatement, and assessment of fish populations and their health. Because of ORSANCO's long history in the field of water resource and quality management, its interstate structure, and its professional staff, it is ideally suited to address many of the basin-scale problems in the region, especially those that are interstate in nature.

In addition to their broad constitutional and statutory powers for regulating water quality, state government entities can also play a significant role in basinwide and regional water resource planning and management. Pennsylvania is like 20 other states cited by the NRC Committee on Watershed Management (NRC, 1999) that have organized some of their management activities around watersheds—most importantly for acid mine drainage and rural nonpoint source pollutants. Formulation of the WRAS and commitments of funds discussed earlier in this chapter represent a significant advance toward confronting these problems. At a minimum, PADEP has to monitor and model (see Chapter 5) how much of the wet weather-related pathogen and heavy metal contamination problems in streams flowing through southwestern Pennsylvania are due to upstream sources. Corrective action should be taken to address these sources as well as those within the Pittsburgh region. The CWARP program discussed and recommended in Chapter 5 can be launched by the basinwide authorities that would establish watershed-based information collection and analysis programs to provide the foundation for work at the subbasin, urban area, or more rural (local) levels.

Multicounty/Metropolitan Scale

As noted previously, improved planning and technical assistance programs for water management at the metropolitan regional scale are needed. Large-scale transportation plans developed at that scale can have profound effects on land use and related water supply, stormwater, sanitary sewer services, and other aspects of water resources. Consideration of those effects should be incorporated in regional planning. Related needs at the metropolitan scale include the following:

- examine alternatives to the existing, highly fragmented pattern of water resource services;
- promote improved coordination among regional transportation, economic development, land use, and water resources; and
- provide assistance to small urban centers and rural areas in matters of water supply and wastewater disposal.

At least two options are available to pursue that goal. One is to enhance capabilities of the existing metropolitan planning organization, SPC; the alternative is to create a new organization. The SPC currently derives its authority in large part from federal transportation incentives. If it is to do more than simply design transportation systems that follow existing development trends—and, in particular, if it is to take a leadership role in regional water planning—SPC's regional planning will have to become more comprehensive. Several basic tasks could be conducted beneficially at that scale. Second, water resource considerations should be integrated with land use and transportation planning to determine resource availabilities, development needs, constraints, and environmental consequences of regional development. Plans at that scale should serve as guides for large-scale urban infrastructure investments that guide growth. Other tasks that have been successful in similar settings are technical assistance to local governments and subarea plans for watersheds within the region.

At least two problems potentially limit SPC as an effective leader in regional water planning and management. First, leadership of the organization is limited to elected officials. For the commission to be effective in bringing about cooperation among the region's numerous local governments, it is important that those elected officials be at the table. However, some of the major water-related issues of concern to the region go beyond the sphere of local governments. Participation by other knowledgeable individuals, water management agencies, community and nongovernmental organizations, academia, and other entities is necessary if a metropolitan planning organization is to better capture the benefits of the region's leadership. Second, the commission is not proportionally representative of its 10-county population. Each of the 10 member counties is represented by five members of the governing board. Representation should reflect the fact that those counties with relatively high densities and a large number of intimately interrelated local governments have priorities that are quite different from those of the predominantly rural counties with relatively smaller and more dispersed populations.

The second option is to create a new special-purpose water quality planning and technical assistance organization at the multicounty level. Its principal advantages would be the creation of a strong voice for water quality improvements and development of a specialized staff for both technical matters and public outreach. Its disadvantages include the following: (1) creation of yet another regional organization that would have to raise revenue; (2) possible duplication of SPC's regional database; and (3) a more difficult task of integrating water resource considerations with land use and transportation planning conducted by SPC.

In the judgment of the committee, the SPC is the region's best choice for planning at the multicounty/metropolitan scale if its governance and participation can be modified to address the aforementioned limitations of participation and representation. One option for SPC to broaden participation is to take an active role in establishing and supporting a regional water forum discussed later in this chapter. SPC would also have to enhance staff capability in water planning and management.

Specific tasks that must to be accomplished by SPC (or a new special-purpose organization) include the following:

1. Prepare a regional framework plan for water resources that integrates water and land resource uses and capabilities with its transportation planning responsibilities:

    - identify the extent of need and management alternatives for on-site sewage treatment and disposal system (OSTDS) management in predominantly rural counties;
    - work with PADEP to identify the extent of need and management alternatives for municipal wastewater management in lesser urbanized counties;
    - work with the Ohio River Basin Regional Water Resources Committee created by PADEP under the 2002 WRPA to identify the extent of need and management alternatives for public water supplies in predominantly rural counties; and
    - identify critical water resource areas in need of protection or restoration.

2. Provide a continuing regional forum for discussion of issues and management options for addressing common problems shared by at least a subset of local governments.
3. Provide advice to local governments as appropriate.

Urban Core and Outlying Urban Centers

In the urban core, including much of Allegheny County and portions of Washington, Westmoreland, Butler, and Beaver Counties, the dominant water quality management problems are combined sewer overflows and sanitary sewer overflows (SSOs) resulting from wet weather conditions. These specific problems, however, are inextricably linked to the more general problem of stormwater management. Actions taken to manage stormwater flows in one location may have significant effects over a much larger portion of the network; elimination of some CSOs and SSOs will have spillover effects on separate stormwater conveyances. To some extent, these problems may also exist in smaller, detached urban centers in the region.

Continued fragmentation of the management of the sewer collection-conveyance-treatment system (i.e., maintaining the status quo) is not a satisfactory situation. Not only is it inefficient; it also impedes solutions. A system in which more than 80 municipalities discharge unregulated quantities of stormwater runoff and sanitary sewage into a centralized conveyance-treatment system over which the treatment management agency (ALCOSAN) has insufficient physical and fiscal control is a recipe for maintaining and possibly worsening current water quality conditions. If contributing municipalities persist in independent operations of their collection systems, some form of performance standards and incentives to comply with those standards should be established. The cost to small communities that choose to "go it alone" in satisfying such standards could be prohibitive.

A plan that integrates separate sanitary sewer systems, separate stormwater systems, and combined sewers should be prepared for the region's urban core. Its geographic coverage should include, at a minimum, all of the watersheds that contribute urban stormwater runoff and/or sanitary sewage to the ALCOSAN system, excluding those watersheds upstream on the Allegheny, Monongahela, and Youghiogheny Rivers. Exact delineation of boundaries for such a plan will have to be made pending a more detailed examination of the area, but a first approximation is provided in Figure 5-2.

A recent example of integrated management of separate sanitary sewers, separate storm sewers, and combined sewers is the case of nearby Morgantown, West Virginia, described in Box 6-2. Although Morgantown is a much smaller community than Pittsburgh, it experiences similar water quality problems.

Combined sewer overflows are inherently linked to collection systems for sanitary sewage and stormwater runoff. Although ultimate decisions on governance of these problems may result in a clear delineation of responsibilities, those decisions should be made in light of consequences for all aspects of stormwater and sanitary sewage collection, treatment, and disposal.

Several alternatives should be considered for planning and managing sanitary sewage and stormwater in the region's urban core; principal among them are the following:

- **Option A:** General-purpose metropolitan government—specifically merger of the City of Pittsburgh and surrounding municipalities with Allegheny County
- **Option B:** Creation of a countywide sewage collection organization, with or without authority over stormwater management, either by dedication of sewer systems to Allegheny County or through an administrative arrangement with Allegheny County using authority under Pennsylvania Acts 67 and 68
- **Option C:** Creation of one or more special districts to manage sewer collection with or without authority over stormwater management
- **Option D:** Expansion of the role of ALCOSAN to include sewer collection systems, with or without authority over stormwater management
- **Option E:** Continuation of the decentralized system but with performance standards and voluntary participation in a regional maintenance organization (RMO) provided on a fee-for-service. ALCOSAN would be encouraged to establish the RMO.

BOX 6-2
Integrated Sanitary and Stormwater Sewer Management in Morgantown, West Virginia

Morgantown, West Virginia, located approximately 90 miles south of Pittsburgh, has a permanent population of 25,000 and an additional student population of 25,000 at West Virginia University. The municipality owns and operates the Morgantown Utility Board (MUB) that serves approximately 21,000 potable water customers and 14,000 sanitary sewer customers. Beginning in August 2002, MUB inaugurated a stormwater utility serving 10,000 customers. MUB is governed by a Board of Directors appointed by the City Council.

Until 2002, stormwater services in Morgantown were provided by the city's street department. Spurred in part by the need to obtain a stormwater discharge permit under Phase II of regulations promulgated by EPA, the city moved responsibility to MUB. Factors influencing that move included MUB's greater expertise with state and federal water quality permits and an acknowledgement that the time had come to undertake costly rehabilitation of its stormwater system. Also, two densely populated and growing neighborhoods were experiencing damage from floodwater, much of which originated beyond the city's planning and zoning jurisdiction. MUB's acceptance of this new responsibility acknowledged the inherent interconnectedness of its separate sanitary sewers, separate stormwater management system, and combined sewer system, all of which were subject to state and federal permits.

A key issue in the formation of the stormwater utility was establishing the service area. After much discussion, the area was defined as the watershed from which overland flow is delivered to a receiving stream within the city limits. Minor adjustments to that delineation were necessary to account for a few other practical considerations. Because MUB's stormwater jurisdiction extended beyond the city's jurisdiction and West Virginia law was silent on the issue of municipal stormwater management outside corporate boundaries, state enabling legislation was necessary to address the issue of authority. The legislature passed the enabling legislation through a series of amendments to the West Virginia Code.

After several months of interaction with affected stakeholders, the City Council passed the newly enabled municipal ordinances creating the stormwater utility. A flat rate of $3.63 per month was adopted for single-family residences. For multifamily residences and nonresidential properties, the fee was set at $0.00145 per month per square foot of measured impervious surface. These fees are expected to generate about $750,000 annually.

SOURCE: Timothy L. Ball, MUB, personal communication, 2004.

Whatever option is chosen, management entities would have to meet performance standards to minimize CSOs and SSOs. Performance standards should, at a minimum, conform to EPA's CSO Control Policy (EPA, 1994; see Chapter 5 for further information), the joint memorandum "Enforcement Efforts Addressing Sanitary Sewer Overflows" (Harman and Perciasepe, 1995) from EPA's Office of Enforcement and Compliance Assurance and Office of Water, and "Chapter X: Setting Priorities for Addressing Discharges from Separate Sanitary Sewers" (EPA, 1996) of the National Pollutant Discharge Elimination System (NPDES) Enforcement Management System. EPA's Region IV Capacity, Management, Operation and Maintenance (CMOM) policy could serve as a model (see Herman, 2000). Discussion of options A through E as they pertain to southwestern Pennsylvania is provided below.

Option A

Discussion of consolidation of the City of Pittsburgh and Allegheny County has been raised recently as a possibility by the Allegheny Conference on Community Development in the context of the region's current fiscal woes (McNulty, 2003; see also Chapter 2). Both the mayor and the county executive at that time called for a study of the proposition. The issue was raised more recently at a meeting of the League of a Women Voters in February 2004 (Cohan, 2004) by Mayor Tom Murphy and county Chief Executive Dan Onorato who commented that the issue was definitely on their agendas. A merger of the City of Pittsburgh and Allegheny County would cover a sizable portion of the geographical area of the region's high density urban core. Given the broad array powers delegated to municipal government, the merged government would have a wide array of authority to implement CWARP in that area. However, a merger of city and county government would achieve less than its full potential if the City of Pittsburgh is the only municipality involved. If surrounding municipalities that contribute sanitary sewage, stormwater, or combined sewage to the urban core continue to operate independent collection systems, the need for an additional mechanism for cooperation will remain. With full recognition that the decision will not and should not be made solely on water-related issues and will be politically difficult, the committee recommends the merger of city and county governments as an efficient and effective option for planning and management of water quality. With such a merger, a large portion of the area affected by CSOs, SSOs, and urban stormwater runoff could be brought under a single management entity that can integrate water management with land use. Under this option, ALCOSAN would continue to own and operate interceptors and treatment facilities.

If Option A is chosen, management of water resources in contiguous urban areas outside Allegheny County that contribute flows to Allegheny County would not be addressed. Management in those areas could be either independent or by administrative arrangements with Allegheny County under one of several Pennsylvania statutes enabling intergovernmental cooperation. Options D could be adopted for contiguous urban areas; Options C or E are possibilities for outlying urban areas that are detached from the urban core.

Option B

Option B for managing sanitary sewage and stormwater in the urban core has at least three components, namely, (1) a countywide sewer organization; (2) continuation and possible expansion of 3RWW programs; and (3) continuation of ALCOSAN as the operator of major interceptor sewers and centralized wastewater treatment.

Countywide sewer system management could be established either by B(1) dedication of local sewer systems to the county or by B(2) intergovernmental contracts through an Intergovernmental Cooperative Planning and Implementation Agreement as authorized under Pennsylvania Acts 67 and 68. Either the City of Pittsburgh or Allegheny County could be the operating entity. If Allegheny County took on this responsibility, it would have to establish an administrative unit capable of handling these duties. However, a decision would still have to be made about who would be responsible for stormwater runoff.

If municipal governments chose to surrender management of their sewer collection systems to a countywide entity under either Option B(1) or B(2), that entity would function as an operating agency for all sewer systems services and include the following responsibilities:

- construction, operation, and maintenance of all sewer systems in the county;
- construction, operation, and maintenance of decentralized stormwater treatment required to satisfy the long-term control plan for CSOs; and
- financing through tax revenues or a system of charges to individual residential, commercial, industrial, and governmental contributors to the sewer network.

The second component of Option B includes the 3RWW, and the committee recommends that it be continued or expanded. Its functions are and should include the following:

- conduct a public education program and provide technical assistance to local governments for stormwater and CSO management;
- provide an educational program to local governments for identifying and correcting illicit connections to sewer system; and
- monitor, analyze, and report periodically on the status of stormwater and CSO management in Allegheny county.

ALCOSAN's existing role would continue under Option B, and it would (1) provide conveyance of combined sewers to the treatment plant; (2) provide appropriate treatment for sanitary and storm sewage conveyed to the central treatment plant; and (3) establish and collect fees for those services.

Option C

The third option is to form one or more sewer utilities where groupings of municipalities would be determined on geographic, political, and economic criteria. As in Option B, a decision would have to be made about who would be responsible for stormwater runoff. Option C would be very similar to Option B except that it would be organized under different authority. It would be similar to ALCOSAN and financed through a set of user fees. If its responsibilities were limited to sanitary sewer and CSOs, fees would be assessed on wastewater discharges. If its responsibilities included broader responsibilities for stormwater management, it could also include stormwater utility fees.

It is important to note that the option of special-purpose districts is not limited to a single district such as ALCOSAN. Metropolitan areas with a very large special district serving the central core of the urban area may have smaller districts that offer similar services to outlying areas. The U.S. Census Bureau lists all individual special districts with major financial activity (MFA), defined as those units with either revenues or expenditures in excess of $10 million or debt in excess of $20 million. In the Chicago metropolitan area, for example, where the Metropolitan Water Reclamation District had expenditures of about $650 million in 1997, there are two MFA special districts that deliver sewerage services in Kane and Lake Counties, each with less than $20 million of expenditures in 1997. In the St. Louis metropolitan area, an MFA special district delivers sewer service in St. Charles County. It spent about $27 million in 1997. In Boston, the South Essex Sewerage District, serving portions of the North Shore, spent $87 million in 1997. Similar subregional arrangements could be identified within the region beyond the boundaries of ALCOSAN.

Option D

Instead of creating any new organization to manage sewage collection systems and possibly stormwater management as well, ALCOSAN's existing mandate could be expanded to include collection systems and possibly stormwater. Municipalities could choose to either dedicate their systems to ALCOSAN or retain ownership.

If municipalities chose to retain ownership, ALCOSAN would be charged with the following tasks:

- establishment and enforcement of performance standards for municipal stormwater discharges throughout the county;
- construction, operation, and maintenance of the decentralized stormwater treatment facility required to satisfy a long-term control plan;
- establishment of a rate structure for assessment of management fees;
- construction, operation, and maintenance of drainage systems for member communities; and
- continued contribution to the operation of 3RWW.

If local governments chose to surrender their general stormwater management responsibilities to ALCOSAN, those responsibilities would be added to the preceding list and ALCOSAN should be given authority to charge stormwater utility fees for those services.

As in the case of Option B, the committee recommends continuation or expansion of 3RWW.

Option E

The fifth option is for municipalities to retain local ownership of sewer collection and stormwater facilities. Operators of systems would be required to demonstrate their capacity to operate and maintain those systems in accordance with performance criteria established by PADEP and EPA or to enter into service contracts with a qualified provider. ALCOSAN would be encouraged to develop a sewer services division that would provide requisite operation and maintenance functions on a fee-for-service basis. As in Options A, B, and D the continuation or expansion of 3RWW is recommended.

Summary of Options A-E

Discussion of Option A is already in progress. Given the length of time required to make a decision on a merger and uncertainty about the outcome of that decision, it is recommended that Allegheny County take the lead in the near future to form a task force to consider Options B, C, D, and E. Although the committee is of the opinion that Options B-E are all viable, it prefers Option B. A countywide sewer organization created under Option B could develop an Act 537 plan (see Chapter 5 for further information) that includes all local governments in Allegheny County and portions of neighboring counties that choose to participate. Such a countywide organization would also have the option of contracting with ALCOSAN for selected services as suggested under Option E. Thus, Option B offers advantages of economies of scale in planning, operation, and maintenance; facilitates a systems approach to management; can be managed through existing institutions; and keeps governance of the program close to politically accountable public officials. Specific project planning and development as well as operational coordination would be conducted under the appropriate option identified above. Lastly, the advantages and disadvantages of these five options for planning and managing sanitary sewage and stormwater in the region's urban core are summarized in Box 6-3.

BOX 6-3
Organization Options for Urban Core and Outlying Urban Centers

Option A: General-purpose metropolitan government
Advantages:
- Integration of fragmented infrastructure
- Integration of water management with other local government functions, particularly those related to land use and development processes
- Economies of scale
- Ability to exercise broad range of municipal powers
- Politically accountable elected officials

Disadvantages:
- Loss of local identity and autonomy
- Management in contiguous urban areas outside Allegheny County unresolved

Option B: Countywide sewage collection organization using authority under Pennsylvania Acts 67 and 68
Advantages:
- Integration of fragmented infrastructure
- Economies of scale

Disadvantages:
- Loss of local autonomy
- Land use powers more limited than those of general-purpose government
- Appointed boards less politically accountable to public
- Management in contiguous urban areas outside Allegheny County unresolved

Option C: Creation of one or more special districts to manage sewer collection with or without authority over stormwater management
Advantages:
- Integration of fragmented infrastructure
- Economies of scale
- Geographic coverage more flexible than county-wide system

Disadvantages:
- Loss of local autonomy
- Powers more limited than those of general-purpose government
- Appointed boards less politically accountable to public

Option D: Expansion of the role of ALCOSAN to include sewer collection systems, with or without authority over stormwater management
Advantages:
- Integration of fragmented infrastructure
- Economies of scale
- Geographic coverage more flexible than county-wide system

Disadvantages:
- Any prior decisions, agreements, or disagreements by and with ALCOSAN that would inhibit its ability to achieve efficiency
- Powers more limited than those of general-purpose government
- Appointed boards less politically accountable to public

Option E: Continuation of the decentralized system with performance standards
Advantages:
- Retention of limited local autonomy
- Integration of water management with other local government functions, particularly those related to land use and development processes
- Politically accountable public officials

Disadvantages:
- Less economically efficient
- Less management expertise available to many units of local government
- Less expertise and intelligence about system-wide infrastructure

Rural Areas

As stated previously, the primary rural area problems identified by the committee are inadequate on-site wastewater disposal and water supplies, and the actions recommended in Chapter 5 to address these deficiencies (e.g., register all individual and cluster OSTDSs within each county in southwestern Pennsylvania) should be undertaken cooperatively by several agencies. At the state level, the WRAS program should be expanded to include assessment of effects of inadequate wastewater disposal on water quality. In doing so, PADEP should work closely with local governments having legal authority over such systems. Although the legal authority to control on-site water supply and wastewater disposal rests with municipalities and counties, it is unlikely that individual units of local government will have sufficient resources to support an effective management capability for these functions. The SPC could and should take strong leadership in bringing local governments together to address these issues. In addition to PADEP, SPC should request assistance from EPA and nongovernmental organizations having prior experience with programs of this kind. The Allegheny County experience in these activities should provide a sound foundation for other counties in the region.

Consistent with Chapter 5, the committee recommends that initial funding from a regional user surcharge be provided to initiate such county activities that eventually would be self-sustaining.

Regional Water Forum

Regardless of which management option is chosen at the metropolitan regional scale, there is a need for a continuing forum on water resources and related issues in the region and, more specifically, for general oversight of the CWARP, the progress being made, and the need for further actions toward achievement of clean water. As discussed earlier in this chapter, southwestern Pennsylvania is served by a diversity of water planning and management institutions, both large and small in geographic scope and both public and private in legal structure. These existing entities appear to be reasonably cooperative on issues of common regional interest. An excellent recent success in intersectoral collaboration was the Western Division of the Pennsylvania Economy League's WSIP project and its 2002 report discussed earlier (see also Appendix B).

The WSIP, however, was both ad hoc and focused on a single issue. The WSIP Steering Committee went out of existence with the publication of its report in 2002, although leaders in the effort continue to press for action as evidenced by this study and report. However, this NRC study is also a singular event to consider and provide advice on options. Water resource planning and management for southwestern Pennsylvania should be continuous and multiple-purpose in scope, including at least the following areas of concern:

- water quality—CSOs, SSOs, stormwater, failing OSTDSs, acid mine drainage (AMD), other;
- water supply—household, institutional, industrial, other;
- aquatic and riparian habitat restoration, including fisheries;
- recreation—urban riverfront, water contact; boating, fishing, other;
- flood hazard and drought mitigation; and
- "sense of place" for the Pittsburgh region, (e.g., preservation of historic cultural landscapes and structures relating to the Three Rivers watershed).

An overlying water resource management "umbrella" is needed to incorporate as many areas of water management concern and stakeholders as possible. The "umbrella" would not be an operating agency that could assume all of these functions. Rather, it would be a regional water forum that would offer open access to interested stakeholders (or representatives of different classes of stakeholders) and be directed to the full spectrum of interconnected water resource issues such as those listed above.

The 2002 WSIP report proposed a water management framework involving three functional areas, each under the leadership of an existing institution, to be coordinated by a hypothetical Watershed Alliance for the Three Rivers Region (WATRR). As stated previously, the three-pronged and interrelated functional elements of the proposed framework would include technology and education, regional goal and priority setting, and advocacy.

The committee supports this approach and concurs with the desirability of using existing and well-respected institutions to lead each functional area, in collaboration with interested stakeholders and with each other. The ultimate choice of a lead agency (or agencies) in each area is, of course, a regional determination.

The crucial element of the proposed framework, the coordinating forum they termed WATRR, was not described in detail in the WSIP report except that it should be a "new unincorporated entity…to represent the united front to organizations and entities both inside and outside the region. Staff support for WATRR should be provided by the [Southwest Pennsylvania Commission]" (WSIP, 2002). Particular attention to the scope and organization of such an umbrella entity should be ensured, as described below.

Scope of a Regional Water Forum

The WSIP Steering Committee and this NRC committee were both established to address water quality issues in the Pittsburgh region. Clearly, however, water quality investment and policies are related to the other functions of water resource management listed in Box 6-1, as well as to broader issues of urban sprawl, "smart growth," economic development, and social equity.

A regional water forum, however established, should be charged with a broad mandate to assess priorities for water infrastructure planning, maintenance, and construction in relation to the full spectrum of water resource goals and in relation to the social, economic, and environmental welfare of the region as a whole.

Establishment and Organization

There are several options for the creation and organization of a regional water forum, some of which were listed in the (WSIP, 2002) report, that include the following:

- existing agency takes lead role, such as SPC;
- new statutory entity (unlikely in the region's current legislative and fiscal climate);
- incorporated nongovernmental organization (NGO) for regional water planning, (e.g., 3RWW expanded geographically and functionally); and
- unincorporated network of public and private stakeholders, established by voluntary memoranda of understanding (MOUs).

The last option listed above for a regional water forum is recommended by the committee for careful consideration. This concept would involve an open-access network of public and private organizations interested in water issues that voluntarily agree to participate in meetings, discussions, research projects, and other activities relating to various water planning issues. Coordination of the network would be provided by a coordinating council, a functionally enlarged and ongoing equivalent of the WSIP Steering Committee. The coordinating council would represent major NGO and government stakeholders including EPA, the U.S. Army Corps of Engineers, PADEP, the City of Pittsburgh, local academic experts, ALCOSAN, 3RWW, the Western Division of the Pennsylvania Economy League, and the Pennsylvania Environmental Council (PEC), among others with significant interest in water quality in southwestern Pennsylvania. Ideally, office space, staff, and administration of the network would be provided by one or more of the major NGOs, possibly with funding from Pittsburgh area foundations. Membership and participation in the Three Rivers Regional Water Forum would be open to counties, municipalities, special districts, watershed organizations, academic and research institutions, and other interests. A conceptual diagram representing this arrangement is shown in Figure 6-3. A description of the Chicago Wilderness network is provided in Box 6-4, knowledge of which may be instructive in the development of a regional water forum. Although the membership and exact organizational plan of the forum should be determined locally, the committee recommends that SPC in coordination with ORSANCO design an appropriate structure for the regional water forum, which could be funded by participating members.

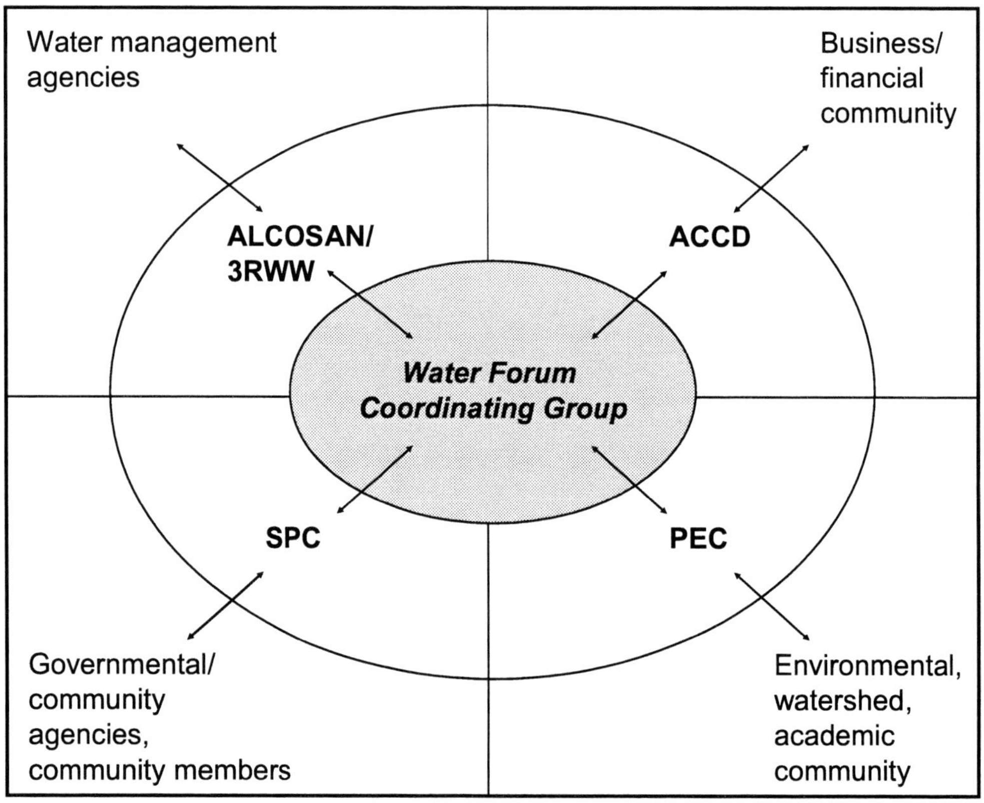

FIGURE 6-3 Concept diagram for a Three Rivers Regional Water Forum.

BOX 6-4
The Chicago Wilderness: A Model for a Three Rivers Regional Water Forum?

The Chicago Wilderness (CW) network, established in the mid-1990s to promote regional biodiversity and environmental education, could be an important model for the development of a Three Rivers Regional Water Forum. While its focus is on biodiversity rather than water, the organization and process of the CW represents an interesting precedent for an open-access, public-private consortium of regional stakeholders around a particular cluster of scientific and policy concerns. Although the CW does not per se take positions on biodiversity issues, its value lies in facilitating collaborative efforts to analyze issues and formulate recommendations for public policy by subgroups of member organizations organized as task forces.

The CW currently includes about 160 member governments, agencies, NGOs, educational institutions, and business corporations. Its geographic area loosely includes the six Illinois counties of the Chicago MSA, Kenosha County in Wisconsin, and Lake and Porter Counties in Indiana. Office space and staff resources are provided by three Chicago area organizations: the Field Museum of Natural History, the Brookfield Zoo, and the Nature Conservancy Chicago Chapter. Startup funding was provided by grants from the EPA, the U.S. Forest Service, and the U.S. Fish and Wildlife Service for research on biodiversity.

Also known as the Chicago Region Biodiversity Council, the CW is "governed" by three leadership entities established under its policies and procedures: (1) an executive council comprising the above three organizations plus additional members that provide resources to CW; (2) a steering committee that includes representation of specified sectors and classes of governments and private interest groups; and (3) a coordinating group established by the steering committee, which holds monthly meetings open to all CW members. The coordinating group (1) implements steering committee decisions, (2) oversees the CW work plan, (3) sets agendas for meetings, and (4) represents CW at professional meetings. The CW corporate council includes participating business firms.

The CW is not incorporated and does not have 501(c)(3) tax-exempt status so it does not compete with its member organizations for funding. The work of CW is carried out through meetings of members, mission-specific task forces, a proposals committee, and a nominating committee. The CW supports ecological restoration activities through a network of citizen volunteers.

In its first few years, the CW has become a respected voice for "ecological citizenship" in the Chicago region. In addition to its web site (*www.chicagowilderness.org*), it has published *Chicago Wilderness, An Atlas of Biodiversity* (Sullivan, 1997), which describes the major ecosystems and selected species with text and graphics directed to the general public. It coordinates environmental education programs for inner city and suburban school systems and is helping to protect and restore habitat sites in the Chicago region. The CW conducts research on biodiversity through various task forces and subgroups. It participates in such related regional initiatives as the Chicago Regional Transportation Plan and the Green Infrastructure Regional Mapping Project and provides speakers for conferences in the region and around the country.

PUBLIC PARTICIPATION

One of the keys to a successful region-wide water quality improvement program is participation by all interested parties. A major function of a regional water forum is to promote participation in planning and decision making across the entire region, but participation should be encouraged by all organizations that share some responsibility for CWARP. Participation by elected officials is a necessity, but other civic leaders, corporate interests, environmental groups, and other groups and individuals should be encouraged to take active roles in the many elements of the programs outlined in Chapters 5 and 6. To be effective, the region should be proactive in seeking participation and keeping participants well informed on current developments. Activities of the proposed regional water forum, the SPC or other regional organization, Allegheny County,

ALCOSAN, and other responsible parties should establish both electronic and traditional communication links readily available to the public. The Internet and e-mail make it possible to provide a full range of data, action proposals, and other information to the public at modest cost.

Participation by elected and appointed representatives of planning, management, and regulatory agencies who make key decisions is essential, but a word of caution is appropriate. Leaders in the drinking water industry in many regions of the country may not be adequately informed as to what the public is willing to pay for improved quality. In the late 1990s, the American Water Works Association conducted a series of surveys that showed the public was far more willing to spend money for drinking water quality improvement than its leaders expected (Reekie, 2000). The history of public votes on water quality improvements is largely positive. The proper manner to judge such support is presentation of specific proposals with full disclosure of benefits and costs to which the public can respond through its elected and appointed representatives.

If specific proposals are to be supported, a greater level of public education and participation in the process would appear to be needed. In November of 2000, the Pennsylvania Economy League sponsored a telephone-based survey (Pennsylvania Economy League, 2000) that produced some revealing results, among them are the following:

- among all public problems, sewerage issues ranked ninth (only 2 percent said this was the most important problem facing the community);
- when asked the degree of concern on a scale of 0 to 100, water contamination, clean drinking water, and water pollution were in the 50 to 60 percent range for sewerage users and owners of OSTDSs;
- about one-fourth thought the waters were extremely polluted—the remainder thought they were not very polluted;
- two-thirds were knowledgeable about their sewerage systems;
- 50 percent did not think sewers were causing contamination;
- 30 percent felt they should pay to solve community problems;
- 16 percent opposed consolidation even if it meant lowering their water bill;
- 30 percent believed their community was contributing to water contamination;
- most significantly favored future support for water quality improvements;
- 80 percent said they would fix an on-site condition if it was contributing to the region's water quality problem;
- 50 percent would pay more to assist low-income, elderly, and OSTDS users; and
- community growth was attractive to two-thirds of users of OSTDSs.

Finally, the survey concluded that providing additional information about important water issues and their relationship to the community was needed and that homeowners would pay more if they were shown to be part of the problem and its solution.

Results of the survey reveal that although water quality issues have not been overlooked, they are perceived to be a major problem by only a few. Those results also indicate that whereas there is a willingness to pay more to solve these problems, it depends on better education of the public as to the sources of problems and their solutions.

FINANCING WATER QUALITY PROGRAMS

Whereas major decisions that will ultimately determine the costs of water quality improvement in the region are yet to be made, enough is known to predict that the costs are likely to be substantial. For example, as discussed in Chapter 5, the estimated cost of the ALCOSAN draft long-term wet weather control plan (LTCP) is in excess of \$3 billion—or \$9,000 per ALCOSAN customer (TPRC, 2002)—and will include expenditures in the following areas:

1. gathering and analyzing information to provide a supportable basis for future investments;
2. capital construction for major wastewater treatment plant expansion, storage and conveyance improvements, and collector system upgrades;
3. administrative costs to facilitate the preceding activities and cooperative actions among local, area-wide and watershed-based organizations responsible for water management;
4. regulatory and technical assistance to encourage, schedule and build improvements to small systems in local communities and in connection with on-site waste disposal management; and
5. information dissemination and stakeholder participation activities to involve interested citizens and elected officials in the development and implementation of future programs.

With the exception of items 2 and 3, expenses for these activities could be funded with a small regional sewer service fee or a grant from a state or federal agency.

Sources of Funds

Numerous state and federal programs provide financial assistance for addressing water quality, often targeted at particular problems. A few have already been mentioned in this chapter and further information is provided below. Foremost among the traditional funding sources for municipalities is EPA's Clean Water State Revolving Fund for the construction of municipal wastewater facilities and implementation of NPS control and estuary projects. The Commonwealth of Pennsylvania's revolving fund is PENNVEST,[8] which provides grants and low interest loans.

In addition to the sources of funding noted above, Pennsylvania voters approved a \$250 million bond issue on water and wastewater infrastructure in May 2004. The Pennsylvania General Assembly passed legislation in November, 2004 establishing the criteria and procedures for spending these funds. The legislation specifies that \$200 million of the \$250 million authorized by the voters is to be used by the Commonwealth Financing Authority for grants and loans for water and sewer projects "which are related to economic development." "Economic development" means "a project which involves the investment of capital in Pennsylvania enterprises and communities or which results in the creation of new or the preservation of existing jobs in this Commonwealth." Up to \$125 million of the total funds here may be used for grants, and the remainder can be used for loans. The remaining \$50 million from the bond issue is allocated to PENNVEST, which is to use it for repairs, modernizations, and expansions to existing water and sewer systems designed to address health or environmental issues. In addition, the legislation authorizes the state to incur an additional

---

[8] Since its establishment in 1988, PENNVEST has served the communities and citizens of Pennsylvania by funding sewer, streamwater, and drinking water throughout the Commonwealth. Further information about PENNVEST is available on-line at *http://www.pennvest.state.pa.us/pennvest/site/*.

$50 million to $100 million in debt (beyond the $250 million authorized in the spring referendum) for PENNVEST to use for the same purposes. There is no restriction on whether these funds can be used as grants or loans.

The U.S. Department of Agriculture's Rural Utility Service (RUS)[9] administers various water and environmental programs, providing loans, grants, and loan guarantees for drinking water, sanitary sewer, solid waste, and storm drainage facilities in rural areas and in cities and towns with populations of 10,000 or less. As noted previously, both PENNVEST and RUS programs have been active in financing infrastructure in southwestern Pennsylvania; funds should continue to be available from these sources for appropriate activities.

Similarly, several state and federal programs are available for addressing AMD. For example, the Clean Streams Program (formerly called the Appalachian Clean Streams Initiative), administered by the U.S. Department of the Interior's Office of Surface Mining, provides for cooperative agreements to nonprofit organizations, especially small watershed groups, that undertake local AMD reclamation projects. Other AMD financing programs are discussed in Chapter 2.

Pennsylvania's Growing Greener program was established in 1999 to preserve farmland and protect open space, eliminate the maintenance backlog in state parks, clean up abandoned mines and restore watersheds, and provide new and upgraded water and sewer systems. As indicated in Table 6-4, some of these sources already have been used to support PADEP's WRAS program.

Although funding has been available from a variety of sources (some of which are discussed here), levels of funding are substantially lower than what was available during the 1970s and early 1980s. Special funding may be made available for special problems of national interest such as homeland security, but despite vigorous and coordinated efforts of the water industry, it is not likely that funding levels will be restored to anywhere near their historic highs in past decades. In the absence of additional intergovernmental aid, the region is left to carry much of the financial burden on its own shoulders or seek direct special appropriations.

Basic options for local governments are general tax revenues, water rates, and various fees and charges. Because fees or charges can go beyond simple cost recovery, they can be used to encourage cooperative actions and promote actions that reduce generation of wastewater, stormwater runoff, pollutant discharges, and water demand.

Water use fees are commonly levied by public and private water supply systems in southwestern Pennsylvania and elsewhere to cover capital, operating, and maintenance costs. Although flat rate structures were once common, charges based on metered water use are now routine for both commercial and residential customers (EPA, 1997) and are commonplace in southwestern Pennsylvania. The desirability of pricing water to recover the costs of water supply systems and encourage the conservation of scarce supplies is relatively noncontroversial. Moreover, because wastewater volumes are correlated positively with water supply, charges on water supply provide indirect incentives to reduce the generation of wastewater.

Fees are also commonly imposed on the discharge of wastewater into public sewers and treatment works in the Pittsburgh region and elsewhere. The purpose is to cover costs, but if the charges are tied to the volume of discharges, then these fees can also provide an incentive to reduce the flow of wastewater delivered into sewers and treatment works. However, whereas metering of water sales to individual customers is routine, metering of wastewater from individual residential and commercial customers is not. Accordingly, fees for connection to public sewers and treatment

---

[9] Further information about the RUS can be found on-line at *http://www.usda.gov/rus/*.

works generally do not provide incentives to limit wastewater flows that are unrelated to use (e.g., wet weather increases).

ALCOSAN's rate structure for sewer services to many of its member communities is based on the volume of water consumed, not on the volume or peak rate at which wastewater is received. This structure fails to provide an incentive for contributing communities to reduce flows from combined sewers and from sanitary sewers with excessive infiltration. It is also inequitable because communities with properly operated and maintained systems that contribute little to combined sewer flows pay the same per unit of water used as communities with high wet weather flows. The committee recommends that this system of charges be phased out and replaced with a rate structure based on volumes and peak rates of discharge into ALCOSAN's interceptor sewers. Devices to measure these flows should be installed at sufficient locations to adequately estimate volumes and peak rates. Such devices could be installed to support real-time control of the collection system as described in Chapter 5.

Discharge fees can also be levied on those who discharge wastes into streams. Many states, including Pennsylvania, assess fees under the Clean Water Act (CWA) NPDES permitting program (see also Chapter 3) to cover administrative costs. Similar to fees on wastewater discharges to sewers, discharge fees can provide an incentive to reduce the flow of polluting discharges to surface waters if the fee is based on volume of discharge and is large enough to alter the behavior of the discharger. However, these fees tend to be small compared to pollution control costs and in many cases are unrelated to discharge volumes (EPA, 2001). With the exception of single residences for which there is no charge, Pennsylvania charges a modest flat fee of $500 for an NPDES permit.

Nonpoint sources of pollution are not required to have permits under the CWA. However, there are a variety of potential mechanisms for extending the "polluter-pays" principle to nonpoint sources in order to generate revenues and create disincentives for NPS pollution (e.g., Shortle and Horan, 2001). For agriculture, these include most notably fees on polluting inputs such as fertilizers and pesticides.

A recent innovation to help municipalities cope with the costs of stormwater management is the stormwater management fee. The fee is typically based on the amount of impervious surface area (i.e., paved areas, areas under hard roofs). Fees based in this way have desirable incentives in that they discourage impervious surface areas that generate high levels of surface runoff. This type of fee may be levied by local governments, and many states have authorized local governments to establish quasi-independent stormwater utilities. These utilities are then given authority to levy a fee or tax to defray the costs of stormwater management. One estimate put the number of stormwater utilities in the United States in 2000 at more than 400 (Kasperson, 2000).

It is important to emphasize that the preceding financing options are not mutually exclusive. All of them (and perhaps others) will have to be considered in developing a financing mechanism for water quality improvement in southwestern Pennsylvania.

A crude estimate of the revenue potential in the Pittsburgh region, with several simplifying conditions, is provided in Table 6-6. According to the 1995 American Housing Survey by the Departments of Housing and Urban Development (HUD) and the U.S. Census Bureau, there were 1,051,700 housing units in the Pittsburgh MSA. Of these, 871,500 were on public water supplies (either publicly or privately owned) and 797,100 were on public sewers (either publicly or privately owned). These calculations assume that there is no growth in the numbers of customers, that increases apply to each monthly water bill and each monthly sewer bill, and that increases in

TABLE 6-6 Revenue Potential and Bond Capacity of Pittsburgh Region

Increase in Monthly Water and Sewer Bill (dollars)	New Revenue (million dollars per year) Water	Sewer	Total	Bond Capacity (billion dollars)
5	52.29	47.83	100.12	1.36
10	104.58	95.65	200.23	2.72
15	156.87	143.48	300.35	4.08
20	209.16	191.30	400.46	5.44

SOURCES: HUD and U.S. Census Bureau, 1995.

rates are uniform across all purveyors in the MSA. Furthermore, bonding capacity[10] is calculated on the basis of 20-year bonds with equal annual payments having a yield of 4.0 percent. It is clear from these first approximations (see also Chapter 5) that fully funding estimated costs of improvements as stated in the TPR report (TPRC, 2002) of ALCOSAN's draft LTCP (ALCOSAN, 1999) could lead to substantial increases in water and sewer rates. Customers could respond to increased rates by lowering rates of use through various water conservation activities, thereby reducing available revenues. Such reductions in use would have some effect on the need for the planned expansion and operation of wastewater treatment facilities. These effects should be examined when the costs of needed facilities are better known and financing packages are formulated.

Affordability

Black & Veatch Corporation, an engineering, consulting, and construction company, periodically publishes results of surveys of water and wastewater rates for cities in the United States. One such study, "Pennsylvania Water/Wastewater Rate Survey—2001," compared the cost of water and wastewater services in 24 different communities throughout the Commonwealth of Pennsylvania. This study compared residential (household) rates at two common quarterly levels of usage: 3,000 cf (cubic feet) of water or about 22,500 gallons used in three months; and 6,000 cf, or about 45,000 gallons. Out of 24 communities in Pennsylvania, Pittsburgh ranked 16 out of 24 for water and sewer charges resulting from use of 22,500 gallons of water in three months and 15 out of 24 for water and sewer charges resulting from use of 45,000 gallons of water in three months.

Table 6-7 presents a comparison of quarterly water or wastewater bills for residential households in Pittsburgh and in cities in nearby states located in the Great Lakes and Atlantic coast regions. This group of cities includes those having populations ranging from less than half to more than twice that of Pittsburgh. Quarterly bills for 3,000 and 6,000 cf of water were nearly the same in Pittsburgh and Akron, the two cities with the highest bills. The lowest bills for cities in this size range were about half of the bills for Pittsburgh and Akron.

Table 6-8 shows the results of the Black & Veatch survey of water and wastewater bills in the 49 largest cities in the United States. Pittsburgh's water and wastewater bills fall between the highest 75th and 90th percentile for both 3,000 and 6,000 cf of water usage. Thus, Pittsburgh's bills already exceed those paid by customers in more than three-fourths of the largest cities.

[10] Bonding capacity = annual revenue generated/annual payment per dollar of bond issued.

TABLE 6-7 Comparison of Residential Quarterly Water and Wastewater Bills for Pittsburgh and Cities in Nearby States

City, State	Population	Year	Bill for 3,000 cf	Bill for 6,000 cf
Pittsburgh, PA	334,563	2001	$157.35	$303.90
Fort Wayne, IN	185,716	2000	$99.60	$174.39
Indianapolis, IN	741,304	2000	$90.09	$155.97
Grand Rapids, MI	185,437	2000	$122.70	$205.20
Syracuse, NY	147,306	2001	$65.04	$130.05
Charlotte, NC	540,828	2001	$100.80	$185.70
Greensboro, NC	223,891	2001	$83.94	$174.75
Akron, OH	215,712	2000	$157.05	$300.36
Cincinnati, OH	336,400	2000	$115.59	$219.27
Cleveland, OH	495,817	2000	$102.00	$212.34
Columbus, OH	632,945	2000	$100.50	$183.30
Toledo, OH	312,174	2000	$72.60	$137.70

SOURCE: Black & Veatch Corporation, 2000 and 2001.

TABLE 6-8 Comparison of Range of Quarterly Residential Water and Wastewater Bills for Pittsburgh and in 49 Largest Cities in United States

Rank or	Usage of 3,000 cf		Usage of 6,000 cf	
Percentile	City, State	Bill	City, State	Bill
Highest	Seattle, WA	$241.59	Seattle, WA	$500.91
90th	San Francisco, CA	$174.96	Atlanta, GA	$337.80
75th	Honolulu, HI	$138.57	San Diego, CA	$250.47
50th	New York City, NY	$111.90	Oakland, CA	$203.04
25th	St. Louis, MO	$93.87	Denver, CO	$169.59
10th	El Paso, TX	$79.41	Sacramento, CA	$123.36
Lowest	Memphis, TN	$40.20	Memphis, TN	$80.37
No rank	Pittsburgh, PA (2001)	$157.35	Pittsburgh, PA (2001)	$303.90

NOTE: This survey included 50 cities, but one did not provide data on wastewater rates.
SOURCE: Black & Veatch Corporation, 2002.

These data indicate that even before the challenge of solving combined sewer overflow problems is addressed, Pittsburgh's water and sewer bills are relatively high compared to cities of similar size. As reported in the 2000 Census, median household income in Allegheny, Washington, and Westmoreland Counties ranged from $37,106 to $38,329. At prices given in Table 6-8, water and sewer bills would run from 3.2 to 3.3 percent of household income before taxes. Distributions of household income are shown in Figure 6-4. It is important to note that approximately one-third of the households had income of less than $25,000 in 2000. The added cost of addressing wet weather water quality problems would be especially onerous to this group of residents. A major challenge will be finding a way to address such problems without generating water and wastewater bills that are unaffordable and out of line with bills in nearby cities.

Challenges

It is clear that developing and implementing effective strategies for funding water quality improvement in southwestern Pennsylvania may well be the greatest challenge for the region. Among the financing issues the region will face are the following:

- Current water, stormwater, and wastewater services repayment capacity exists, but is unevenly distributed across communities and may be limited by affordability. Revenue sharing may be essential to achieve progress across the region.
- The current fragmented structure of local government in southwestern Pennsylvania tends to increase the cost of all aspects of wastewater management. Once opportunities to exploit economies of scale are forgone, the cost of financing is increased.
- Access to new and innovative technology that can reduce labor costs is expensive and requires expertise that may be beyond the reach of small independent communities, but in the long run, it can significantly reduce costs and improve performance (NDWAC Affordability Work Group, 2003).

Current Financial Developments and Committee Recommendations

As this study was being conducted, a number of analyses were being prepared on local water and wastewater rates in the region's urban core. ALCOSAN and other utilities have adopted new rates to reflect rising costs, such as the control of infiltration and inflow as part of compliance with recently adopted consent decrees. The 3RWW conducted a rate survey that was completed as this report was in review in the fall of 2004. The development of a sound regional program that equitably shares future costs—considering relative contributions of wastewater flows, repayment capacity, and efficient use of debt obligations—would go a long way to making the future priority investments in affordable water quality improvement. Affordability is a significant issue in southwestern Pennsylvania, and many of the recommendations and conclusions in the National Drinking Water Advisory Council (NDWAC. 2003) report to EPA on its small systems affordability

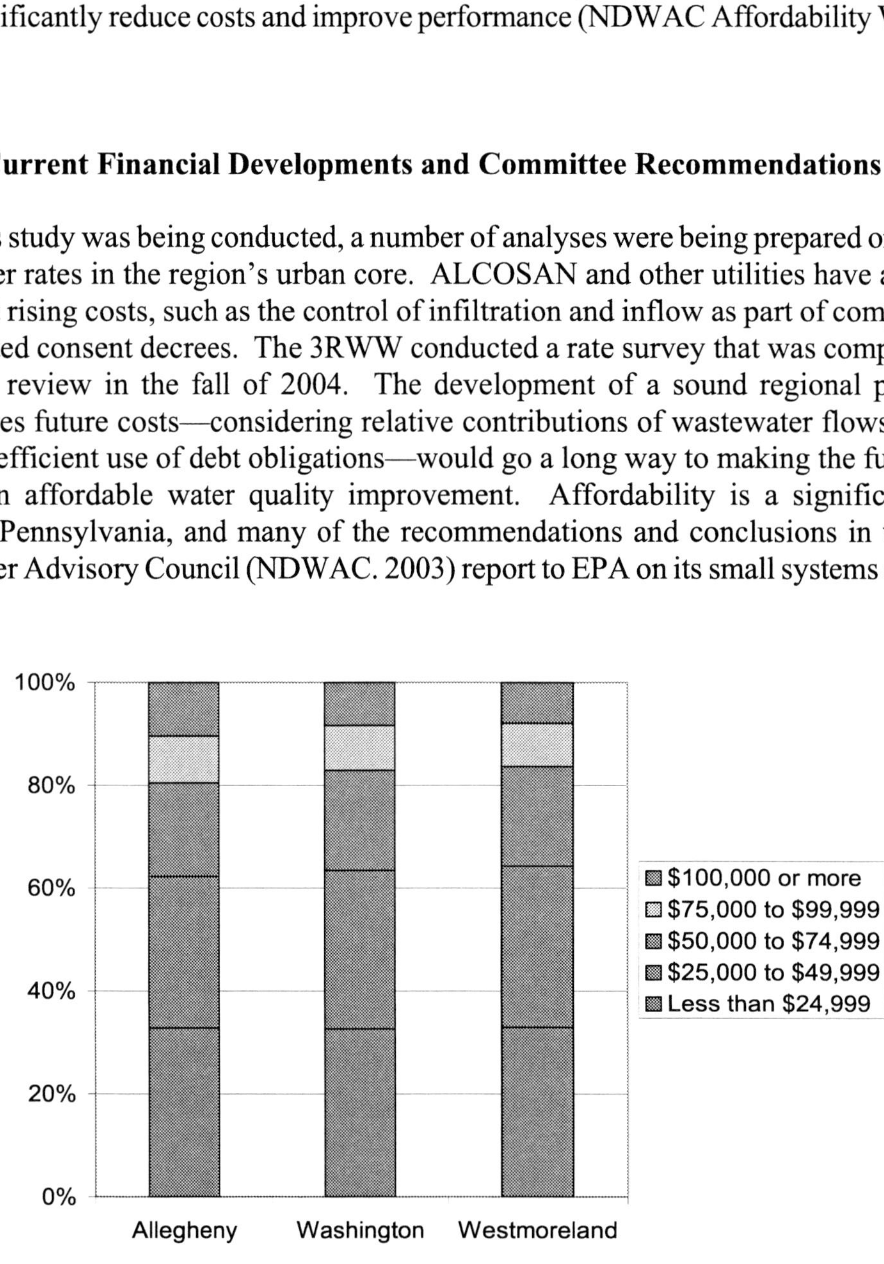

FIGURE 6-4 Distribution of household income in select southwestern Pennsylvania counties in 2000. SOURCE: U.S. Census Bureau, http://factfinder.census.gov.

criteria can be considered for wastewater systems in the region. The committee recommends the following actions toward a framework for a new regional financial approach:

- Develop and implement a sewer and/or water user surcharge, as recently proposed in Maryland, to fund the next five years of planning and data gathering under CWARP or a similar program. Ideally, the charge would be in addition to wastewater and water bills throughout the Three Rivers basin or, as a minimum, in the region's urban core (see Figure 6-1).
- Initiate a flow-based repayment system for ALCOSAN and other regional wastewater treatment providers that reflects, to the extent practicable, the actual contributions of flow into sewerage systems.
- Select one or more forms of regional governance that have the necessary legal authority and administrative expertise to finance capital improvements and operating and maintenance expenses of management programs. Such authorities should include the power to incur debt for capital projects, establish user charges, and collect revenues necessary to pay for all expenses except those financed by intergovernmental grants.
- Continue efforts to increase regional assistance through PENNVEST and other sources of funds that can generate support for specific programs such as development of county-based management programs for on-site waste disposal and AMD control.
- To the extent that assistance is not available, continuing studies are needed regarding the efficient application of current local taxes and user charges to cover the start-up efforts identified above, with the goal of creating repayment mechanisms based on an equitable regional user charge system. Ultimately the system would generate sufficient revenues to repay debt obligations that will be necessary to fund priority facilities.

SUMMARY: CONCLUSIONS AND RECOMMENDATIONS

Water management in southwestern Pennsylvania is highly fragmented among federal and state governments as well as 11 counties, 595 municipalities, and 492 water and sewer providers. Water planning in southwestern Pennsylvania has to be addressed on a regional scale and should be holistic rather than focused on particular goals; it should consider water quality, water supply, flood hazard mitigation, aquatic and riparian habitat protection and restoration, and recreation. In choosing an appropriate organization or set of organizations to address these concerns, the following three factors should be considered:

1. water resource management functions for which improvements are necessary or desirable;
2. the level of government or private sector enterprise to which management functions should be entrusted and to which legal authority should be delegated by the legislature; and
3. the geographic scale that is appropriate to achieve efficiency by exploiting economies of scale and making significant regional interdependencies internal to the planning area.

Consistent with the CWARP approach recommended in Chapter 5, changes are necessary at the following geographic scales: river basins and interstate river basins and watersheds; metropolitan region scale (multi-county areas); metropolitan urban core areas; and rural areas outside of the urban core.

Some problems, particularly those related to long-distance transport of potentially pathogenic microorganisms, heavy metals, and persistent toxic chemicals, transcend regional and state boundaries. Basinwide planning is needed to address these issues. ORSANCO and PADEP are the appropriate agencies to establish the necessary monitoring, modeling, and formulation of management strategies at that scale. Management of water quality at that scale should be integrated with activities mandated under WRPA 2002 and previously undertaken by PADEP's Unified Watershed Assessment and the WRAS program. Basinwide plans should assess and, where appropriate, adopt programs to address agricultural and other nonpoint pollution sources.

Planning for transportation systems is occurring at the multicounty, regional scale by the SPC. These plans can have significant effects on regional land use and water-related services. Concerns about land use and associated water supplies, wastewater disposal, and stormwater management should be incorporated into planning at that scale. The SPC is probably the region's best choice for accomplishing this goal, but its present representative structure and lack of water resource expertise limit its capacity to do so. Its regional databases on land use, transportation, and economic development are its strengths relative to water resource planning. Its effectiveness may be limited by the absence of regional leadership other than elected officials from local governments and may also by the makeup of its current membership where each of its counties and the City of Pittsburgh are equally represented. Rather, representation should reflect the fact that those counties with relatively high densities and a large number of intimately interrelated local governments have priorities that are quite different from those of more rural counties with more dispersed populations.

An important step that SPC in coordination with ORSANCO could take to broaden representation and advance public education on regional water resources would be to establish a Three Rivers Regional Water Forum as conceptually illustrated in Figure 6-2. The forum should be charged with a broad mandate to assess priorities for water infrastructure planning, maintenance, and construction as these activities are related to regional transportation, land use, and economic development. The Forum should include elected and appointed officials of local governments, regional leaders in the private sector, academia, environmental organizations, and other NGOS, and participation should be encouraged by all organizations that share some responsibility for the proposed CWARP. Although there are several options for the creation and organization of a regional water forum, an unincorporated network of public and private stakeholders established by voluntary MOUs is recommended for careful consideration. However, the participants and exact organization plan should be determined locally.

The dominant water quality management problems in Pittsburgh's urban core are overflows from combined and separate sanitary sewers resulting from wet weather conditions. Continued fragmented management of the sewer collection-conveyance-treatment system (i.e., maintaining the status quo) is not a satisfactory situation. Planning and management of sanitary and combined sewers should be integrated with stormwater management. There are at least five viable organizational arrangements to serve that purpose, designated as the following options: (A) merger of city and county government; (B) establishment of county-wide management either by dedication of the systems to Allegheny County or through an administrative arrangement with Allegheny County using authority under Pennsylvania Acts 67 and 68; (C) creation of one or more special districts to manage sewer collection with or without authority over stormwater management; (D) expansion of the role of ALCOSAN to include sewer collection systems, with or without authority over stormwater management; and (E) continuation of the decentralized system but with performance standards and voluntary participation in an RMO provided on a fee-for-service basis. ALCOSAN would be encouraged to establish the RMO.

All five options are viable, and discussions of Option A between Allegheny County and the City of Pittsburgh have already occurred. It is recommended that Allegheny County take a leadership role in search of a consensus on one of the four remaining options. A merger of city and county government, though politically difficult, is desirable from the perspective of water quality management. The committee also prefers Option B (establishment of county-wide management) to Options C, D, or E because it captures economies of scale in planning and management; facilitates the use of a systems approach; and keeps decision making closer to politically accountable public officials. Whatever option is ultimately chosen, management entities would have to meet performance standards to minimize CSOs and SSOs.

ORSANCO, with its prior experience with similar problems in the Ohio River basin, can be of valuable assistance in reaching a consensus on these options. In both scenarios, 3RWW should be continued or expanded to conduct public education programs for stormwater and CSO management; to provide technical assistance to local governments for stormwater and CSO management; to provide education to local governments on identifying and correcting illicit connections to sewer systems; and to monitor, analyze, and report on the status of stormwater and CSO management in Allegheny County.

Additional steps are also needed to systematically address wastewater disposal problems and inadequate water supply in rural and small urban areas outside the region's metropolitan urban core. The actions recommended to address OSTDS deficiencies in Chapter 5 should be undertaken cooperatively by several agencies. At the state level, the WRAS program should be expanded to include assessment of the effects of inadequate wastewater disposal on water quality. In doing so, the PADEP should work closely with local governments having legal authority over such systems. The SPC could and should take strong leadership in bringing local governments together to address these issues. In addition to PADEP, the SPC should request assistance from EPA and nongovernmental organizations having prior experience with programs of this kind. The Allegheny County experience in these activities should provide a sound foundation for other counties in the region.

Financing programs to support the water quality improvement activities discussed in this chapter include the following: a new regional surcharge, changing the basis for current wastewater charges, expansion of debt repayment capacity, increased assistance, and studies of regional equity through user charge systems. Most of the funding will have to come from local sources. A first approximation to the financing capacity of the metropolitan urban core suggests that water and sewer fees and stormwater fees would have to be increased substantially to pay for outlays estimated in ALCOSAN's draft LTCP. Furthermore, ALCOSAN's current pricing structure fails to send the right signal to municipalities that contribute flow to the ALCOSAN system. Instead of charging communities on the basis of water consumption, ALCOSAN should be charging on the basis of actual wastewater flows entering its interceptors. Charges should be established for total volume and peak rates of flow. Flow-measuring devices should be installed as needed to implement a revised rate structure.

It is clear from these preliminary figures that the cost of the program as outlined in the draft LTCP could have a significant effect on costs of water and sewer services for residents of the region. The committee reached no conclusion as to whether achievement of water quality goals in the region is financially feasible. This must await more detailed estimates of costs for each phase of the program and a more careful analysis of impacts on residential, commercial, and industrial consumers of regional water and sewer services. Use of an adaptive implementation strategy within CWARP as discussed in Chapter 5 would permit judgments to be made at each step of the process. Committee support of that process should not be interpreted as a strategy for indefinite postponement of major investment decisions. As long as the region is making good- faith efforts, as each step of the process is completed the region can make cost-effective decisions based on evaluation of the outcomes of prior actions. The committee believes that the involvement of key stakeholder groups should be a significant feature of each of the activities undertaken by the groups identified in this chapter. In addition, information that exists today, as well as that developed under all elements of CWARP, should be made readily available to the public. Among this information are sources of water quality problems, their significance, appropriate solutions, costs, and social impacts.

REFERENCES

ALCOSAN (Allegheny County Sanitary Authority). 1999. Draft Combined Sewer Overflow Program Phase I Activity Report: Regional Long Term Wet Weather Control Concept Plan. Pittsburgh, PA: ALCOSAN.

American Rivers, NRDC (National Resources Defense Council), Smart Growth America. 2002. Paving Our Way to Water Shortages: How Sprawl Aggravates the Effects of Drought. Washington, DC.

Black & Veatch Corporation. 2000. Indiana Water/Wastewater Rate Survey—2000; Michigan Water/Wastewater Rate Survey—2000; Ohio Water/Wastewater Rate Survey—2000. Kansas City, MO: Black & Veatch.

Black & Veatch Corporation. 2001. New York Water/Wastewater Rate Survey—2001; North Carolina Water/Wastewater Rate Survey—2001; North Carolina Water/Wastewater Rate Survey—2001; Pennsylvania Water/Wastewater Rate Survey—2001. Kansas City, MO: Black & Veatch.

Black & Veatch Corporation, 2002. 50 Largest Cities Water/Wastewater Rate Survey—2002. Kansas City, MO: Black & Veatch.

Cohan, J. 2004. Crisis spawns opportunity: Pittsburgh, Allegheny County talking openly about a merger. Pittsburgh Post Gazette. Available on-line at *http://www.post-gazette.com/pg/04053/276148.stm*. Accessed May 26, 2004.

EPA (U.S. Environmental Protection Agency). 1994. Combined Sewer Overflow (CSO) Control Policy. FRL-4732-7. Federal Register 59(75). Available on-line at *http:// www.epa.gov/npdes/pubs/owm0111.pdf*. Accessed March 29, 2004.

EPA. 1996. The Enforcement Management System: National Pollutant Discharge Elimination System (Clean Water Act): Chapter X: Setting Priorities for Addressing Discharges from Separate Sanitary Sewers. Washington, DC: Office of Regulatory Enforcement.

EPA. 1997. Community Water System Survey, Volume I: Overview. 815-R-97-001a. Washington, DC: Office of Water.

EPA. 2001. The United States Experience with Economic Incentives for Protecting the Environment. 240-R-01-001. Washington, DC: Office of the Administrator.

General Assembly of Pennsylvania. 2002. House Bill No. 2302 Amending Title 27 (Environmental Resources) of the Pennsylvania Consolidated Statutes. Chapter 31 Water Resources Planning. Available on-line at *http://www.legis.state.pa.us/wu01/li/bi/bt/ 2001/0/hb2302 p4697.htm*. Accessed May 26, 2004.

Herman, S. 2000. Memorandum: Compliance and Enforcement Strategy Addressing Combined Sewer Overflows ad Sanitary Sewer Overflows. Washington, DC: EPA Office of Enforcement and Compliance Assurance.

Herman, S. and R. Perciasepe. 1995. Memorandum: Enforcement Efforts Addressing Sanitary Sewer Overflows. Washington, DC: EPA Office of Enforcement and Compliance Assurance and Office of Water.

HUD (Department of Housing and Urban Development) and the U.S. Census Bureau. 1995. Current Housing Reports H170/95-13: American Housing Survey for the Pittsburgh Metropolitan Area in 1995. Washington, DC: HUD and U.S. Census Bureau.

Kasperson, J. 2000. The Stormwater Utility: Will it Work in Your Community? Available on-line at *http://www.forester.net/sw_0011_utility.html*. Accessed November 18, 2004.

McNulty, T. 2003. Study urged on merger of Pittsburgh, Allegheny County. Pittsburgh Post-Gazette. Available on-line at *http://www.post-gazette.com/localnews/20030916conference0916p1.asp*. Accessed May 26, 2004.

NDWAC (National Drinking Water Advisory Council) Affordability Work Group. 2003. Recommendations of the National Drinking Water Advisory Council to the U.S. EPA on Its National Small Systems Affordability Criteria. Available on-line at *http://www.epa.gov/safewater/ndwac/pdfs/report_ndwac_affordabilitywg_final_.08-08-03.pdf*. Accessed August 18, 2004.

NRC (National Research Council). 1999. New Strategies for America's Watersheds. Washington, DC: National Academy Press.

PADEP (Pennsylvania Department of Environmental Protection). 2002a. 2002 Pennsylvania Water Quality Assessment 305(b) Report. Available on-line at *http://www.dep.state.pa.us/dep/deputate/watermgt/Wqp/WQStandards/305_wq2002_narr.pdf*. Accessed April 5, 2004.

PADEP. 2002b. 2002 Section 303(d) List of Impaired Waterbodies. Available on-line *http://www.dep.state.pa.us/dep/deputate/watermgt/wqp/wqstandards/303-2002/303d-Report.htm#2002_List*. Accessed April 6, 2004.

Pennsylvania Economy League. 2000. Southwestern Pennsylvania Wastewater Survey. Pittsburgh, PA: Pennsylvania Economy League.

Reekie, L. 2000. Attitudes and perceptions of drinking water customers. AWWA Research Foundation Presentation at H2Obiettivo Symposium, Torino, Italy, May.

Rusk, D. 1995. Cities Without Suburbs. Washington, DC: Woodrow Wilson Center Press.

Rusk, D. 2000. Growth Management: The Core Regional Issue in Reflections on Regionalism, B. Katz (ed.). Washington, DC: Brookings Institution Press.

Shortle, J. and R. Horan. 2001. The economics of nonpoint pollution control. Journal of Economic Surveys 15:255-290.

Sullivan, J. 1997. Chicago Wilderness, An Atlas of Biodiversity. Chicago, IL: Chicago Region Biodiversity Council.

TPRC (Third Party Review Committee). 2002. Third Party Review of the ALCOSAN Regional Long Term Wet Weather Control Concept Plan. Pittsburgh, PA: ALCOSAN.

WSIP (Southwestern Pennsylvania Water and Sewer Infrastructure Project Steering Committee). 2002. Investing in Clean Water: A Report from the Southwestern Pennsylvania Water and Sewer Infrastructure Project Steering Committee. Pittsburgh, PA: Campaign for Clean Water.

Yaro, R. 2000. Growing and governing smart: A case study of the New York region. In Reflections on Regionalism, B. Katz (ed.). Washington, DC: Brookings Institution Press.

7

Implications Beyond Southwestern Pennsylvania

In the course of this study of water quality improvement in southwestern Pennsylvania, the committee gained knowledge and insights on several technical, policy, and institutional issues that have broader implications and might be considered useful by others responsible for national efforts to protect and enhance water quality. These implications are organized into six areas and discussed below.

INFORMATION SYSTEMS

Programs and organizations that gather water quality data in southwestern Pennsylvania are identified in Chapters 3 and 4 of this report. Each agency that collects such data does so under a federal, state, or local (sometimes private utility) mandate and budget. Although some states such as California with its basin plans, and Pennsylvania with its recent water resources legislation (e.g., Pennsylvania's Water Resources Planning Act [WRPA] of 2002; see Chapter 6 for further information), have moved toward a framework for integrated data gathering, such coordinated development of water quality data is the exception rather than the rule. To change this paradigm and avoid the collection of redundant or useless data, coordinated and efficient monitoring efforts must be developed by appropriate entities for every watershed that is the subject of significant assessment and/or improvement.

Databases should be thoughtfully integrated, and modeling efforts coordinated. Frequently, data are collected by agencies with single-purpose directives, such as flow measurements by the U.S. Army Corps of Engineers (USACE) and/or the U.S. Geological Survey (USGS); various water quality measurements by water utilities, wastewater dischargers, USGS, and others; and biological measurements by various state fishery and environmental agencies. Each state should give consideration to establishing a comprehensive water quality data management program, with a single "clearinghouse" designated to coordinate data collection and analysis for each (sub)watershed.

HEALTH AND ECOLOGICAL IMPACTS OF WATER QUALITY

Prudent investments in sewerage infrastructure require the establishment of reasonably direct relationships between causes of water quality degradation and their effects. Historically, these relationships have not generally been established. For example, it is clear that the excessive concentrations of microorganisms measured during and immediately after wet weather periods in the Three Rivers in Allegheny County have not been correlated with outbreaks of gastrointestinal illness. Further, although the impacts of acid mine drainage on stream habitat and ecology are typically measured by water pH and heavy metal concentrations, downstream effects are more subtle, difficult, and expensive to identify (e.g., through bioassessments). Thus, current water quality planning, regulation, and action to protect both public health and the environment are based largely on indirect criteria, without the benefit of human health or ecological assessments. This national condition also applies in southwestern Pennsylvania. Therefore, when considering present and potential investments and improvement projects for wastewater management and drinking water treatment, a significant investment to improve environmental and health data to demonstrate the actual impact of water quality conditions on the environment and public health would seem justified.

POTENTIAL FEDERAL POLICY CONFLICTS WITH REGIONAL OPTIMIZATION

An important lesson learned from this examination of water quality management in southwestern Pennsylvania is that long-term control plans for combined sewer overflows (CSOs) should be designed so as to permit creative and flexible solutions that can be adjusted with knowledge and changing conditions. Current federal CSO policy could be interpreted to maximize delivery of such overflows to secondary wastewater treatment plants. Decisions to adopt such action as part of a control plan should be made in the context of a comprehensive plan for a specific management area. The comprehensive plan should consider a broad array of options that may be complementary to or alternatives of maximization of flow to secondary treatment plants. The preferred management strategy should be programmed over a reasonable time horizon to facilitate feedback from less capital-intensive elements to better inform decisions about more capital-intensive ones.

To one degree or another, existing federal and state policies support regional cooperative water quality management programs. However, most financial assistance has been directed toward projects that provide specific benefits, but which may not be optimal in a regional context. To the extent that funds available to a Comprehensive Watershed Assessment and Response Plan ("CWARP-type") program can be applied with some flexibility, they should be directed toward achieving optimal regional benefits.

STAKEHOLDER REPRESENTATION AND PARTICIPATION

In the conduct of public affairs related to environmental quality, elected officials, their administrators, and private utility operators must consider a multitude of individual and collective concerns about strategies, costs, and local projects. In many parts of the country, these disparate concerns have delayed, significantly changed, or vetoed water quality improvement projects. The committee's recommendation to establish a Three Rivers Regional Water Forum is an attempt to address this issue in southwestern Pennsylvania and, in concept, may be beneficial to other areas of the United States.

A 2000 survey conducted by the Pennsylvania Economy League and related national surveys (e.g., American Water Works Association Research Foundation) of public attitudes regarding water and investment indicate that there is more potential support for expenditures for water quality improvement than is generally believed by those responsible for water matters.

Although water quality was not a major issue in southwestern Pennsylvania in 2000, it was found to be of concern to about 50 percent of the people surveyed who believe that public health is threatened by sewage.

If the water quality improvement projects likely to result from a CWARP approach are to be successful, responsible officials must recognize that there is latent support among the public for significant investment in improvements. The only way to verify this support is to present programs and projects to the public for approval even though they will likely result in increased water or wastewater bills. This model may be applicable nationally, and to the extent possible, should address the concerns of special interests and provide long-range solutions to water quality problems.

PAYING FOR WATER QUALITY IMPROVEMENTS

The significant issue in water quality improvements needed and proposed for southwestern Pennsylvania—and, in large part, the reason this study was commissioned by the Allegheny Conference on Community Development—is their potentially high costs. As discussed in Chapters 5 and 6, past federal grants have been a substantial part of the solution to water quality problems for many communities nationwide, thus reducing the local burden, but in the committee's judgment such federal grants are no longer likely to be available. Thus, it is likely that most future costs will be borne by the businesses and residents of southwestern Pennsylvania, or perhaps the Commonwealth of Pennsylvania. Since solutions will likely result in significant increases in costs of wastewater services, the region should seek ways to minimize the impact on those who are least able to pay.

More broadly, water and wastewater utilities will have to identify creative financing solutions as costs of water quality improvements increase. Although utilities can expect federal and perhaps state assistance in areas such as security, little more will be provided except in the case of smaller utilities with utility-wide affordability issues. Those water and wastewater systems that require asset upgrades to maintain their useful lives (e.g., the U.S. Environmental Protection Agency's [EPA's] Capacity, Management, Operation, and Maintenance [CMOM] requirements for wastewater collection systems) and improvements to meet regulations will be faced with high costs. One option that should be considered is the creation of regional financing mechanisms. This could be accomplished by regional sharing of the burden for individuals with limited ability to pay. Accounts that fund lifeline or similar rates are one way to mitigate increasing monthly bills of those who can least afford them. This report includes a discussion of the importance of assessing the economic benefits of water quality improvement. A reliable assessment of such benefits would go a long way in evaluating the feasibility of new, larger investments.

REGIONALIZATION AND COOPERATION

According to an August 8, 2004, article in the *Pittsburgh Post-Gazette*,[1] Allegheny County contains the greatest number of municipalities per capita in the nation and this character generally typifies the 11-county study area. Thus, it should be no surprise that the region

[1] Available on-line at *http://www.post-gazette.com/pg/04221/357841.stm*.

potentially can achieve tremendous advantages through cooperative actions in planning and implementing water quality improvements. Such advantages would include economies of scale, access to increasingly needed technology, and improved reliability and security. Nonetheless, there may be some resistance to regionalization in southwestern Pennsylvania and elsewhere, because some communities and utilities in the United States have a strong attachment to self-sufficiency and maintaining the status quo. This may be due to desires to control rate structures and use revenues for other municipal functions or to fears of job loss or of dealing with an unresponsive separate party with newly assigned responsibilities. It will be an important challenge in southwestern Pennsylvania and elsewhere to find ways to coordinate, consolidate, cooperate, and regionalize *appropriate* individual functions in a manner that preserves local identity while gaining the improved performance to achieve multiple water resource management objectives. As discussed in Chapter 6, the recommended creation of a regional water forum constitutes an opportunity to facilitate this development in southwestern Pennsylvania.

Acronyms

3R2N	Three Rivers Second Nature
3RWW	Three Rivers Wet Weather Demonstration Program
ABAG	Association of Bay Area Governments
ACCD	Allegheny Conference on Community Development
ACHD	Allegheny County Health Department
AGI	Acute gastrointestinal illness
ALCOSAN	Allegheny County Sanitary Authority
AMD	Acid mine drainage
AML	Abandoned mine land
ATSDR	Agency for Toxic Substances and Disease Registry
AWQS	Ambient water quality standards
BMP	Best management practices
BOD	Biological oxygen demand
CAA	Clean Air Act
CATS	Chicago Area Transit Study
CDC	Centers for Disease Control and Prevention
CMOM	Capacity, Management, Operations, and Maintenance
CMU	Carnegie Mellon University
CSO	Combined sewer overflow
CVMP	Citizens' Volunteer Monitoring Program
CW	Chicago Wilderness
CWA	Clean Water Act
CWARP	Comprehensive Watershed Assessment and Response Plan
DDT	Dichlorodiphenyltrichloroethane
DHHS	U.S. Department of Health and Human Services
EBMUD	East Bay Municipal Utility District
EMAP	Environmental Monitoring and Assessment Program
EPA	U.S. Environmental Protection Agency
ERC	Extended reactive control
EWGCC	East-West Gateway Coordinating Council
FBM	Flow balancing method
FDA	U.S. Food and Drug Administration
GIS	Geographic information system
GWR	Ground Water Rule (proposed)
HUC	Hydrologic unit code
IARC	International Agency for Research on Cancer
IAWPR	International Association on Water Pollution Research and Control
IAWQ	International Association on Water Quality
IBI	Index of Biotic Integrity
ICR	Information Collection Rule
I/I	Infiltration and inflow

ISTEA	Intermodal Surface Transportation Efficiency Act
JLA	Joint local agency
KIPDA	Kentuckyiana Regional Planning and Development Agency
LRC	Local reactive control
LTCP	Long-term wet weather control plan
MCL	Maximum contaminant level
MFA	Major financial activity
MMSD	Milwaukee Metropolitan Sewer District
MOU	Memorandum of understanding
MPN	Most probable number
MPO	Metropolitan planning organization
MSA	Metropolitan statistical area
MSD	Metropolitan sewer district
MWRD	Metropolitan Water Reclamation District
NAPRA	National agricultural pesticide risk assessment
NAWQA	National Water Quality Assessment Program
NGO	Nongovernmental organization
NIPC	Northeastern Illinois Planning Commission
NMR	Nine Mile Run
NMR-GP	Nine Mile Run Greenway Project
NODP	National Onsite Demonstration Project
NPDES	National Pollutant Discharge Elimination System
NPS	Nonpoint source
NRC	National Research Council
NRDC	Natural Resources Defense Council
OGP	Optimal global predictive
ORSANCO	Ohio River Valley Water Sanitation Commission
OSM	Office of Surface Mining, Reclamation, and Enforcement
OSTDS	On-site sewage treatment and disposal system
PABS	Pennsylvania Biological Survey
PADEP	Pennsylvania Department of Environmental Protection
PADER	Pennsylvania Department of Environmental Resources
PADOH	Pennsylvania Department of Health
PASEO	Pennsylvania Association of Sewage Enforcement Officers
PCB	Polychlorinated biphenyl
PEC	Pennsylvania Environmental Council
PEL	Probable effect level
PENNVEST	Pennsylvania Infrastructure Investment Authority
PFBC	Pennsylvania Fish and Boat Commission
PFU	Plaque forming unit
POWR	Pennsylvania Organization for Watersheds and Rivers
PRWA	Pennsylvania Rural Water Association
PWSA	Pittsburgh Water and Sewer Authority
RBP	Rapid Bioassessment Protocol
RCAP	Rural Community Assistance Program
RMO	Regional Maintenance Organization

RTA	Regional Transportation Authority
RTC	Real-time control
RUS	Rural Utility Service
RWRC	Regional Water Resources Council
SDWA	Safe Drinking Water Act
SEWRPC	Southeastern Wisconsin Regional Planning Commission
SLMSD	Saint Louis Metropolitan Sewer District
SMCL	Secondary Maximum Contaminant level
SMCRA	Surface Mining Control and Reclamation Act
SPC	Southwestern Pennsylvania Commission
SSO	Sanitary sewer overflow
STEP	Septic tank effluent pump
SWAP	Source Water Assessment and Protection
SWRC	Statewide Water Resources Committee
TCMC	Twin Cities Metro Council
TDS	Total dissolved solids
TMDL	Total maximum daily load
TPR	Third party review
TPRC	Third Party Review Committee
TSS	Total suspended solids
USACE	U.S. Army Corps of Engineers
USDA	U.S. Department of Agriculture
USGS	U.S. Geological Survey
USPHS	U.S. Public Health Service
VOC	Volatile organic compound
WATRR	Watershed Alliance for the Three Rivers Region
WPC	Western Pennsylvania Conservancy
WQIF	Water Quality Improvement Fund
WQS	Water quality standard
WRAS	Watershed Restoration Action Strategy
WRPA	Water Resources Planning Act
WSIP	Southwestern Pennsylvania Water and Sewer Infrastructure Project
WSTB	Water Science and Technology Board
WWTP	Wastewater treatment plant

Appendix A

Southwestern Pennsylvania Local Resource Panel

Convened Summer 2002

Mr. Wilder Bancroft
Manager, retired, Environmental Quality Programs
Allegheny County Department of Health

Ms. Ann Bargerstock
Director
Greene County Planning Commission

Mr. David Borneman
Director, Engineering and Construction
Allegheny County Sanitary Authority

The Honorable David Coder
Chairman, Board of Commissioners
Greene County

Dr. Jared Cohon
Chair, Southwestern Pennsylvania Water and Sewer Infrastructure Project Steering Committee
President, Carnegie Mellon University

Mr. Tim Collins
Project Director
3 Rivers 2nd Nature

Dr. Anthony DiGioia
Chairman, Board of Directors
GAI Consultants, Inc.

Mr. Charles Duritsa
Regional Director
Pennsylvania Department of Environmental Protection

Dr. David Dzombak
Professor, Department of Civil and Environmental Engineering
Carnegie Mellon University

Mr. Larry Gasparato
Project Specialist
(Pennsylvania Infrastructure Investment Authority)

Mr. Art Gazdik
Member, PADEP of Water Resources Advisory Committee
Engineer, Ross Township

Mr. Alex Graziani
Executive Director, Smart Growth Partnership of Westmoreland County
University of Pittsburgh at Greensburg

Mr. James Hassinger
Executive Director
Southwestern Pennsylvania Commission

Mr. David Johnston
Executive Director
Butler County Planning Commission

Mr. Kenneth Komoroski
Attorney
Kirkpatrick & Lockhart LLP

Ms. Barbara McMillen
Rural Development Specialist
U.S. Department of Agriculture

Mr. Louis Oliva
Senior Vice President, Industrial Services Group
Grubb & Ellis

Ms. Jan Oliver
Manager of Wet Weather Programs
Allegheny County Sanitary Authority

Ms. Ruthann Omer
President
The Gateway Engineers, Inc.

Mr. Richard Palilla
Executive Director
Armstrong County Department of Planning and Development

Mr. Dewitt Peart
Executive Director
Regional Development Consortium

Mr. Greg Phillips
Manager
Westmoreland County Conservation District

Mr. Robert Polczynski
Public Services Officer, Western Region
First Commonwealth Bank

Ms. Rita Pollock
Consultant, Southwestern Pennsylvania Commission

Mr. Tony Russo
General Manager
Wilkinsburg Penn Joint Water Authority

Mr. John Schombert
Executive Director
Three Rivers Wet Weather Demonstration Program

Mr. Larry Schweiger
President and Chief Executive Officer
Western Pennsylvania Conservancy

Mr. Gary Sheppard
County Extension Director
Pennsylvania State Cooperative Extension Service

Mr. Bob Softcheck
General Manager
North Fayette County Municipal Authority

Mr. Paul Spence
Vice President
GAI Consultants, Inc.

Mr. Ronald Tarquinio
President
Tarquin Real Estate, Inc.

Allegheny Conference on Community Development Staff

Mr. Harold Miller
President

Ms. Jan Lauer
Senior Vice President

Mr. Ken Zapinski
Vice President

Mr. Joshua Donner
Senior Policy Analyst

Appendix B

Summary of Select Reports Concerning Water and Wastewater Quality Problems of Southwestern Pennsylvania

***Investing in Clean Water: A Report from the Southwestern Pennsylvania Water and Sewer Infrastructure Project Steering Committee* (WSIP, 2002)**

This report considers the pervasive water quality and wastewater issues facing the southwestern Pennsylvania region by examining current problems (e.g., combined sewer overflows, aging infrastructure, communities without sewage treatment) and exploring possible solutions to such problems. The authoring steering committee concludes that a key to dealing with wastewater pollution is adopting a regional approach to protecting, treating, and delivering the region's water. The steering committee acknowledges that urban and rural areas face different problems, but believes that the needs of the watershed and its infrastructure transcend political and economic boundaries. If municipalities, authorities, and homeowners work together to address and resolve these problems, the burden of cost will be spread out and diminished accordingly. The report recommends three regional strategies to improve the water and wastewater systems of southwestern Pennsylvania:

1. plan and prioritize water and wastewater investments;
2. help communities find the most cost-effective solutions and educate the public; and
3. advocate for legislative and regulatory action and for state and federal funding.

***Water Quality in the Allegheny and Monongahela River Basins: Pennsylvania, West Virginia, New York, and Maryland, 1996-98* (Anderson et al., 2000)**

This report was generated under the U.S. Geological Survey's National Water Quality Assessment (NAWQA) Program and summarizes the major findings about water quality in the Allegheny and Monongahela River Basins from 1996 to 1998. The report states that the major influences on stream and river water quality were abandoned coal mines; maintenance of navigation channels; increased urban development; and reductions in agricultural, industrial, and coal production activities. Major factors affecting groundwater quality included coal mining, pesticide and fertilizer use, gasoline and oxygenate use, and naturally occurring concentrations of radon.

***Plumbing the Future: Sewage Infrastructure and Sustainability in Western Pennsylvania* (Environmental Law Institute, 1999)**

At the request of the Heinz Endowments, the Environmental Law Institute studied the relationship between sewage infrastructure decisions in western Pennsylvania and the effects of those decisions on urban, suburban, and rural landscapes of the region. The report examines ways in which the laws and policies affecting sewage infrastructure investment can be utilized to repair problems with aging infrastructure without increasing pressure for sprawl. It also seeks to assure regional decision makers that new development supported by sewage infrastructure investments is sustainable and consistent with regional goals. The report focuses on the demographic, environmental, and technical issues affecting current infrastructure; identifies the legal and institutional issues affecting decision makers; examines financial alternatives and opportunities for infrastructure improvements or replacements; and identifies promising new approaches including new uses for existing laws.

***Draft Combined Sewer Overflow Program Phase I Activity Report: Regional Long Term Wet Weather Control Concept Plan* (ALCOSAN, 1999)**

The Allegheny County Sanitary Authority (ALCOSAN) operates under a National Pollution Discharge Elimination System permit (see also Box 1-1) which is administered by the Pennsylvania Department of Environmental Protection (PADEP). Under this permit, ALCOSAN was required to prepare a long term control plan (LTCP) for combined sewer systems—although ALCOSAN also incorporated sanitary sewer overflow (SSO) planning into its wet weather planning efforts—and published this report in March of 1999. This draft plan advocates the use of a "presumption approach" as allowed by the U.S. Environmental Protection Agency (EPA) (see Chapter 5 for further information) to address Allegheny County's combined sewer overflow (CSO) problem. The LTCP calls for the elimination of SSOs by providing for secondary treatment of flow equivalent to the flow from separate sewered areas.

***Third Party Review of the ALCOSAN Regional Long Term Wet Weather Control Concept Plan* (TPRC, 2002)**

This report is an independent review of ALCOSAN's 1999 LTCP (summarized above) and was conducted by the Third Party Review Committee (TPRC). The TPRC held information gathering meetings and completed limited water quality modeling based on existing data. The report provides data, observations, and findings relevant to the continued refinement and development of a cost-effective LTCP for ALCOSAN.

***Southwestern Pennsylvania Citizens' Vision for Smart Growth: Strengthening Communities and Regional Economy* (Sustainable Pittsburgh, 2003)**

This report is the result of dialogue among concerned citizens and organizations of the Pittsburgh area regarding regional planning efforts and the application of "smart-growth" principles in an effort to promote sustainable development as well as economic growth. In regard to water issues, the report asserts that sewers and water systems may foster sprawl because water infrastructure is necessary for new development. Thus, sewer service providers and water suppliers have de facto authority to influence the density and location of new developments. The report also urges regional cooperation in an effort to steer development to existing communities by repairing and upgrading water systems that are already in place.

REFERENCES

ALCOSAN (Allegheny County Sanitary Authority). 1999. Draft Combined Sewer Overflow Program Phase I Activity Report: Regional Long Term Wet Weather Control Concept Plan. Pittsburgh, PA: ALCOSAN.

Anderson, R., K. Beer, T. Buckwalter, M. Clark, S. McAuley, J. Sams, and D. Williams. 2000. Water Quality in the Allegheny and Monongahela River Basins: Pennsylvania, West Virginia, New York, and Maryland (1996-98). Denver, CO: U.S. Geologic Survey.

ELI (Environmental Law Institute). 1999. Plumbing the Future: Sewerage and Sustainability in Western Pennsylvania. Washington, DC: ELI.

Sustainable Pittsburgh. 2003. Southwestern Pennsylvania Citizens' Vision for Smart Growth: Strengthening Communities and Regional Economy. Pittsburgh, PA: Sustainable Pittsburgh.

TPRC (Third Party Review Committee). 2002. Third Party Review of the ALCOSAN Regional Long Term Wet Weather Control Concept Plan. Pittsburgh, PA: ALCOSAN.

WSIP (Southwestern Pennsylvania Water and Sewer Infrastructure Project Steering Committee). 2002. Investing in Clean Water: A Report from the Southwestern Pennsylvania Water and Sewer Infrastructure Project Steering Committee. Pittsburgh, PA: Campaign for Clean Water.

Appendix C

Glossary

SOURCES: Adapted from 25 PA Code § 73; EPA, 1997 and 2002; NSFC, 2001; PADEP, 2003 and 2004a, b, c.

Abandoned well—A well whose use has been permanently discontinued or that is in a state of such disrepair cannot be used for its intended purpose.

Absorption area—A component of an individual or community sewage system where liquid from a treatment tank seeps into the soil; it consists of an aggregate-filled area containing piping for the distribution of liquid and the soil or sand-soil combination located beneath the aggregate.

Acid mine drainage (AMD)—Drainage of water from areas that have been mined for coal or other mineral ores. The water has a low pH because of its contact with sulfur-bearing material and is harmful to aquatic organisms.

Aeration—A process that promotes biological degradation of organic matter in water. The process may be passive (e.g., when waste is exposed to air) or active (e.g., when a mixing or bubbling device introduces the air).

Alternate sewage system—Method of demonstrated on-site sewage treatment and disposal not described in 25 PA Code § 73. Such systems may be considered for individual or community on-site use to solve an existing pollution or public health problem, overcome site suitability deficiencies, overcome engineering problems related to the site or its proposed use, or utilize a successful experimental design under varying site conditions.

Assimilative capacity—The ability of a natural body of water to receive wastewaters or toxic materials without deleterious effects and without damage to aquatic life or humans who consume the water.

Benefit-cost analysis—An economic method for assessing the benefits and costs of achieving alternative health-based standards at given levels of health protection.

Biological oxygen demand—A measure of the amount of oxygen consumed in the biological processes that break down organic matter in water. The greater the BOD, the greater is the degree of pollution.

Biological contaminants—Living organisms or derivatives (e.g., viruses, bacteria, fungi, mammal and bird antigens) that can cause harmful health effects when inhaled, swallowed, or otherwise taken into the body.

Borehole disposal—Individual or community systems, discharging to a borehole, abandoned water well, drywell, ventilation shaft, or other subterranean structure.

Cesspool—An outdated (nineteenth to mid-twentieth century) method of sewage disposal that is not permitted in modern regulations. A cesspool may be described as an "igloo-like" structure, built of loose (without mortar) rock or building blocks, that is buried underground. Cesspools are not watertight and allow the sewage entering them to drain into the surrounding area. Unlike septic tanks, cesspools provide very little treatment to sewage before releasing it to the environment, and unlike holding tanks, cesspools do not retain sewage for treatment elsewhere.

Chemical stressors—Chemicals released to the environment through industrial waste, auto emissions, pesticides, and other human activity that can cause illness and even death in plants and animals.

Chlorination—The application of chlorine to drinking water, sewage, or industrial waste to disinfect or oxidize undesirable compounds.

Cistern—Small tank or storage facility used to store water for a home or farm; often used to store rainwater.

Clarification—Clearing action that occurs during wastewater treatment when solids settle out, it is often aided by centrifugal action and chemically induced coagulation in wastewater.

Cluster OSTDS—An on-site sewage treatment and disposal system under some form of common ownership and management that provides treatment and dispersal or discharge of wastewater from two or more homes or buildings but less than an entire community. (*See* On-site sewage treatment and disposal system)

Coliform index—A rating of water purity based on a count of fecal bacteria.

Coliform organism—Microorganisms found in the intestinal tract of humans and animals, their presence in water indicates fecal pollution and potentially adverse contamination by pathogens.

Collector sewers—Pipes used to collect and carry wastewater from individual sources to an interceptor sewer that will carry it to a treatment facility.

Combined sewer overflows—Discharge of a mixture of stormwater and domestic waste when the flow capacity of a sewer system is exceeded during rainstorms.

Combined sewers—A sewer system that carries both sewage and stormwater runoff. Normally, its entire flow goes to a waste treatment plant, but during a heavy storm, the volume of water may be so great as to cause overflows of untreated mixtures of stormwater and sewage into receiving waters. Stormwater runoff may also carry toxic chemicals from industrial areas or streets into the sewer system.

Community water system—A public water system that serves at least 15 service connections used by year-round residents or regularly serves at least 25 year-round residents.

Consent order—A legal document, also known as a consent decree, signed by the U.S. Environmental Protection Agency (EPA) and an individual, business, or other entity, committing that entity to take corrective action or refrain from an activity. The consent order describes the actions to be taken and can be enforced in court.

Contaminant—Any physical, chemical, biological, or radiological substance or matter that has an adverse effect on air, water, or soil. (*See also* pollutant, these two terms are used interchangeably in this report.)

Conventional pollutants—Statutorily listed pollutants that are well understood by scientists. These may be in the form of organic waste, sediment, acid, bacteria, viruses, nutrients, oil and grease, or heat.

Conventional sewage system—System employing the use of demonstrated on-site sewage treatment and disposal technology in a manner specifically recognized by 25 PA Code § 73; includes septic tank or gravity absorption trenches, in-ground seepage bed, aerobic treatment system, pressure distribution absorption system, subsurface sand filter, elevated sand mound, and recycling-incinerating-composting toilets.

Corrosion—The dissolution and wearing away of metal caused by a chemical reaction—for example, between water and the pipes, chemicals touching a metal surface, or contact between two metals.

Criteria—Descriptive factors taken into account by EPA in setting standards for various pollutants. These factors are used to determine limits on allowable concentration levels and to limit the number of violations per year. When issued by EPA, the criteria provide guidance to the states on how to establish their standards.

Design capacity—The average daily flow that a treatment plant or other facility is designed to accommodate.

Designated uses—Those water uses identified in state water quality standards that must be achieved and maintained as required under the federal Clean Water Act. Uses can include public water supply, aquatic life use, and contact recreation, among others.

Direct discharger—A municipal or industrial facility that introduces pollution through a defined conveyance or system such as outlet pipes; a point source.

Direct runoff—Water that flows over the ground surface or through the ground directly into streams, rivers, and lakes.

Discharge—The flow of surface water in a stream or canal or the outflow of groundwater from a flowing artesian well, ditch, or spring; it can also apply to discharge of liquid effluent from a facility.

Dissolved oxygen (DO)—The oxygen freely available in water, vital to fish and other aquatic life and for the prevention of odors. DO levels are considered an important indicator of a waterbody's ability to support desirable aquatic life. Secondary and advanced wastewater treatment are generally designed to ensure adequate DO in waste-receiving waters.

Dissolved solids—Disintegrated organic and inorganic material in water. Excessive amounts make water unfit to drink or use in industrial processes.

Drainage—Improving the productivity of agricultural land by removing excess water from the soil by means such as ditches or subsurface drainage tiles.

Drainage basin—The area of land that drains water, sediment, and dissolved materials to a common outlet at some point along a stream channel.

Effluent—Wastewater, treated or untreated, that flows out of a treatment plant, sewer, or industrial outfall. Generally refers to wastes discharged into surface waters.

Effluent limitation—Restrictions established by a state or the EPA on quantities, rates, and concentrations in wastewater discharges.

EMAP data—Environmental monitoring data collected under the auspices of EPA's Environmental Monitoring and Assessment Program. All EMAP data share the common attribute of being of known quality, having been collected in the context of explicit data quality objectives (DQOs), and being subject to a consistent quality assurance program.

Environmental indicator—A measurement, statistic, or value that provides a proximate gauge or evidence of the effects of environmental management programs or of the state or condition of the environment.

Eutrophication—The slow aging process induced by higher levels of nutritive compounds (e.g., nitrogen and phosphorus) during which a waterbody can evolve into a bog or marsh and eventually disappear. Human activities can accelerate the process.

Experimental sewage system—Method of on-site sewage treatment and disposal not described in the Pennsylvania Code that is proposed for the purpose of testing and observation. These systems may be considered for individual or community on-site use to solve an existing pollution or public health problem; overcome site suitability deficiencies; overcome engineering problems related to the site or its proposed use; evaluate new concepts or technologies applicable to on-site disposal; or evaluate the applicability to on-site disposal of established concepts or technologies having successful use in comparable applications in the field of engineering.

Facilities plans—Plans and studies related to the construction of treatment works necessary to comply with the federal Clean Water Act. A facilities plan investigates needs and provides information on the cost-effectiveness of alternatives; a recommended plan; an environmental assessment of the recommendations; and descriptions of the treatment works, costs, and a completion schedule.

Fecal coliform bacteria—Bacteria found in and emanating from the intestinal tracts of mammals, including humans. Their presence in water or sludge is an indicator of microbial pollution and possible contamination by pathogens.

Finished water—The condition of water when it has passed through all the processes in a water treatment plant and is ready to be delivered to consumers through a distribution system for consumption or contact use.

Flocculation—Process by which clumps of solids in water or sewage aggregate through biological or chemical action so they can be separated from water or sewage.

Groundwater—Freshwater found beneath the earth's surface, usually in aquifers, that supplies wells and springs. Because groundwater is a major source of drinking water, there is growing concern over contamination from leaching agricultural or industrial pollutants or leaking underground storage tanks.

Heavy metals—Metallic elements with high atomic weights (e.g., mercury, chromium, cadmium, arsenic, lead) that can damage living things at low concentrations and tend to accumulate in the food chain.

Holding tank—A tank, whether permanent or temporary, to which sewage is conveyed by a water-carrying system; a watertight receptacle that receives and retains sewage and is designed and constructed to facilitate ultimate disposal of sewage at another site. Holding tanks do not treat sewage; they merely store sewage that will be treated at another location. Unlike septic tanks, holding tanks have no outlet to a soil absorption area.

Indicator—In biology, any biological entity, processes, or community whose characteristics show the presence of specific environmental conditions. In chemistry, a substance that demonstrates a visible change, usually of color, at a desired point in a chemical reaction.

Industrial waste—Unwanted materials from an industrial operation; that may be liquid, sludge, solid, or hazardous waste.

Infiltration—The penetration of water through the ground surface into subsurface soil or the penetration of water from the soil into sewer or other pipes through defective joints, connections, or manhole walls.

Influent—Water, wastewater, or other liquid flowing into a reservoir, basin, or treatment plant.

Innovative technologies—New or inventive methods to treat effectively hazardous waste and reduce risks to human health and the environment.

Interceptor sewers—Large sewer lines that, in a combined system, control the flow of sewage to the treatment plant. In a storm, they allow some of the sewage to flow directly into a receiving stream, thus keeping it from overflowing onto the streets. Also used in separate sewer systems to collect the flows from main and trunk sewers and carry them to treatment points.

Land application—Application of wastewater onto the ground for subsequent treatment or reuse.

Lateral sewers—Privately owned pipes that run under city streets and receive the sewage from homes and businesses, as opposed to domestic feeders and main trunk lines.

Leachate—Water that collects contaminants as it trickles through wastes, pesticides, or fertilizers. Leaching may occur in farming areas, feedlots, and landfills and may result in hazardous substances entering surface water, groundwater, or soil.

Leachate collection system—A system that gathers leachate and pumps it to the surface for treatment.

Maximum contaminant level (MCL)—The maximum permissible level of a contaminant in water delivered to any user of a public system. MCLs are enforceable standards.

Municipal sewage—Waste (mostly liquid) originating from a community; it may be composed of domestic wastewaters and/or industrial discharges.

Nonpoint source (NPS)—Diffuse pollution source (i.e., without a single point of origin or not introduced into a receiving stream from a specific outlet). The pollutants are generally carried off the land by stormwater or through the air. Common NPs are agriculture, forestry, urban, mining, construction, land disposal, and city streets.

On-site sewage treatment and disposal system (OSTDS)—A system relying on natural processes and/or mechanical components that is used to collect, treat, and disperse or discharge wastewater from single dwellings or buildings.

Outfall—The physical location at which effluent is discharged into receiving waters.

Overflow rate—One of the guidelines for the design of settling tanks and clarifiers in a wastewater treatment plant; it is used by plant operators to determine if tanks and clarifiers are over- or underused.

Package plant—A small sewage treatment plant of compact, prefabricated design to reduce capital costs, utilizing mechanical and/or aerobic treatment; used for sewage flows of greater than 0.002 million gallons per day (mgd) but less than 0.1 mgd; commonly privately owned; requires regular operation and maintenance by professional operators.

Pathogens—Microorganisms (bacteria, viruses, or parasites) that can cause disease in humans.

Permit—An authorization, license, or equivalent control document issued by the EPA or a state agency to implement the requirements of an environmental regulation (e.g., a permit to operate a wastewater treatment plant or a facility that may generate harmful emissions).

Point source—A stationary location or fixed facility from which pollutants are discharged; any single identifiable source of pollution, (e.g., a pipe, ditch, ship, ore pit, factory smokestack).

Pollutant—Generally, any substance introduced into the environment that adversely affects the usefulness of a resource or the health of humans, animals, or ecosystems. (*See also* contaminant; these two terms are used interchangeably in this report)

Potable water—Water that is safe for drinking and cooking.

Primary drinking water regulation—Rule that applies to public water systems and specifies a contaminant level, that, in the judgment of EPA, will not adversely affect human health.

Primary waste treatment—First steps in wastewater treatment in which screens and sedimentation tanks are used to remove most materials that float or will settle. Primary treatment removes about 30 percent of carbonaceous biochemical oxygen demand from domestic sewage.

Privy vault—A hole in the ground to receive waste, underlying an outhouse.

Public water system (PWS)—A system that provides piped water for human consumption to at least 15 service connections or regularly serves 25 individuals.

Pumping station—Mechanical device installed in sewer or water systems or other liquid-carrying pipelines to move the liquids to a higher level.

Raw sewage—Untreated wastewater and its contents.

Raw water—Intake water prior to any treatment or use.

Real-time monitoring—Monitoring and measuring environmental developments with technology and communications systems that provide time-relevant information to the public in an easily understood format that people can use in day-to-day decision making about their health and the environment.

Receiving waters—A river, lake, ocean, stream, or other watercourse into which wastewater or treated effluent is discharged.

River basin—The land area drained by a river and its tributaries.

Runoff—That part of precipitation, snow melt, or irrigation water that runs off the land into streams or other surface water.

Sanitary sewers—Underground pipes that carry off only domestic or industrial waste, not storm- water.

Sanitary water—Water discharged from sinks, showers, kitchens, or other nonindustrial operations, but not from commodes.

Screening—Use of screens to remove coarse floating and suspended solids from sewage.

Secondary drinking water regulations—Nonenforceable regulations applying to public water systems and specifying the maximum contamination levels that, in the judgment of EPA, are required to protect the public welfare. They apply to any contaminants that may adversely affect the odor or appearance of such water and consequently may cause people served by the system to discontinue its use.

Secondary treatment—The second step in most publicly owned waste water treatment systems in which bacteria consume the organic parts of the waste. This is accomplished by bringing together waste, bacteria, and oxygen in trickling filters or in the activated sludge process. The treatment removes floating and settleable solids and about 90 percent of the oxygen-demanding substances and suspended solids. Disinfection is the final stage of secondary treatment.

Seepage—Percolation of water through the soil from unlined canals, ditches, laterals, watercourses, or water storage facilities.

Septic system—An OSTDS designed to treat and dispose of domestic sewage. A typical septic system consists of a tank that receives waste from a residence or business and an absorption area for subsequent disposal of the liquid effluent (sludge) that remains after decomposition of the solids by bacteria in the tank, which must be pumped out periodically.

Septic tank—An underground storage tank for wastes from homes not connected to a sewer line. Waste goes directly from the home to the tank.

Settling tank—A holding area for wastewater, in which heavier particles sink to the bottom for removal and disposal.

Sewage—The waste and wastewater produced by residential and commercial sources and discharged into sewers.

Sewer—A channel or conduit that carries wastewater and stormwater runoff from the source to a treatment plant or receiving stream "sanitary" sewers carry household, industrial, and commercial waste, "storm" sewers carry runoff from rain or snow, "combined" sewers handle both.

Sewerage—The entire system of sewage collection, treatment, and disposal infrastructure.

Sludge—A semisolid residue from any of a number of water treatment processes; it can be regulated as hazardous waste.

Storm sewer—A system of pipes (separate from sanitary sewers) that carries water runoff from buildings and land surfaces.

Straight pipe—One that discharges untreated or partially treated sewage from a single dwelling structure onto the ground surface, or into a ditch, storm sewer, adjacent waterbody, or other water.

Stressors—Physical, chemical, or biological entities that can induce adverse effects on ecosystems or human health.

Surface runoff—Precipitation, snow melt, or irrigation water in excess of the amount that can infiltrate the soil surface and be stored in small surface depressions; a major transporter of nonpoint source pollutants in rivers, streams, and lakes.

Synthetic organic chemicals— Man-made organic chemicals, some are volatile, others tend to stay dissolved in water instead of evaporating.

Total suspended solids—A measure of the suspended solids in wastewater, effluent, or waterbodies, determined by tests for "total suspended nonfilterable solids."

Toxic substance—A chemical or mixture that may present an unreasonable risk of injury to health or the environment.

Toxicant—A harmful substance or agent that may injure an exposed organism.

Toxicity—The degree to which a substance or mixture of substances can harm humans or animals. *Acute toxicity* involves harmful effects in an organism through a single or short-term exposure. *Chronic toxicity* is the ability of a substance or mixture of substances to cause harmful effects over an extended period, usually upon repeated or continuous exposure, sometimes lasting for the entire life of the exposed organism. *Subchronic* *toxicity* is the ability of the substance to cause effects for more than one year but less than the lifetime of the exposed organism.

Treated wastewater—Wastewater that has been subjected to one or more physical, chemical, and biological processes to reduce its potential of being a health hazard.

Urban runoff—Stormwater from city streets and adjacent domestic or commercial properties that carries pollutants of various kinds into the sewer systems and receiving waters.

Waste load allocation—(1) The maximum load of pollutants that each discharger of waste is allowed to release into a particular waterway. Discharge limits are usually required for each specific water quality criterion being, or expected to be, violated. (2) The portion of a stream's total assimilative capacity assigned to an individual discharge.

Wastewater—The spent or used water from a home, community, farm, or industry that contains dissolved or suspended matter.

Wastewater treatment plant—A facility containing a series of tanks, screens, filters, and other processes by which pollutants are removed from wastewater.

Water quality criteria—Levels of water quality expected to render a body of water suitable for its designated use. Criteria are based on specific levels of pollutants that would make the water harmful if used for drinking, swimming, farming, fish production, or industrial processes.

Water quality standards—State-adopted and EPA-approved ambient standards for waterbodies. The standards prescribe the use of the waterbody and establish the water quality criteria that must be met to protect designated uses.

Watershed—The land area that drains into a stream; the watershed for a major river may encompass a number of smaller watersheds that ultimately combine at a common point.

Watershed approach—A coordinated framework for environmental management that focuses public and private efforts on the highest priority problems within hydrologically defined geographic areas by taking systematically into consideration both ground- and surface water flow.

Wildcat sewer—Community straight pipe or collection system (community sewer) serving more than one equivalent dwelling and discharging untreated or partially treated sewage to the surface of the ground, storm sewers, or other water.

REFERENCES

25 PA Code § 73. Standards for On-lot (site) Sewage Treatment Facilities. Available on-line at *http://www.pacode.com/secure/data/025/chapter73/chap73toc.html.* Accessed August 18, 2004.

EPA (U.S. Environmental Protection Agency). 1997. Terms of Environment. Available on-line at *http://www.epa.gov/OCEPAterms/.* Accessed August 18, 2004.

EPA. 2002. Onsite Wastewater Treatment Systems Manual. Available on-line at *http://www.epa.gov/ORD/NRMRL/Pubs/625R00008/html/625R00008gloss.htm.* Accessed August 18, 2004.

NSFC (National Small Flows Clearinghouse). 2001. Poster: Wastewater Collection and Treatment Systems for Small Communities. Morgantown, WV: NSFC.

PADEP (Pennsylvania Department of Environmental Protection). 2003. Experimental Systems Guidance. 362-0300-008. Harrisburg, PA: Bureau of Water Supply and Wastewater Management.

PADEP. 2004a. Alternate Systems Guidance. 362-0300-007. Harrisburg, PA: Bureau of Water Supply and Wastewater Management.

PADEP. 2004b. Act 537: Pennsylvania Sewage Facilities Act with Index. Harrisburg, PA: Bureau of Water Supply and Wastewater Management.

PADEP. 2004c. Understanding Holding Tanks. Fact Sheet 3800-FS-DEP2807. Harrisburg, PA: Office of Water Management.

Appendix D

WATER SCIENCE AND TECHNOLOGY BOARD

R. RHODES TRUSSELL, *Chair*, Trussell Technologies, Inc., Pasadena, California
RICHARD G. LUTHY,* *Chair,* Stanford University, Stanford, California
JOAN B. ROSE,* *Vice Chair,* Michigan State University, East Lansing
RICHELLE M. ALLEN-KING,* State University of New York, Buffalo
GREGORY B. BAECHER, University of Maryland, College Park
MARY JO BAEDECKER, U.S. Geological Survey (retired), Reston, Virginia
KENNETH R. BRADBURY,* Wisconsin Geological and Natural History Survey, Madison
JAMES CROOK,* Water Reuse Consultant, Norwell, Massachusetts
JOAN G. EHRENFELD, Rutgers University, New Brunswick, New Jersey
DARA ENTEKHABI, Massachusetts Institute of Technology, Cambridge
EFI FOUFOULA-GEORGIOU,* University of Minnesota, Minneapolis
GERALD GALLOWAY, Titan Corporation, Reston, Virginia
PETER GLEICK, Pacific Institute for Studies in Development, Environment, and Security, Oakland, California
CHARLES N. HAAS, Drexel University, Philadelphia, Pennsylvania
KAI N. LEE, Williams College, Williamstown, Massachusetts
JOHN LETEY, JR.,* University of California, Riverside
CHRISTINE L. MOE, Emory University, Atlanta, Georgia
ROBERT PERCIASEPE, National Audubon Society, Washington, D.C.
JERALD L. SCHNOOR, University of Iowa, Iowa City
LEONARD SHABMAN, Resources for the Future, Washington, D.C.
KARL K. TUREKIAN, Yale University, New Haven, Connecticut
HAME M. WATT, Independent Consultant, Washington, D.C.
CLAIRE WELTY, University of Maryland, Baltimore County, Baltimore
JAMES L. WESCOAT, JR., University of Illinois at Urbana-Champaign

Staff

STEPHEN D. PARKER, Director
LAURA J. EHLERS, Senior Program Officer
MARK C. GIBSON, Senior Program Officer
JEFFREY W. JACOBS, Senior Program Officer
WILLIAM S. LOGAN, Senior Program Officer
LAUREN E. ALEXANDER, Program Officer
STEPHANIE E. JOHNSON, Program Officer
M. JEANNE AQUILINO, Financial and Administrative Associate
ELLEN A. DE GUZMAN, Research Associate
PATRICIA JONES KERSHAW, Senior Program Associate

ANITA A. HALL, Program Associate
DOROTHY K. WEIR, Senior Program Assistant

* Terms expired June 30, 2004.

Appendix E

Committee and Staff Biographical Information

Jerome B. Gilbert (NAE), *Chair*, is a consulting professional engineer who provides policy and technical advice on water and wastewater management, water transfers and rights, and regulatory compliance. Mr. Gilbert was general manager and chief engineer (1981-1991) for the East Bay Municipal Utility District in the Oakland, California region and served as president of the American Academy of Environmental Engineers (1991-1992) and the American Water Works Association (1979-1980). He served on the Executive Committee of the International Water Association from 1998 to 2004. Mr. Gilbert was a founding member of the National Research Council (NRC) Water Science and Technology Board (WSTB; 1982-1986) and is a member of the National Academy of Engineering (NAE). He served on several NRC committees, most recently the Committee on Privatization of Water Services in the United States. Mr. Gilbert holds B.S. and M.S. degrees in civil engineering from the University of Cincinnati and Stanford University, respectively.

Brian J. Hill was senior vice president for watersheds for the Pennsylvania Environmental Council (PEC) and also served as director for the PEC French Creek Project based in Meadville, Pennsylvania prior to accepting a position in May 2004 in the Policy Office of the Governor in Harrisburg, Pennsylvania, and resigning from the committee. The purpose of the project and the council's overall watershed protection program is to spur grassroots efforts to reduce point and nonpoint source water pollution. Previously Mr. Hill served as director of the Western Pennsylvania Office of the PEC in Pittsburgh for six years; he has developed a wide variety of environmental educational programs including workshops and seminars on solid and hazardous waste management, the reuse of industrial sites, and land use policy in Pennsylvania. Mr. Hill is a past chairman of the Citizens Advisory Council to the Pennsylvania Department of Environmental Protection (PADEP) and has served as a member of the PADEP's Environmental Quality Board. He received his B.S. in environmental science from Allegheny College and his M.S. in natural resource management from the University of New Hampshire.

Jeffrey M. Lauria is a vice president of Malcolm Pirnie, Inc., a century-old New York-based firm of civil and environmental engineers and scientists specializing in water issues. In this position, Dr. Lauria directs large-scale program management and engineering master plans for wastewater, wet weather, watershed, and water quality projects. He also has comprehensive national and international experience in wastewater, drinking water, and stormwater treatment processes and related expertise in hydraulic, hydrologic, water quality, and mathematical modeling to support decision optimization at more than 200 project locations. Dr. Lauria has also served as a technical adviser to several state and local governments and on scientific and managerial councils from the private sector. He received a B.E. in civil engineering from Manhattan College, and an M.E. and Ph.D. in environmental engineering from Manhattan College and Polytechnic University, respectively.

Gary S. Logsdon recently retired as a senior consultant for Black & Veatch Corporation, a worldwide engineering, consulting, and construction company based in Kansas City, Missouri. Previously, Dr. Logsdon served for more than 25 years with the U.S. Public Health Service and the U.S. Environmental Protection Agency (EPA). At Black & Veatch, he provided oversight for studies of drinking water treatment and worked with water utilities to optimize their operations. Dr. Logsdon has a wide range of experience in water treatment technology development and application; he has conducted research on water filtration for removal of waterborne intestinal parasite cysts, bacteria, and turbidity and on the modification of water quality for corrosion control in water distribution systems. He is a former WSTB member and served on the NRC Committee on Small Water Supply Systems. Dr. Logsdon received his B.S. and M.S. degrees in civil and sanitary engineering from the University of Missouri at Columbia, and a D.Sc. in environmental engineering from Washington University.

Perry L. McCarty (NAE) is Silas H. Palmer Professor Emeritus in the Department of Civil and Environmental Engineering at Stanford University. Dr. McCarty's research interests include aerobic and anaerobic biological processes for water quality control, advanced wastewater treatment processes, and movement, fate, and control of hazardous chemicals in groundwater. Dr. McCarty is a member of the NAE, and has served on many NAE and NRC panels, committees, boards, and commissions. Dr. McCarty received his B.S. degree from Wayne State University and his M.S. and Sc.D. degrees in sanitary engineering from the Massachusetts Institute of Technology.

Patricia Miller is a senior training coordinator and hydrogeologist at Tetra Tech, Inc., in Cincinnati, Ohio. She previously worked as an extension specialist at Michigan State University and, until recently, at West Virginia University. Her earlier work experience with the Virginia Department of Conservation and Recreation and the Virginia Department of Health involved watershed, total maximum daily load (TMDL), and decentralized wastewater programs. Dr. Miller's research and extension activities include environmental health, drinking water, and surface and groundwater protection, especially as related to septic systems and other on-site and small community wastewater treatment systems. She received her B.S. in geology from Tulane University, her M.S. in geology and mineralogy from Ohio State University, and her Ph.D. in environmental sciences from the University of Texas at Dallas.

David H. Moreau is professor and prior chair of the Department of City and Regional Planning at the University of North Carolina, Chapel Hill. Dr. Moreau's research interests include analysis, planning, financing, and evaluation of water resources and related environmental programs and he is actively involved in water resources planning at the local, state, and federal levels. He has chaired or served on several NRC committees, most recently as a member of the Panel on Peer Review of the Committee to Assess the U.S. Army Corps of Engineers Methods of Analysis and Peer Review for Water Resources Project Planning. Dr. Moreau serves as chairman of the North Carolina Environmental Management Commission, the state's regulatory commission for water quality, air quality, and water allocation. Dr. Moreau received a B.S. and M.S. in civil engineering from Mississippi State University and North Carolina State University, respectively, and a Ph.D. in water resources from Harvard University.

Nelson P. Moyer is a senior scientist with the Cadmus Group, Inc., and he holds an adjunct professor appointment in the University of Iowa College of Public Health. He is a diplomate of the Board of Medical and Molecular Biology with certification in medical and public health microbiology. Prior to joining the Cadmus Group, Inc., in August 2002, Dr. Moyer served for 28 years as chief microbiologist in the State Public Health Laboratories in Oklahoma and Iowa. His research interests include molecular epidemiology, application of indicator organisms to pollution monitoring, and bacterial colonization of potable water systems. He has served on numerous advisory committees supporting the EPA, the U.S. Food and Drug Administration, the American Water Works Association Research Foundation, and the Public Health Foundation. Dr. Moyer received his B.S. degree in bacteriology-chemistry from the Florida State University, and his Ph.D. degree in microbiology-biochemistry from the Louisiana State University.

Rutherford H. Platt is a professor of geography and planning law at the University of Massachusetts at Amherst. Dr. Platt specializes in federal, state, and regional policies concerning land and water resource management and natural disasters. Among many books and other publications, he is the author of *Land Use and Society: Geography, Law, and Public Policy—Revised Edition* published in 2004. He is director of the Ecological Cities Project based at the University of Massachusetts, Amherst, to share experience and research on the protection, restoration, and management of urban greenspaces across the United States and elsewhere. Dr. Platt is the past chair of the NRC Roundtable on Natural Disasters and a former member of the WSTB. He has previously served on eight committees of the NRC, twice as chair, and was recently appointed a national associate of the National Academies. He received a B.S. in political science from Yale University, and a Ph.D. in geography from the University of Chicago; he and holds a J.D. from the University of Chicago Law School.

Stuart S. Schwartz is director of the Center for Environmental Science, Technology, and Policy at Cleveland State University (CSU). Before joining CSU, Dr. Schwartz served as associate director of the Water Resources Research Institute of the University of North Carolina. Previously, Dr. Schwartz served as an associate hydrologic engineer at the Hydrologic Research Center in San Diego, California, and directed the Section for Cooperative Water Supply Operations on the Potomac at the Interstate Commission on the Potomac River Basin. Dr. Schwartz's research and professional interests are in the application of probabilistic hydrologic forecasting and multiobjective decision making in risk-based water resources management, watershed management, and water supply systems operations. He currently serves on the NRC Committee on USGS Water Resources Research. He received his B.S. and M.S. in biology-geology from the University of Rochester and his Ph.D. in systems analysis from the Johns Hopkins University.

James S. Shortle is distinguished professor of agricultural and environmental economics in the Department of Agricultural Economics and Rural Sociology at Pennsylvania State University. His recent work focuses on the design of incentive-based approaches for reducing pollution from agriculture and other nonpoint pollution sources; measuring and predicting relationships between the environment and the economy; integrating economic and environmental information for water quality decision making; and decision making to mitigate impacts of climate change. Dr. Shortle received a B.S. and M.A. in economics from the University of New Mexico and a Ph.D. in economics from Iowa State University.

Joel A. Tarr is the Richard S. Caliguiri University Professor of History and Policy at Carnegie Mellon University. His research areas include the history of the urban environment and of urban technological systems. One of his specialties is the history of the Pittsburgh region, and in 2003 his book *Devastation and Renewal: An Environmental History of Pittsburgh and Its Region* was published. Dr. Tarr served previously on two NRC committees, most recently the Committee on International Comparison of National Policies and Expectations Affecting Public Transit. He received his B.S. and M.A. degrees from Rutgers University and his Ph.D. in American history from Northwestern University.

Jeanne M. VanBriesen is an assistant professor in the Department of Civil and Environmental Engineering and the Department of Biomedical Engineering at Carnegie Mellon University. Her primary research interests are in biological processes in aquatic environmental systems, including biological treatment processes for wastewater and drinking water and microbiological stability in drinking water. She also conducts research in bioremediation of recalcitrant organic compounds, modeling environmental systems involving complex biogeochemistry, and treatment and remediation of mixed wastes in aquatic surface and subsurface systems. Dr. VanBriesen received a B.S. in education (chemistry) and an M.S. and Ph.D. in civil (environmental) engineering from Northwestern University.

Paul F. Ziemkiewicz is director of the National Mine Land Reclamation Center at West Virginia University and the West Virginia Water Research Institute. Dr. Ziemkiewicz is also a research professor in the Division of Plant and Soil Sciences at West Virginia University. His current research focuses on acid mine drainage, land reclamation, coal combustion by-products, and water quality impacts relating to the coal industry. In addition to his research activities, he currently serves on both state and federal policy advisory committees focusing on reclamation and acid mine drainage. Dr. Ziemkiewicz received his B.S. in biology and M.S. in range ecology from Utah State University and his Ph.D. in forest ecology from the University of British Columbia.

STAFF

Mark C. Gibson is a senior program officer at the NRC's Water Science and Technology Board and was responsible for the completion of this report. Since joining the NRC in 1998, he has served as study director for six committees, including the Committee on Drinking Water Contaminants that released three reports, the Committee on Assessing and Valuing the Services of Aquatic and Related Terrestrial Ecosystems, and the Committee on Indicators for Waterborne Pathogens. Mr. Gibson received his B.S. in biology from Virginia Polytechnic Institute and State University and his M.S. in environmental science and policy in biology from George Mason University.

Dorothy K. Weir is a senior program assistant with the Water Science and Technology Board. She received a B.S. in biology from Rhodes College in Memphis, Tennessee and an M.S. degree in environmental science and policy from Johns Hopkins University. Ms. Weir joined the NRC in 2003.